INFORMATION THEORY

BY

ROBERT B. ASH

University of Illinois
Urbana, Illinois

DOVER PUBLICATIONS, INC.
NEW YORK

Published in Canada by General Publishing Company, Ltd., 30 Lesmill Road, Don Mills, Toronto, Ontario.

This Dover edition, first published in 1990, is an unabridged and corrected republication of the work originally published by Interscience Publishers (a division of John Wiley & Sons), New York, 1965.

Manufactured in the United States of America
Dover Publications, Inc., 31 East 2nd Street, Mineola, N.Y. 11501

Library of Congress Cataloging-in-Publication Data

Ash, Robert B.
Information theory / by Robert B. Ash.
p. cm.
Includes bibliographical references and index.
ISBN 0-486-66521-6 (pbk.)
1. Information theory. I. Title.
Q360.A8 1990
003′.54—dc20 90-45415
CIP

PREFACE

Statistical communication theory is generally regarded as having been founded by Shannon (1948) and Wiener (1949), who conceived of the communication situation as one in which a signal chosen from a specified class is to be transmitted through a channel, but the output of the channel is not determined by the input. Instead, the channel is described statistically by giving a probability distribution over the set of all possible outputs for each permissible input. At the output of the channel, a received signal is observed, and then a decision is made, the objective of the decision being to identify as closely as possible some property of the input signal.

The Shannon formulation differs from the Wiener approach in the nature of the transmitted signal and in the type of decision made at the receiver. In the Shannon model, a randomly generated message produced by a source of information is "encoded," that is, each possible message that the source can produce is associated with a signal belonging to a specified set. It is the encoded message which is actually transmitted. When the output is received, a "decoding" operation is performed, that is, a decision is made as to the identity of the particular signal transmitted. The objectives are to increase the size of the vocabulary, that is, to make the class of inputs as large as possible, and at the same time to make the probability of correctly identifying the input signal as large as possible. How well one can do these things depends essentially on the properties of the channel, and a fundamental concern is the analysis of different channel models. Another basic problem is the selection of a particular input vocabulary that can be used with a low probability of error.

In the Wiener model, on the other hand, a random signal is to be communicated directly through the channel; the encoding step is absent. Furthermore, the channel model is essentially fixed. The channel is generally taken to be a device that adds to the input signal a randomly generated "noise." The "decoder" in this case operates on the received signal to produce an estimate of some property of the input. For example, in the prediction problem the decoder estimates the value of the input at some future time. In general, the basic objective is to design a decoder, subject to a constraint of physical realizability, which makes the best estimate, where the closeness of the estimate is measured by an appropriate criterion. The problem of realizing and implementing an optimum decoder is central to the Wiener theory.

I do not want to give the impression that every problem in communication theory may be unalterably classified as belonging to the domain of either Shannon or Wiener, but not both. For example, the radar reception problem contains some features of both approaches. Here one tries to determine whether a signal was actually transmitted, and if so to identify which signal of a specified class was sent, and possibly to estimate some of the signal parameters. However, I think it is fair to say that this book is concerned entirely with the Shannon formulation, that is, the body of mathematical knowledge which has its origins in Shannon's fundamental paper of 1948. This is what "information theory" will mean for us here.

The book treats three major areas: first (Chapters 3, 7, and 8), an analysis of channel models and the proof of coding theorems (theorems whose physical interpretation is that it is possible to transmit information reliably through a noisy channel at any rate below channel capacity, but not at a rate above capacity); second, the study of specific coding systems (Chapters 2, 4, and 5); finally, the study of the statistical properties of information sources (Chapter 6). All three areas were introduced in Shannon's original paper, and in each case Shannon established an area of research where none had existed before.

The book has developed from lectures and seminars given during the last five years at Columbia University; the University of California, Berkeley; and the University of Illinois, Urbana. I have attempted to write in a style suitable for first-year graduate students in mathematics and the physical sciences, and I have tried to keep the prerequisites modest. A course in basic probability theory is essential, but measure theory is not required for the first seven chapters. All random variables appearing in these chapters are discrete and take on only a finite number of possible values. For most of Chapter 8, the random variables, although continuous, have probability density functions, and therefore a knowledge of basic probability should suffice. Some measure and Hilbert space theory is helpful for the last two sections of Chapter 8, which treat time-continuous channels. An appendix summarizes the Hilbert space background and the results from the theory of stochastic processes that are necessary for these sections. The appendix is not self-contained, but I hope it will serve to pinpoint some of the specific equipment needed for the analysis of time-continuous channels.

Chapters 1 and 3 are basic, and the concepts developed there appear throughout the book. Any of Chapters 4 through 8 may be read immediately after Chapters 1 and 3, although the reader should browse through the first five sections of Chapter 4 before looking at Chapter 5. Chapter 2 depends only on Chapter 1.

In Chapter 4, the exposition is restricted to binary codes, and the generalization to codes over an arbitrary finite field is sketched at the end of the chapter. The analysis of cyclic codes in Chapter 5 is carried out by a matrix development rather than by the standard approach, which uses abstract algebra. The matrix method seems to be natural and intuitive, and will probably be more palatable to students, since a student is more likely to be familiar with matrix manipulations than he is with extension fields.

I hope that the inclusion of some sixty problems, with fairly detailed solutions, will make the book more profitable for independent study.

The historical notes at the end of each chapter are not meant to be exhaustive, but I have tried to indicate the origins of some of the results.

I have had the benefit of many discussions with Professor Aram Thomasian on information theory and related areas in mathematics. Dr. Aaron Wyner read the entire manuscript and supplied helpful comments and criticism. I also received encouragement and advice from Dr. David Slepian and Professors R. T. Chien, M. E. Van Valkenburg, and L. A. Zadeh.

Finally, my thanks are due to Professor Warren Hirsch, whose lectures in 1959 introduced me to the subject, to Professor Lipman Bers for his invitation to publish in this series, and to the staff of Interscience Publishers, a division of John Wiley and Sons, Inc., for their courtesy and cooperation.

Urbana, Illinois
July, 1965

Robert B. Ash

CONTENTS

CHAPTER ONE

A Measure of Information

1.1. Introduction

Information theory is concerned with the analysis of an entity called a "communication system," which has traditionally been represented by the block diagram shown in Fig. 1.1.1. The source of messages is the person or machine that produces the information to be communicated. The encoder associates with each message an "object" which is suitable for transmission over the channel. The "object" could be a sequence of binary digits, as in digital computer applications, or a continuous waveform, as in radio communication. The channel is the medium over which the coded message is transmitted. The decoder operates on the output of the channel and attempts to extract the original message for delivery to the destination. In general, this cannot be done with complete reliability because of the effect of "noise," which is a general term for anything which tends to produce errors in transmission.

Information theory is an attempt to construct a mathematical model for each of the blocks of Fig. 1.1.1. We shall not arrive at design formulas for a communication system; nevertheless, we shall go into considerable detail concerning the theory of the encoding and decoding operations.

It is possible to make a case for the statement that information theory is essentially the study of one theorem, the so-called "fundamental theorem of information theory," which states that "it is possible to transmit information through a noisy channel at any rate less than channel capacity with an arbitrarily small probability of error." The meaning of the various terms "information," "channel," "noisy," "rate," and "capacity" will be clarified in later chapters. At this point, we shall only try to give an intuitive idea of the content of the fundamental theorem.

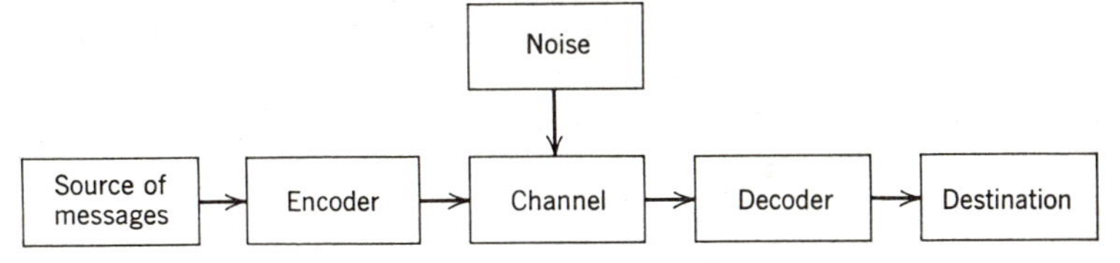

Fig. 1.1.1. Communication system.

Imagine a "source of information" that produces a sequence of binary digits (zeros or ones) at the rate of 1 digit per second. Suppose that the digits 0 and 1 are equally likely to occur and that the digits are produced independently, so that the distribution of a given digit is unaffected by all previous digits. Suppose that the digits are to be communicated directly over a "channel." The nature of the channel is unimportant at this moment, except that we specify that the probability that a particular digit

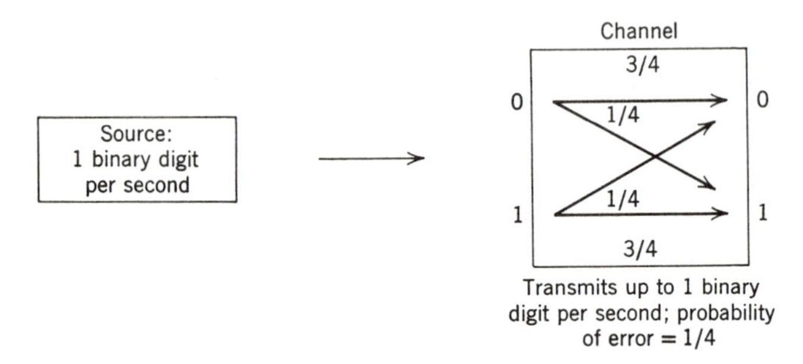

Fig. 1.1.2. Example.

is received in error is (say) 1/4, and that the channel acts on successive inputs independently. We also assume that digits can be transmitted through the channel at a rate not to exceed 1 digit per second. The pertinent information is summarized in Fig. 1.1.2.

Now a probability of error of 1/4 may be far too high in a given application, and we would naturally look for ways of improving reliability. One way that might come to mind involves sending the source digit through the channel more than once. For example, if the source produces a zero at a given time, we might send a sequence of 3 zeros through the channel; if the source produces a one, we would send 3 ones. At the receiving end of the channel, we will have a sequence of 3 digits for each one produced by the source. We will have the problem of *decoding* each sequence, that is, making a decision, for each sequence received, as to the identity of the source digit. A "reasonable" way to decide is by means of a "majority selector," that is, a rule which specifies that if more ones than zeros are received, we are to decode the received sequence as a "1"; if more zeros than ones appear, decode as a "0." Thus, for example, if the source produces a one, a sequence of 3 ones would be sent through the channel. The first and third digits might be received incorrectly; the received sequence would then be 0 1 0; the decoder would therefore declare (incorrectly) that a zero was in fact transmitted.

We may calculate the probability that a given source digit is received in error; it is the probability that at least 2 of a sequence of 3 digits will be received incorrectly, where the probability of a given digit's being incorrect is 1/4 and the digits are transmitted independently. Using the standard formula for the distribution of successes and failures in a sequence of Bernoulli trials, we obtain

$$\binom{3}{2}\left(\frac{1}{4}\right)^2\frac{3}{4}+\left(\frac{1}{4}\right)^3=\frac{10}{64}<\frac{1}{4}.$$

Thus we have lowered the probability of error; however, we have paid a price for this reduction. If we send 1 digit per second through the channel, it now takes 3 seconds to communicate 1 digit produced by the source, or three times as long as it did originally. Equivalently, if we want to synchronize the source with the channel, we must slow down the rate of the source to $\frac{1}{3}$ digit per second while keeping the channel rate fixed at 1 digit per second. Then during the time (3 seconds) it takes for the source to produce a single digit, we will be able to transmit the associated sequence of 3 digits through the channel.

Now let us generalize this procedure. Suppose that the probability of error for a given digit is $\beta < 1/2$, and that each source digit is represented by a sequence of length $2n + 1$; a majority selector is used at the receiver. The effective transmission rate of the source is reduced to $1/(2n + 1)$ binary digits per second while the probability of incorrect decoding is

$$p(e) = P\{n + 1 \text{ or more digits in error}\} = \sum_{k=n+1}^{2n+1}\binom{2n+1}{k}\beta^k(1-\beta)^{2n+1-k}.$$

Since the expected number of digits in error is $(2n + 1)\beta < n + 1$, the weak law of large numbers implies that $p(e) \to 0$ as $n \to \infty$. (If S_{2n+1} is the number of digits in error, then the sequence $S_{2n+1}/(2n + 1)$ converges in probability to β, so that

$$p(e) = P\{S_{2n+1} \geq n + 1\} = P\left\{\frac{S_{2n+1}}{2n+1} \geq \frac{n+1}{2n+1}\right\}.$$

$$= P\left\{\frac{S_{2n+1}}{2n+1} \geq \beta + \varepsilon\right\} \to 0 \quad \text{as} \quad n \to \infty.)$$

Thus we are able to reduce the probability of error to an arbitrarily small figure, at the expense of decreasing the effective transmission rate toward zero.

The essence of the fundamental theorem of information theory is that in order to achieve arbitrarily high reliability, *it is not necessary to reduce the transmission rate to zero*, but only to a number called the *channel capacity*.

The means by which these results are obtained is called *coding*. The process of coding involves the insertion of a device called an "encoder" between the source and the channel; the encoder assigns to each of a specified group of source messages a sequence of symbols called a *code word* suitable for transmission through the channel. In the above example, we have just seen a primitive form of coding; we have assigned to the source digit 0 a sequence of zeros, and to the source digit 1 a sequence of ones. The received sequence is fed to a decoder which attempts to determine the identity of the original message. In general, to achieve reliability without sacrificing speed of transmission, code words are not assigned to single digits but instead to long blocks of digits. In other words, the encoder waits for the source to produce a block of digits of a specified length, and then assigns a code word to the entire block. The decoder examines the received sequence and makes a decision as to the identity of the transmitted block. In general, encoding and decoding procedures are considerably more elaborate than in the example just considered.

The discussion is necessarily vague at this point; hopefully, the concepts introduced will eventually be clarified. Our first step in the clarification will be the construction of a mathematical measure of the information conveyed by a message. As a preliminary example, suppose that a random variable X takes on the values 1, 2, 3, 4, 5 with equal probability. We ask how much information is conveyed about the value of X by the statement that $1 \leq X \leq 2$. Originally, if we try to guess the value of X, we have probability 1/5 of being correct. After we know that X is either 1 or 2, we have a higher probability of guessing the right answer. In other words, there is less *uncertainty* about the second situation. Telling us that $1 \leq X \leq 2$ has reduced the uncertainty about the actual value of X. It thus appears that if we could pin down the notion of uncertainty, we would be able to measure precisely the transfer of information. Our approach will be to set up certain requirements which an uncertainty function should "reasonably" satisfy; we shall then prove that there is only one function which meets all the requirements. We must emphasize that it is not important *how* we arrive at the measure of uncertainty. The axioms of uncertainty we choose will probably seem reasonable to most readers, but we definitely will not base the case for the measure of uncertainty on intuitive grounds. The usefulness of the uncertainty measure proposed by Shannon lies in its operational significance in the construction of codes. Using an appropriate notion of uncertainty we shall be able to define the information transmitted through a channel and establish the existence of coding systems which make it possible to transmit at any rate less than channel capacity with an arbitrarily small probability of error.

1.2. *Axioms for the uncertainty measure*

Suppose that a probabilistic experiment involves the observation of a discrete random variable X. Let X take on a finite number of possible values $x_1, x_2, \ldots, x_M$, with probabilities $p_1, p_2, \ldots, p_M$, respectively. We assume that all p_i are strictly greater than zero. Of course $\sum_{i=1}^{M} p_i = 1$. We now attempt to arrive at a number that will measure the uncertainty associated with X. We shall try to construct two functions h and H. The function h will be defined on the interval $(0, 1]$; $h(p)$ will be interpreted as the uncertainty associated with an event with probability p. Thus if the event $\{X = x_i\}$ has probability p_i, we shall say that $h(p_i)$ is the uncertainty associated with the event $\{X = x_i\}$, or the uncertainty removed (or information conveyed) by revealing that X has taken on the value x_i in a given performance of the experiment. For each M we shall define a function H_M of the M variables $p_1, \ldots, p_M$ (we restrict the domain of H_M by requiring all p_i to be > 0, and $\sum_{i=1}^{M} p_i = 1$). The function $H_M(p_1, \ldots, p_M)$ is to be interpreted as the average uncertainty associated with the events $\{X = x_i\}$; specifically, we require that $H_M(p_1, \ldots, p_M) = \sum_{i=1}^{M} p_i h(p_i)$. [For simplicity we write $H_M(p_1, \ldots, p_M)$ as $H(p_1, \ldots, p_M)$ or as $H(X)$.] Thus $H(p_1, \ldots, p_M)$ is the average uncertainty removed by revealing the value of X. The function h is introduced merely as an aid to the intuition; it will appear only in this section. In trying to justify for himself the requirements which we shall impose on $H(X)$, the reader may find it helpful to think of $H(p_1, \ldots, p_M)$ as a weighted average of the numbers $h(p_1), \ldots, h(p_M)$.

Now we proceed to impose requirements on the functions H. In the sequel, $H(X)$ will be referred to as the "uncertainty of X"; the word "average" will be understood but will, except in this section, generally not be appended. First suppose that all values of X are equally probable. We denote by $f(M)$ the average uncertainty associated with M equally likely outcomes, that is, $f(M) = H(1/M, 1/M, \ldots, 1/M)$. For example, $f(2)$ would be the uncertainty associated with the toss of an unbiased coin, while $f(8 \times 10^6)$ would be the uncertainty associated with picking a person at random in New York City. We would expect the uncertainty of the latter situation to be greater than that of the former. In fact, our first requirement on the uncertainty function is that

$f(M) = H(1/M, \ldots, 1/M)$ *should be a monotonically increasing function of* M; *that is,* $M < M'$ *implies* $f(M) < f(M')$ $(M, M' = 1, 2, 3, \ldots)$.

Now consider an experiment involving two independent random variables X and Y. Let X take on the values $x_1, x_2, \ldots, x_M$ with equal probability, and let Y take on the values $y_1, y_2, \ldots, y_L$, also with equal probability.

Thus the joint experiment involving X and Y has ML equally likely outcomes, and therefore the average uncertainty of the joint experiment is $f(ML)$. If the value of X is revealed, the average uncertainty about Y should not be affected because of the assumed independence of X and Y. Hence we expect that the average uncertainty associated with X and Y together, minus the average uncertainty removed by revealing the value of X, should yield the average uncertainty associated with Y. Revealing the value of X removes, on the average, an amount of uncertainty equal to $f(M)$, and thus the second requirement on the uncertainty measure is that

$$f(ML) = f(M) + f(L) \qquad (M, L, = 1, 2, \ldots).$$

At this point we remove the restriction of equally likely outcomes and turn to the general case. We divide the values of a random variable X into two groups, A and B, where A consists of $x_1, x_2, \ldots, x_r$ and B consists of $x_{r+1}, x_{r+2}, \ldots, x_M$. We construct a compound experiment as follows. First we select one of the two groups, choosing group A with probability $p_1 + p_2 + \cdots + p_r$ and group B with probability $p_{r+1} + p_{r+2} + \cdots + p_M$. Thus the probability of each group is the sum of the probabilities of the values in the group. If group A is chosen, then we select x_i with probability $p_i/(p_1 + \cdots + p_r)$ $(i = 1, \ldots, r)$, which is the

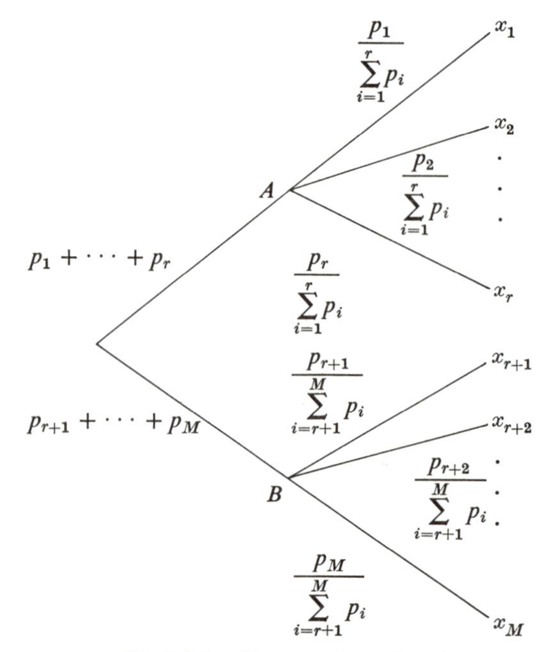

Fig. 1.2.1. Compound experiment.

conditional probability of x_i given that the value of X lies in group A. Similarly, if group B is chosen, then x_i is selected with probability $p_i/(p_{r+1} + \cdots + p_M)$ $(i = r + 1, \ldots, M)$. The compound experiment is diagrammed in Fig. 1.2.1. It is equivalent to the original experiment associated with X. For if Y is the result of the compound experiment, the probability that $Y = x_1$ is

$$\begin{aligned} P\{Y = x_1\} &= P\{A \text{ is chosen and } x_1 \text{ is selected}\} \\ &= P\{A \text{ is chosen}\}P\{x_1 \text{ is selected} \mid A \text{ is chosen}\} \\ &= \left(\sum_{i=1}^{r} p_i\right) \frac{p_1}{\sum_{i=1}^{r} p_i} = p_1. \end{aligned}$$

Similarly, $P\{Y = x_i\} = p_i (i = 1, 2, \ldots, M)$ so that Y and X have the same distribution. Before the compound experiment is performed, the average uncertainty associated with the outcome is $H(p_1, \ldots, p_M)$. If we reveal which of the two groups A and B is selected, we remove on the average an amount of uncertainty $H(p_1 + \cdots + p_r, p_{r+1} + \cdots + p_M)$. With probability $p_1 + \cdots + p_r$ group A is chosen and the remaining uncertainty is

$$H\left(\frac{p_1}{\sum_{i=1}^{r} p_i}, \frac{p_2}{\sum_{i=1}^{r} p_i}, \ldots, \frac{p_r}{\sum_{i=1}^{r} p_i}\right);$$

with probability $p_{r+1} + \cdots + p_M$, group B is chosen, and the remaining uncertainty is

$$H\left(\frac{p_{r+1}}{\sum_{i=r+1}^{M} p_i}, \frac{p_{r+2}}{\sum_{i=r+1}^{M} p_i}, \ldots, \frac{p_M}{\sum_{i=r+1}^{M} p_i}\right).$$

Thus *on the average* the uncertainty remaining after the group is specified is

$$(p_1 + \cdots + p_r) H\left(\frac{p_1}{\sum_{i=1}^{r} p_i}, \ldots, \frac{p_r}{\sum_{i=1}^{r} p_i}\right) + (p_{r+1} + \cdots + p_M) H\left(\frac{p_{r+1}}{\sum_{i=r+1}^{M} p_i}, \ldots, \frac{p_M}{\sum_{i=r+1}^{M} p_i}\right).$$

We expect that the average uncertainty about the compound experiment minus the average uncertainty removed by specifying the group equals the average uncertainty remaining after the group is specified. Hence, the third requirement we impose on the uncertainty function is that

$$H(p_1, \ldots, p_M) = H(p_1 + \cdots + p_r, p_{r+1} + \cdots + p_M) + (p_1 + \cdots + p_r)H\left(\frac{p_1}{\sum_{i=1}^{r} p_i}, \ldots, \frac{p_r}{\sum_{i=1}^{r} p_i}\right) + (p_{r+1} + \cdots + p_M)H\left(\frac{p_{r+1}}{\sum_{i=r+1}^{M} p_i}, \ldots, \frac{p_M}{\sum_{i=r+1}^{M} p_i}\right).$$

As a numerical example of the above requirement, we may write

$$H(\underbrace{\tfrac{1}{2}, \tfrac{1}{4}}_{A}, \underbrace{\tfrac{1}{8}, \tfrac{1}{8}}_{B}) = H(\tfrac{3}{4}, \tfrac{1}{4}) + \tfrac{3}{4}H(\tfrac{2}{3}, \tfrac{1}{3}) + \tfrac{1}{4}H(\tfrac{1}{2}, \tfrac{1}{2}).$$

Finally, we require for mathematical convenience that $H(p, 1 - p)$ be a continuous function of p. (Intuitively we should expect that a small change in the probabilities of the values of X will correspond to a small change in the uncertainty of X.)

To recapitulate, we assume the following four conditions as axioms:

1. $H(1/M, 1/M, \ldots, 1/M) = f(M)$ is a monotonically increasing function of M ($M = 1, 2, \ldots$).

2. $f(ML) = f(M) + f(L) \qquad (M, L = 1, 2, \ldots).$

3. $$H(p_1, \ldots, p_M) = H(p_1 + \cdots + p_r, p_{r+1} + \cdots + p_M) + (p_1 + \cdots + p_r)H\left(\frac{p_1}{\sum_{i=1}^{r} p_i}, \ldots, \frac{p_r}{\sum_{i=1}^{r} p_i}\right) + (p_{r+1} + \cdots + p_M)H\left(\frac{p_{r+1}}{\sum_{i=r+1}^{M} p_i}, \ldots, \frac{p_M}{\sum_{i=r+1}^{M} p_i}\right)$$
$$(r = 1, 2, \ldots, M - 1).$$

(Axiom 3 is called the *grouping axiom.*)

4. $H(p, 1 - p)$ is a continuous function of p.

The four axioms essentially determine the uncertainty measure. More precisely, we prove the following theorem.

Theorem 1.2.1. The only function satisfying the four given axioms is

$$H(p_1, \ldots, p_M) = -C \sum_{i=1}^{M} p_i \log p_i, \tag{1.2.1}$$

where C is an arbitrary positive number and the logarithm base is any number greater than 1.

Proof. It is not difficult to verify that the function (1.2.1) satisfies the four axioms. We shall prove that any function satisfying Axioms 1 through 4 must be of the form (1.2.1). For convenience we break the proof into several parts.

[We assume that all logarithms appearing in the proof are expressed to some fixed base greater than 1. No generality is lost by doing this; since $\log_a x = \log_a b \log_b x$, changing the logarithm base is equivalent to changing the constant C in (1.2.1).]

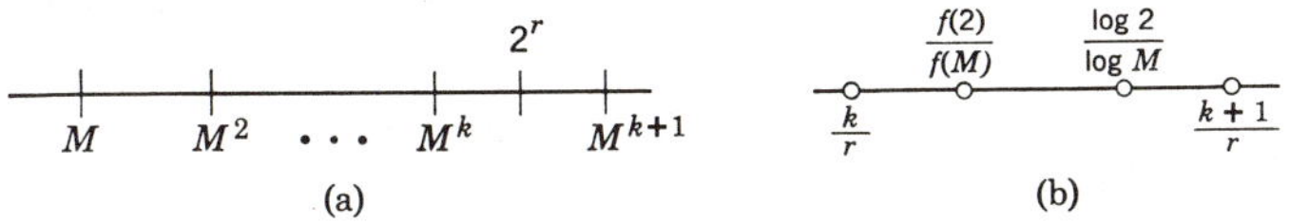

(a) (b)

Fig. 1.2.2. Proof of Theorem 1.2.1.

a. $f(M^k) = kf(M)$ for all positive integers M and k. This is readily established by induction, using Axiom 2. If M is an arbitrary but fixed positive integer, then (a) is immediately true for $k = 1$. Since $f(M^k) = f(M \cdot M^{k-1}) = f(M) + f(M^{k-1})$ by Axiom 2, the induction hypothesis that (a) is true for all integers up to and including $k - 1$ yields $f(M^k) = f(M) + (k - 1)f(M) = kf(M)$, which completes the argument.

b. $f(M) = C \log M$ $(M = 1, 2, \ldots)$ where C is a positive number. First let $M = 1$. We have $f(1) = f(1 \cdot 1) = f(1) + f(1)$ by Axiom 2, and hence $f(1) = 0$ as stated in (b). This agrees with our intuitive feeling that there should be no uncertainty associated with an experiment with only one possible outcome. Now let M be a fixed positive integer greater than 1. If r is an arbitrary positive integer, then the number 2^r lies somewhere between two powers of M, that is, there is a nonnegative integer k such that $M^k \leq 2^r < M^{k+1}$. (See Fig. 1.2.2a.) It follows from Axiom 1 that $f(M^k) \leq f(2^r) < f(M^{k+1})$, and thus we have from (a) that $kf(M) \leq rf(2) < (k + 1)f(M)$, or $k/r \leq f(2)/f(M) < (k + 1)/r$. The logarithm is a montone increasing function (as long as the base is greater than 1) and hence $\log M^k \leq \log 2^r < \log M^{k+1}$, from which we obtain $k \log M \leq r \log 2 < (k + 1) \log M$, or $k/r \leq (\log 2)/(\log M) < (k + 1)/r$. Since $f(2)/f(M)$ and $(\log 2)/(\log M)$ are both between k/r and $(k + 1)/r$, it follows that

$$\left| \frac{\log 2}{\log M} - \frac{f(2)}{f(M)} \right| < \frac{1}{r}. \qquad \text{(See Fig. 1.2.2b.)}$$

Since M is fixed and r is arbitrary, we may let $r \to \infty$ to obtain

$$(\log 2)/(\log M) = f(2)/f(M)$$

or $f(M) = C \log M$ where $C = f(2)/\log 2$. Note that C must be positive since $f(1) = 0$ and $f(M)$ increases with M.

c. $H(p, 1 - p) = -C[p \log p + (1 - p) \log (1 - p)]$ if p is a rational number. Let $p = r/s$ where r and s are positive integers. We consider

$$f(s) = H\Bigl(\underbrace{\frac{1}{s}, \ldots, \frac{1}{s}}_{r}, \underbrace{\frac{1}{s}, \ldots, \frac{1}{s}}_{s-r}\Bigr)$$

$$= H\left(\frac{r}{s}, \frac{s-r}{s}\right) + \frac{r}{s} f(r) + \frac{s-r}{s} f(s-r)$$

(by the grouping axiom 3).

Using (b) we have

$$C \log s = H(p, 1 - p) + Cp \log r + C(1 - p) \log (s - r).$$

Thus

$$\begin{aligned} H(p, 1 - p) &= -C[p \log r - \log s + (1 - p) \log (s - r)] \\ &= -C[p \log r - p \log s + p \log s \\ &\qquad - \log s + (1 - p) \log (s - r)] \\ &= -C\left[p \log \frac{r}{s} + (1 - p) \log \frac{s - r}{s}\right] \\ &= -C[p \log p + (1 - p) \log (1 - p)]. \end{aligned}$$

d. $H(p, 1 - p) = -C[p \log p + (1 - p) \log (1 - p)]$ for all p. This is an immediate consequence of (c) and the continuity axiom 4. For if p is any number between 0 and 1, it follows by continuity that

$$H(p, 1 - p) = \lim_{p' \to p} H(p', 1 - p').$$

In particular, we may allow p' to approach p through a sequence of rational numbers. We then have

$$\begin{aligned} \lim_{p' \to p} H(p', 1 - p') &= \lim_{p' \to p} [-C(p' \log p' + (1 - p') \log (1 - p'))] \\ &= -C[p \log p + (1 - p) \log (1 - p)]. \end{aligned}$$

e. $H(p_1, \ldots, p_M) = -C \sum_{i=1}^{M} p_i \log p_i (M = 1, 2, \ldots)$. We again proceed by induction. We have already established the validity of the above formula for $M = 1$ and 2. If $M > 2$, we use Axiom 3, which yields

$$\begin{aligned} H(p_1, \ldots, p_M) &= H(p_1 + \cdots + p_{M-1}, p_M) \\ &\quad + (p_1 + \cdots + p_{M-1}) H\left(\frac{p_1}{\sum_{i=1}^{M-1} p_i}, \ldots, \frac{p_{M-1}}{\sum_{i=1}^{M-1} p_i}\right) + p_M H(1). \end{aligned}$$

Assuming the formula valid for positive integers up to $M - 1$, we obtain:

$$
\begin{aligned}
H(p_1, \ldots, p_M) &= -C[(p_1 + \cdots + p_{M-1}) \log (p_1 + \cdots + p_{M-1}) \\
&\qquad + p_M \log p_M] \\
&\quad - C(p_1 + \cdots + p_{M-1}) \left[\frac{p_1}{\sum_{i=1}^{M-1} p_i} \log \left(\frac{p_1}{\sum_{i=1}^{M-1} p_i} \right) \right. \\
&\qquad \left. + \cdots + \frac{p_{M-1}}{\sum_{i=1}^{M-1} p_i} \log \left(\frac{p_{M-1}}{\sum_{i=1}^{M-1} p_i} \right) \right] + p_M(0) \\
&= -C \left[\left(\sum_{i=1}^{M-1} p_i \right) \log \left(\sum_{i=1}^{M-1} p_i \right) + p_M \log p_M \right] \\
&\quad - C \left[\sum_{i=1}^{M-1} p_i \log p_i - \left(\sum_{i=1}^{M-1} p_i \right) \log \sum_{i=1}^{M-1} p_i \right] \\
&= -C \sum_{i=1}^{M} p_i \log p_i.
\end{aligned}
$$

The proof is complete.

Unless otherwise specified, we shall assume $C = 1$ and take logarithms to the base 2. The units of H are sometimes called *bits* (a contraction of *binary digits*). Thus the units are chosen so that there is one bit of uncertainty associated with the toss of an unbiased coin. Biasing the coin tends to decrease the uncertainty, as seen in the sketch of $H(p, 1 - p)$ (Fig. 1.2.3).

We remark in passing that the average uncertainty of a random variable X does not depend on the values the random variable assumes, or on anything else except the probabilities associated with those values. The average uncertainty associated with the toss of an unbiased coin is not changed by adding the condition that the experimenter will be shot if the coin comes up tails.

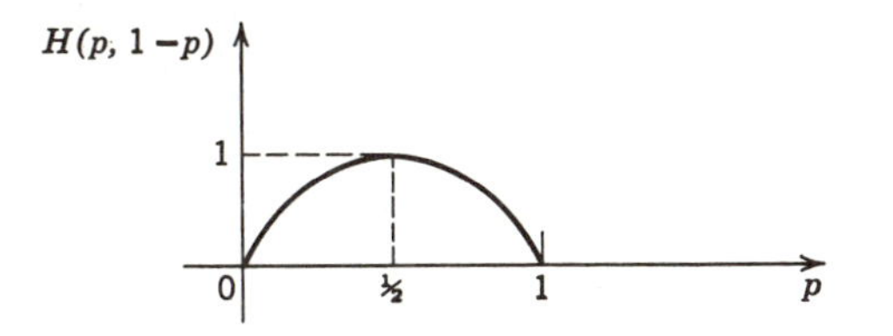

Fig. 1.2.3. $H(p, 1 - p) = -p \log p - (1 - p) \log (1 - p)$.

We also remark that the notations

$$-\sum_{i=1}^{M} p_i \log p_i \quad \text{and} \quad -\sum_{i=1}^{M} p(x_i) \log p(x_i)$$

will be used interchangeably.

Finally, we note that Theorem 1.2.1 implies that the function h must be of the form $h(p) = -C \log p$, provided we impose a requirement that h be continuous (see Problem 1.11).

An alternate approach to the construction of an uncertainty measure involves a set of axioms for the function h. If A and B are independent events, the occurrence of A should not affect the odds that B will occur. If $P(A) = p_1, P(B) = p_2$, then $P(AB) = p_1p_2$ so that the uncertainty associated with AB is $h(p_1p_2)$. The statement that A has occurred removes an amount of uncertainty $h(p_1)$, leaving an uncertainty $h(p_2)$ because of the independence of A and B. Thus we might require that

$$h(p_1p_2) = h(p_1) + h(p_2), \qquad 0 < p_1 \leq 1, \qquad 0 < p_2 \leq 1.$$

Two other reasonable requirements on h are that $h(p)$ be a decreasing function of p, and that h be continuous.

The only function satisfying the three above requirements is $h(p) = -C \log p$ (see Problem 1.12).

1.3. Three interpretations of the uncertainty function

As inspection of the form of the uncertainty function reveals, $H(p_1, \ldots, p_M)$ is a weighted average of the numbers $-\log p_1, \ldots, -\log p_M$, where the weights are the probabilities of the various values of X. This suggests that $H(p_1, \ldots, p_M)$ may be interpreted as the expectation of a random variable $W = W(X)$ which assumes the value $-\log p_i$ with probability p_i. If X takes the value x_i, then $W(X) = -\log P\{X = x_i\}$. Thus the values of $W(X)$ are the negatives of the logarithms of the probabilities associated with the values assumed by X. The expected value of $W(X)$ is $-\sum_{i=1}^{M} P\{X = x_i\} \log P\{X = x_i\} = H(X)$. An example may help to clarify the situation. If $p(x_1) = 2/3, p(x_2) = 1/6, p(x_3) = p(x_4) = 1/12$, then the random variable $W(X)$ has the following distribution:

$$P\left\{W = \log \frac{3}{2}\right\} = \frac{2}{3}; \qquad P\left\{W = \log 6\right\} = \frac{1}{6};$$

$$P\left\{W = \log 12\right\} = \frac{1}{12} + \frac{1}{12} = \frac{1}{6}.$$

The expectation of W is

$$\frac{2}{3} \log \frac{3}{2} + \frac{1}{6} \log 6 + \frac{1}{6} \log 12,$$

which is

$$H\left(\frac{2}{3}, \frac{1}{6}, \frac{1}{12}, \frac{1}{12}\right).$$

There is another interpretation of $H(X)$ which is closely related to the construction of codes. Suppose X takes on five values x_1, x_2, x_3, x_4, x_5 with probabilities .3, .2, .2, .15, and .15, respectively. Suppose that the value of X is to be revealed to us by someone who cannot communicate except by means of the words "yes" and "no." We therefore try to

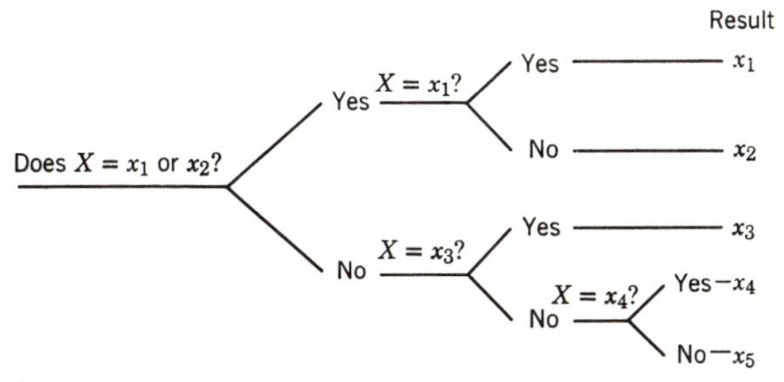

Fig. 1.3.1. A scheme of "yes or no" questions for isolating the value of a random variable.

arrive at the correct value by a sequence of "yes" and "no" questions, as shown in Fig. 1.3.1. If in fact X equals x_4, we receive a "no" answer to the question "Does $X = x_1$ or x_2?" We then ask "Does $X = x_3$?" and receive a "no" answer. The answer to "Does $X = x_4$?" is then "yes" and we are finished. Notice that the number of questions required to specify the value of X is a random variable that equals 2 whenever $X = x_1$, x_2, or x_3, and equals 3 when $X = x_4$ or x_5. The average number of questions required is

$$(0.3 + 0.2 + 0.2)2 + (0.15 + 0.15)3 = 2.3.$$

The essential content of the "noiseless coding theorem," to be proved in Chapter 2, is that the average number of "yes or no" questions needed to specify the value of X can never be less than the uncertainty of X. In this example, $H(X) = -0.3 \log 0.3 - 0.4 \log 0.2 - 0.3 \log 0.15 = 2.27$. We shall see in Chapter 2 that no scheme can be devised which on the average uses fewer questions than the one described above. We can, however, improve the efficiency of our guesswork if we assume that the experiment associated with X is performed independently n times and that we are allowed to wait until all n observations are recorded and then guess all n results simultaneously. For example, if X takes on two values, x_1 and x_2, with probabilities .7 and .3 respectively, one question is needed to specify the value of X. If, however, we are allowed to make guesses about the outcome of a joint experiment involving two independent observations of X, we may use the scheme shown in Fig. 1.3.2, which uses $0.49 + 2(0.21) + 3(0.21 + 0.09) = 1.81$ questions on the average, or 0.905 questions *per value of X*. We show later that by making guesses about longer and longer blocks, the average number of questions per value of X may be made to approach $H(X)$ as closely as desired. In no case can the

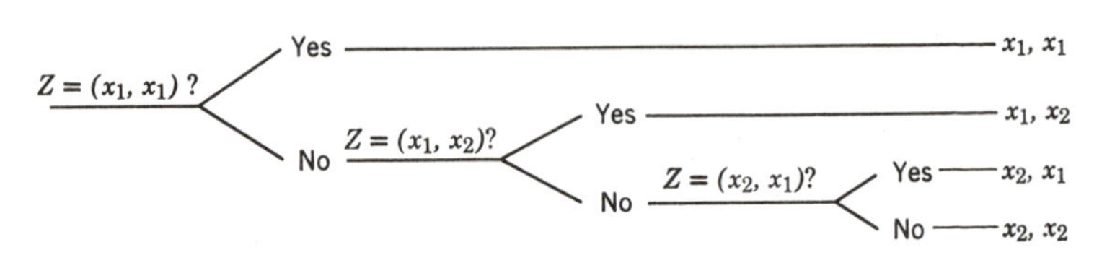

Fig. 1.3.2. "Yes or no" questions associated with two independent observations of X.

average number of questions be less than $H(X)$. Thus we may state the second interpretation of the uncertainty measure:

$H(X)$ = the minimum average number of "yes or no" questions required to determine the result of one observation of X.

There is a third interpretation of uncertainty which is related to the asymptotic behavior of a sequence of independent, identically distributed random variables. Let X be a random variable taking on the values $x_1, \ldots, x_M$ with probabilities $p_1, \ldots, p_M$ respectively. Suppose that the experiment associated with X is performed independently n times. In other words, we consider a sequence $X_1, \ldots, X_n$ of independent, identically distributed random variables, each having the same distribution as X. Let $f_i = f_i(X_1, \ldots, X_n)$ be the number of times that the symbol x_i occurs in the sequence $X_1, \ldots, X_n$; then f_i has a binomial distribution with parameters n and p_i. Given $\varepsilon > 0$ choose any positive number k such that $1/k^2 < \varepsilon/M$; fix ε and k for the remainder of the discussion. Let $\alpha = (\alpha_1, \ldots, \alpha_n)$ be a sequence of symbols, each α_i being one of the elements $x_1, \ldots, x_M$. We say that the sequence α is *typical* if

$$\left| \frac{f_i(\alpha) - np_i}{\sqrt{np_i(1 - p_i)}} \right| < k \quad \text{for all} \quad i = 1, 2, \ldots, M.$$

(The definition of a typical sequence depends on k, and hence on ε; however, the results we are seeking will be valid for any ε and appropriate k.) Thus in a typical sequence, each symbol x_i occurs approximately with its expected frequency np_i, the difference between the actual and expected frequency being of the order of $\sqrt{n}$ and hence small in comparison with n. We now show that, for large n, there are approximately 2^{Hn} typical sequences of length n, each with a probability of approximately 2^{-Hn}, where $H = H(X)$ is the uncertainty of X. More precisely, we prove the following result.

Theorem 1.3.1. 1. The set of nontypical sequences of length n has a total probability $<\varepsilon$.

2. There is a positive number A such that for each typical sequence α of length n,

$$2^{-nH-A\sqrt{n}} < P\{(X_1, \ldots, X_n) = \alpha\} < 2^{-nH+A\sqrt{n}}.$$

3. The number of typical sequences of length n is $2^{n(H+\delta_n)}$ where $\lim_{n\to\infty} \delta_n = 0$.

Proof. The probability that $(X_1, \ldots, X_n)$ is not a typical sequence is

$$P\left\{\left|\frac{f_i - np_i}{\sqrt{np_i(1-p_i)}}\right| \geq k \text{ for at least one } i\right\} \leq \sum_{i=1}^{M} P\left\{\left|\frac{f_i - np_i}{\sqrt{np_i(1-p_i)}}\right| \geq k\right\}.$$

But by Chebyshev's inequality,

$$P\left\{\left|\frac{f_i - np_i}{\sqrt{np_i(1-p_i)}}\right| \geq k\right\} = P\{|f_i - np_i| \geq k\sqrt{np_i(1-p_i)}\}$$

$$\leq \frac{E[|f_i - np_i|^2]}{k^2 np_i(1-p_i)} = \frac{1}{k^2} < \frac{\varepsilon}{M}$$

since f_i has a binomial distribution with parameters n and p_i, and hence has variance $np_i(1 - p_i)$. Thus

$$P\{(X_1, \ldots, X_n) \text{ is not typical}\} < \varepsilon,$$

proving the first part of the theorem. Now let $\alpha = (\alpha_1, \ldots, \alpha_n)$ be a typical sequence. We then have

$$np_i - k\sqrt{np_i(1-p_i)} < f_i(\alpha) < np_i + k\sqrt{np_i(1-p_i)}, \qquad i = 1, \ldots, M. \tag{1.3.1}$$

Also $P\{(X_1, \ldots, X_n) = \alpha\} = p_1^{f_1(\alpha)} p_2^{f_2(\alpha)} \cdots p_M^{f_M(\alpha)}$ because of the independence of the X_i. Writing $p(\alpha)$ for $P\{(X_1, \ldots, X_n) = \alpha\}$ we have $-\log p(\alpha) = -\sum_{i=1}^{M} f_i(\alpha) \log p_i$ or, using (1.3.1),

$$-\sum_{i=1}^{M}(np_i \log p_i - k\sqrt{np_i(1-p_i)} \log p_i) < -\log p(\alpha)$$

$$< -\sum_{i=1}^{M}(np_i \log p_i + k\sqrt{np_i(1-p_i)} \log p_i). \tag{1.3.2}$$

If we let $A = -k \sum_{i=1}^{M} \sqrt{p_i(1 - p_i)} \log p_i > 0$, then (1.3.2) yields

$$nH - A\sqrt{n} < -\log p(\alpha) < nH + A\sqrt{n}$$

or

$$2^{-nH-A\sqrt{n}} < p(\alpha) < 2^{-nH+A\sqrt{n}},$$

proving part 2. Finally, let S be the set of typical sequences. We have proved (part 1) that $1 - \varepsilon < P\{(X_1, \ldots, X_n) \in S\} \leq 1$ and (part 2) that for each $\alpha \in S$,

$$2^{-nH-A\sqrt{n}} < P\{(X_1, \ldots, X_n) = \alpha\} < 2^{-nH+A\sqrt{n}}.$$

Now if S is a subset of a finite sample space with a probability that is known to be between (say) $\frac{3}{4}$ and 1, and each point of S has a probability of between (say) $\frac{1}{16}$ and $\frac{3}{16}$, then the number of points of S is at least $\frac{3}{4}/\frac{3}{16} = 4$ and at most $1/\frac{1}{16} = 16$; for example, the fewest number of points would be obtained if S had the minimum probability $\frac{3}{4}$, and each point of S has the maximum probability $\frac{3}{16}$. By similar reasoning, the number of typical sequences is at least

$$(1 - \varepsilon)2^{nH-A\sqrt{n}} = 2^{nH-A\sqrt{n}+\log(1-\varepsilon)} = 2^{n[H-An^{-1/2}+n^{-1}\log(1-\varepsilon)]}$$

and at most $2^{nH+A\sqrt{n}} = 2^{n(H+An^{-1/2})}$; part 3 follows.

1.4. *Properties of the uncertainty function; joint and conditional uncertainty*

In this section we derive several properties of the uncertainty measure $H(X)$ and introduce the notions of *joint uncertainty* and *conditional uncertainty*.

We first note that $H(X)$ is always nonnegative since $-p_i \log p_i \geq 0$ for all i. We then establish a very useful lemma.

Lemma 1.4.1. Let $p_1, p_2, \ldots, p_M$ and $q_1, q_2, \ldots, q_M$ be arbitrary positive numbers with $\sum_{i=1}^{M} p_i = \sum_{i=1}^{M} q_i = 1$.

Then $-\sum_{i=1}^{M} p_i \log p_i \leq -\sum_{i=1}^{M} p_i \log q_i$, with equality if and only if $p_i = q_i$ for all i. For example, let

$$p_1 = \tfrac{1}{2}, \quad p_2 = p_3 = \tfrac{1}{4}, \quad q_1 = \tfrac{1}{3}, \quad q_2 = \tfrac{4}{9}, \quad q_3 = \tfrac{2}{9}.$$

Then

$$-\tfrac{1}{2}\log\tfrac{1}{2} - \tfrac{1}{4}\log\tfrac{1}{4} - \tfrac{1}{4}\log\tfrac{1}{4} = 1.5$$

and

$$-\tfrac{1}{2}\log\tfrac{1}{3} - \tfrac{1}{4}\log\tfrac{4}{9} - \tfrac{1}{4}\log\tfrac{2}{9} = 1.63.$$

Proof. For convenience we use natural logarithms instead of logarithms to the base 2. Since $\log_2 x = \log_2 e \log_e x$, the statement of the lemma is unaffected by this change. The logarithm is a convex function; in other words, $\ln x$ always lies below its tangent. By considering the tangent at $x = 1$, we obtain $\ln x \leq x - 1$ with equality if and only if $x = 1$ (see

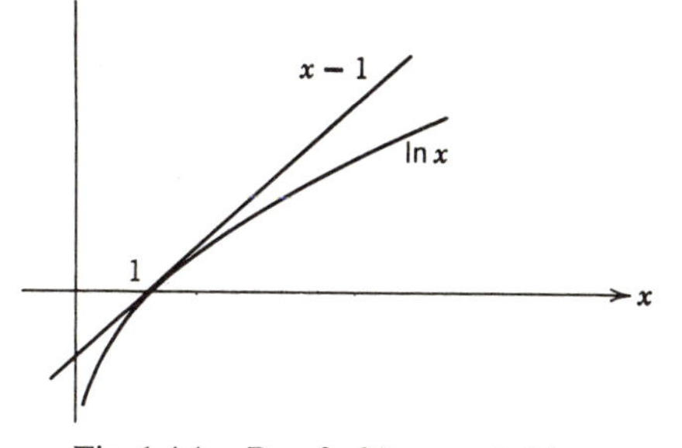

Fig. 1.4.1. Proof of Lemma 1.4.1.

Fig. 1.4.1). Thus $\ln (q_i/p_i) \leq q_i/p_i - 1$ with equality if and only if $p_i = q_i$. Multiplying the inequality by p_i and summing over i we obtain

$$\sum_{i=1}^{M} p_i \ln \frac{q_i}{p_i} \leq \sum_{i=1}^{M} (q_i - p_i) = 1 - 1 = 0$$

with equality if and only if $p_i = q_i$ for all i. Thus

$$\sum_{i=1}^{M} p_i \ln q_i - \sum_{i=1}^{M} p_i \ln p_i \leq 0$$

which proves the lemma.

We would expect that a situation involving a number of alternatives would be most uncertain when all possibilities are equally likely. The uncertainty function $H(X)$ does indeed have this property, as shown by the following theorem.

Theorem 1.4.2. $H(p_1, p_2, \ldots, p_M) \leq \log M$, with equality if and only if all $p_i = 1/M$.

Proof. The application of Lemma 1.4.1 with all $q_i = 1/M$ yields

$$-\sum_{i=1}^{M} p_i \log p_i \leq -\sum_{i=1}^{M} p_i \log \frac{1}{M} = \log M \sum_{i=1}^{M} p_i = \log M$$

with equality if and only if $p_i = q_i = 1/M$ for all i.

We turn now to the problem of characterizing the uncertainty associated with more than one random variable. Suppose X and Y are two discrete random variables associated with the same experiment. Let X and Y have a joint probability function

$$p(x_i, y_j) = P\{X = x_i \text{ and } Y = y_j\} = p_{ij} \qquad (i = 1, \dots, M;\ j = 1, \dots, L).$$

We therefore have an experiment with ML possible outcomes; the outcome $\{X = x_i,\ Y = y_j\}$ has probability $p(x_i, y_j)$. It is natural to define the *joint uncertainty* of X and Y as

$$H(X, Y) = -\sum_{i=1}^{M}\sum_{j=1}^{L} p(x_i, y_j) \log p(x_i, y_j).$$

Similarly, we define the joint uncertainty of n random variables $X_1, X_2, \dots, X_n$ as

$$H(X_1, X_2, \dots, X_n) = -\sum_{x_1, x_2, \dots, x_n} p(x_1, x_2, \dots, x_n) \log p(x_1, \dots, x_n),$$

where $p(x_1, x_2, \dots, x_n) = P\{X_1 = x_1, X_2 = x_2, \dots, X_n = x_n\}$ is the joint probability function of $X_1, X_2, \dots, X_n$. A connection between joint uncertainty and individual uncertainty is established by the following theorem.

Theorem 1.4.3. $H(X, Y) \le H(X) + H(Y)$, with equality if and only if X and Y are independent.

Proof. Since $p(x_i) = \sum_{j=1}^{L} p(x_i, y_j)$ and $p(y_j) = \sum_{i=1}^{M} p(x_i, y_j)$, we may write:

$$H(X) = -\sum_{i=1}^{M} p(x_i) \log p(x_i) = -\sum_{i=1}^{M}\sum_{j=1}^{L} p(x_i, y_j) \log p(x_i)$$

and

$$H(Y) = -\sum_{j=1}^{L} p(y_j) \log p(y_j) = -\sum_{i=1}^{M}\sum_{j=1}^{L} p(x_i, y_j) \log p(y_j).$$

Thus

$$\begin{aligned} H(X) + H(Y) &= -\sum_{i=1}^{M}\sum_{j=1}^{L} p(x_i, y_j)[\log p(x_i) + \log p(y_j)] \\ &= -\sum_{i=1}^{M}\sum_{j=1}^{L} p(x_i, y_j) \log p(x_i)p(y_j) \\ &= -\sum_{i=1}^{M}\sum_{j=1}^{L} p_{ij} \log q_{ij} \end{aligned}$$

where $q_{ij} = p(x_i)p(y_j)$.

We have $H(X, Y) = -\sum_{i=1}^{M}\sum_{j=1}^{L} p_{ij} \log p_{ij}$. We may now apply Lemma 1.4.1 to obtain:

$$-\sum_{i=1}^{M}\sum_{j=1}^{L} p_{ij} \log p_{ij} \le -\sum_{i=1}^{M}\sum_{j=1}^{L} p_{ij} \log q_{ij}$$

with equality if and only if $p_{ij} = q_{ij}$ for all i, j. The fact that we have a double instead of a single summation is of no consequence since the double sum can be transformed into a single sum by reindexing. We need only check that the hypothesis of the lemma is satisfied, that is,

$$\sum_{i=1}^{M}\sum_{j=1}^{L} q_{ij} = \sum_{i=1}^{M} p(x_i)\sum_{j=1}^{L} p(y_j) = 1\cdot 1 = 1.$$

Thus, $H(X, Y) \leq H(X) + H(Y)$ with equality if and only if $p(x_i, y_j) = p(x_i)p(y_j)$ for all i, j; that is, if and only if X and Y are independent.

An argument identical to that used above may be used to establish the following results.

Corollary 1.4.3.1. $H(X_1, \ldots, X_n) \leq H(X_1) + \cdots + H(X_n)$, with equality if and only if $X_1, \ldots, X_n$ are independent.

Corollary 1.4.3.2. $H(X_1, \ldots, X_n, Y_1, \ldots, Y_m) \leq H(X_1, \ldots, X_n) + H(Y_1, \ldots, Y_m)$, with equality if and only if the "random vectors" $X = (X_1, \ldots, X_n)$ and $Y = (Y_1, \ldots, Y_m)$ are independent, that is, if and only if

$$P\{X_1 = \alpha_1, \ldots, X_n = \alpha_n, Y_1 = \beta_1, \ldots, Y_m = \beta_m\}$$
$$= P\{X_1 = \alpha_1, \ldots, X_n = \alpha_n\}P\{Y_1 = \beta_1, \ldots, Y_m = \beta_m\}$$

for all $\alpha_1, \alpha_2, \ldots, \alpha_n, \beta_1, \beta_2, \ldots, \beta_m$.

We turn now to the idea of conditional uncertainty. Let two random variables X and Y be given. If we are given that $X = x_i$, then the distribution of Y is characterized by the set of conditional probabilities $p(y_j \mid x_i)$ $(j = 1, 2, \ldots, L)$. We therefore define the *conditional uncertainty of Y given that $X = x_i$* as

$$H(Y \mid X = x_i) = -\sum_{j=1}^{L} p(y_j \mid x_i) \log p(y_j \mid x_i).$$

We define the *conditional uncertainty of Y given X* as a weighted average of the uncertainties $H(Y \mid X = x_i)$, that is,

$$H(Y \mid X) = p(x_1)H(Y \mid X = x_1) + \cdots + p(x_M)H(Y \mid X = x_M)$$
$$= -\sum_{i=1}^{M} p(x_i)\sum_{j=1}^{L} p(y_j \mid x_i) \log p(y_j \mid x_i).$$

Using the fact that $p(x_i, y_j) = p(x_i)p(y_j \mid x_i)$ we have

$$H(Y \mid X) = -\sum_{i=1}^{M}\sum_{j=1}^{L} p(x_i, y_j) \log p(y_j \mid x_i).$$

We may define conditional uncertainties involving more than two random variables in a similar manner. For example,

$$H(Y, Z \mid X) = -\sum_{i,j,k} p(x_i, y_j, z_k) \log p(y_j, z_k \mid x_i)$$
$$= \text{the uncertainty about } Y \text{ and } Z \text{ given } X.$$

$$H(Z \mid X, Y) = -\sum_{i,j,k} p(x_i, y_j, z_k) \log p(z_k \mid x_i, y_j)$$
$$= \text{the uncertainty about } Z \text{ given } X \text{ and } Y.$$

$$H(Y_1, \ldots, Y_m \mid X_1, \ldots, X_n) =$$
$$-\sum_{x_1, \ldots, x_n, y_1, \ldots, y_m} p(x_1, \ldots, x_n, y_1, \ldots, y_m) \log p(y_1, \ldots, y_m \mid x_1, \ldots, x_n)$$
$$= \text{the uncertainty about } Y_1, \ldots, Y_m \text{ given } X_1, \ldots, X_n.$$

We now establish some intuitively reasonable properties of conditional uncertainty. If two random variables X and Y are observed, but only the value of X is revealed, we hope that the remaining uncertainty about Y is $H(Y \mid X)$. This is justified by the following result.

Theorem 1.4.4. $H(X, Y) = H(X) + H(Y \mid X) = H(Y) + H(X \mid Y)$.

Proof. The theorem follows directly from the definition of the uncertainty function. We write

$$H(X, Y) = -\sum_{i=1}^{M}\sum_{j=1}^{L} p(x_i, y_j) \log p(x_i, y_j)$$
$$= -\sum_{i=1}^{M}\sum_{j=1}^{L} p(x_i, y_j) \log p(x_i) p(y_j \mid x_i)$$
$$= -\sum_{i=1}^{M}\sum_{j=1}^{L} p(x_i, y_j) \log p(x_i) - \sum_{i=1}^{M}\sum_{j=1}^{L} p(x_i, y_j) \log p(y_j \mid x_i)$$
$$= -\sum_{i=1}^{M} p(x_i) \log p(x_i) + H(Y \mid X)$$
$$= H(X) + H(Y \mid X).$$

Similarly, we prove that $H(X, Y) = H(Y) + H(X \mid Y)$.

A corresponding argument may be used to establish various identities involving more than two random variables. For example,

$$H(X, Y, Z) = H(X) + H(Y \mid X) + H(Z \mid X, Y)$$
$$= H(X, Y) + H(Z \mid X, Y)$$
$$= H(X) + H(Y, Z \mid X);$$

$$H(X_1, \ldots, X_n, Y_1, \ldots, Y_m) = H(X_1, \ldots, X_n)$$
$$+ H(Y_1, \ldots, Y_m \mid X_1, \ldots, X_n).$$

It is also reasonable to hope that the revelation of the value of X should not increase the uncertainty about Y. This fact is expressed by Theorem 1.4.5.

Theorem 1.4.5. $H(Y \mid X) \leq H(Y)$, with equality if and only if X and Y are independent.

Proof. By Theorem 1.4.4, $H(X, Y) = H(X) + H(Y \mid X)$.

By Theorem 1.4.3, $H(X, Y) \leq H(X) + H(Y)$, with equality if and only if X and Y are independent. Theorem 1.4.5 then follows.

Similarly, it is possible to prove that

$$H(Y_1, \ldots, Y_m \mid X_1, \ldots, X_n) \leq H(Y_1, \ldots, Y_m)$$

with equality if and only if the random vectors

$$(X_1, \ldots, X_n) \quad \text{and} \quad (Y_1, \ldots, Y_m)$$

are independent.

A joint or conditional uncertainty may be interpreted as the expectation of a random variable, just as in the case of an individual uncertainty. For example,

$$H(X, Y) = -\sum_{i,j} p(x_i, y_j) \log p(x_i, y_j) = E[W(X, Y)]$$

where

$$W(X, Y) = -\log p(x_i, y_j) \quad \text{whenever } X = x_i \text{ and } Y = y_j$$

$$H(Y \mid X) = -\sum_{i,j} p(x_i, y_j) \log p(y_j \mid x_i) = E[W(Y \mid X)]$$

where

$$W(Y \mid X) = -\log p(y_j \mid x_i) \quad \text{whenever } X = x_i \text{ and } Y = y_j.$$

1.5. *The measure of information*

Consider the following experiment. Two coins are available, one unbiased and the other two-headed. A coin is selected at random and tossed twice, and the number of heads is recorded. We ask how much information is conveyed about the identity of the coin by the number of heads obtained. It is clear that the number of heads does tell us something about the nature of the coin. If less than 2 heads are obtained, the unbiased coin must have been used; if both throws resulted in heads, the evidence favors the two-headed coin. In accordance with the discussion at the beginning of this chapter, we decide to measure information as a *reduction in uncertainty*. To be specific, let X be a random variable that has the value 0 or 1 according as the unbiased or the two-headed coin is chosen. Let Y be the number of heads obtained in two tosses of the chosen coin. A diagram representing the experiment, together with various associated probability distributions, is shown in Fig. 1.5.1. The initial uncertainty about the identity of the coin is $H(X)$. After the number of heads is revealed, the uncertainty is $H(X \mid Y)$. We therefore define the *information conveyed about X by Y* as

$$I(X \mid Y) = H(X) - H(X \mid Y). \tag{1.5.1}$$

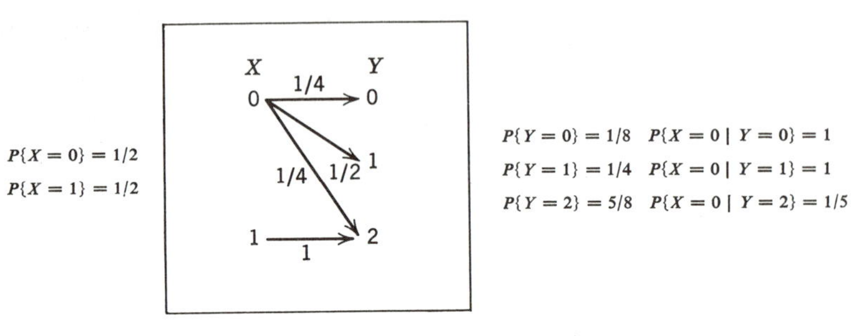

Fig. 1.5.1. A coin-tossing experiment.

The formula (1.5.1) is our general definition of the information conveyed about one random variable by another. In this case the numerical results are:

$$\begin{aligned}
H(X) &= \log 2 = 1, \\
H(X \mid Y) &= P\{Y = 0\}H(X \mid Y = 0) + P\{Y = 1\}H(X \mid Y = 1) \\
&\qquad + P\{Y = 2\}H(X \mid Y = 2) \\
&= \tfrac{1}{8}(0) + \tfrac{1}{4}(0) - \tfrac{5}{8}(\tfrac{1}{5}\log\tfrac{1}{5} + \tfrac{4}{5}\log\tfrac{4}{5}) \\
&= 0.45; \\
I(X \mid Y) &= 0.55.
\end{aligned}$$

The information measure may be interpreted as the expectation of a random variable, as was the uncertainty measure. We may write

$$\begin{aligned}
I(X \mid Y) &= H(X) - H(X \mid Y) \\
&= -\sum_{i=1}^{M}\sum_{j=1}^{L} p(x_i, y_j)\log p(x_i) + \sum_{i=1}^{M}\sum_{j=1}^{L} p(x_i, y_j)\log p(x_i \mid y_j) \\
&= -\sum_{i=1}^{M}\sum_{j=1}^{L} p(x_i, y_j)\log\frac{p(x_i)}{p(x_i \mid y_j)}.
\end{aligned}$$

Thus $I(X \mid Y) = E[U(X \mid Y)]$ where $X = x_i$, $Y = y_j$ implies $U(X \mid Y) = -\log[p(x_i)/p(x_i \mid y_j)]$; we may write

$$U(X \mid Y) = W(X) - W(X \mid Y)$$

where $W(X)$ and $W(X \mid Y)$ are as defined in Sections 1.3 and 1.4.

A suggestive notation that is sometimes used is

$$H(X) = E[-\log p(X)], \qquad H(X \mid Y) = E[-\log p(X \mid Y)],$$

$$I(X \mid Y) = E\left[-\log \frac{p(X)}{p(X \mid Y)}\right];$$

the expression $-\log p(X)$ is the random variable $W(X)$ defined previously, and similarly for the other terms.

The information conveyed about X by Y may also be interpreted as the difference between the minimum average number of "yes or no" questions required to determine the result of one observation of X before Y is observed and the minimum average number of such questions required after Y is observed. However, the fundamental significance of the information measure is its application to the reliable transmission of messages through noisy communication channels. We shall discuss this subject in great detail beginning with Chapter 3. At this point we shall be content to derive a few properties of $I(X \mid Y)$. By Theorem 1.4.5, $H(X \mid Y) \leq H(X)$ with equality if and only if X and Y are independent. Hence $I(X \mid Y) \geq 0$ with equality if and only if X and Y are independent. By Theorem 1.4.4, $H(X \mid Y) = H(X, Y) - H(Y)$; thus $I(X \mid Y) = H(X) + H(Y) - H(X, Y)$. But $H(X, Y)$ is the same as $H(Y, X)$, and therefore

$$I(X \mid Y) = I(Y \mid X).$$

The information measure thus has a surprising symmetry; the information conveyed about X by Y is the same as the information conveyed about Y by X. For the example of Fig. 1.5.1, we compute

$$H(Y) = -\tfrac{1}{8}\log\tfrac{1}{8} - \tfrac{1}{4}\log\tfrac{1}{4} - \tfrac{5}{8}\log\tfrac{5}{8} = 1.3;$$

$$H(Y \mid X) = P\{X = 0\}H(Y \mid X = 0) + P\{X = 1\}H(Y \mid X = 1)$$

$$= \tfrac{1}{2}H(\tfrac{1}{4}, \tfrac{1}{2}, \tfrac{1}{4}) + \tfrac{1}{2}H(1) = 0.75.$$

$$I(Y \mid X) = H(Y) - H(Y \mid X) = 0.55, \qquad \text{as before.}$$

When the conditional probabilities $p(y_j \mid x_i)$ are specified at the beginning of a problem, as they are here, it is usually easier to compute the information using $H(Y) - H(Y \mid X)$ rather than $H(X) - H(X \mid Y)$.

More generally, we may define the information conveyed about a set of random variables by another set of random variables. If $X_1, \ldots, X_n, Y_1, \ldots, Y_m$ are random variables, the *information conveyed about* $X_1, \ldots, X_n$ *by* $Y_1, \ldots, Y_m$ is defined as

$$I(X_1, \ldots, X_n \mid Y_1, \ldots, Y_m)$$
$$= H(X_1, \ldots, X_n) - H(X_1, \ldots, X_n \mid Y_1, \ldots, Y_m).$$

Proceeding as before, we obtain

$$\begin{aligned} I(X_1, \ldots, X_n \mid Y_1, \ldots, Y_m) &= H(X_1, \ldots, X_n) \\ &\quad + H(Y_1, \ldots, Y_m) - H(X_1, \ldots, X_n, Y_1, \ldots, Y_m) \\ &= I(Y_1, \ldots, Y_m \mid X_1, \ldots, X_n). \end{aligned}$$

1.6. Notes and remarks

The axioms for the uncertainty measure given in the text are essentially those of Shannon (1948). A weaker set of axioms which determine the same uncertainty function has been given by Fadiev (1956); Fadiev's axioms are described in Feinstein (1958). The weakest set of axioms known at this time may be found in Lee (1964). The properties of the uncertainty and information measures developed in Sections 1.4 and 1.5 are due to Shannon (1948); a similar discussion may be found in Feinstein (1958). The interpretation of $H(X)$ in terms of typical sequences was discovered by Shannon (1948). The notion of typical sequence is used by Wolfowitz (1961) as the starting point for his proofs of coding theorems; Wolfowitz's term "π-sequence" corresponds essentially to our "typical sequence."

We have thus far required that the arguments $p_1, \ldots, p_M$ of H be strictly positive. It is convenient, however, to extend the domain so that zero values are allowed; we may do this by writing

$$H(p_1, \ldots, p_M) = -\sum_{i=1}^{M} p_i \log p_i \qquad \left(\text{all } p_i \geq 0, \quad \sum_{i=1}^{M} p_i = 1\right)$$

with the proviso that an expression which appears formally as $0 \log 0$ is defined to be 0. This convention preserves the continuity of H. In Lemma 1.4.1, we may allow some of the p_i or q_i to be 0 if we interpret $0 \log 0$ as 0 and $-a \log 0$ as $+\infty$ if $a > 0$.

The quantity $H(X)$, which we have referred to as the "uncertainty of X," has also been called the "entropy" or "communication entropy" of X.

PROBLEMS

1.1 The inhabitants of a certain village are divided into two groups A and B. Half the people in group A always tell the truth, three-tenths always lie, and two-tenths always refuse to answer. In group B, three-tenths of the people are truthful, half are liars, and two-tenths always refuse to answer. Let p be the probability that a person selected at random will belong to group A. Let $I = I(p)$ be the information conveyed about a person's truth-telling status by specifying his group membership. Find the maximum possible value of I and the percentage of people in group A for which the maximum occurs.

1.2 A single unbiased die is tossed once. If the face of the die is 1, 2, 3, or 4, an unbiased coin is tossed once. If the face of the die is 5 or 6, the coin is tossed twice. Find the information conveyed about the face of the die by the number of heads obtained.

1.3 Suppose that in a certain city, $\frac{3}{4}$ of the high-school students pass and $\frac{1}{4}$ fail. Of those who pass, 10 percent own cars, while 50 percent of the failing students own cars. All of the car-owning students belong to fraternities, while 40 percent of those who do not own cars but pass, as well as 40 percent of those who do not own cars but fail, belong to fraternities.

a. How much information is conveyed about a student's academic standing by specifying whether or not he owns a car?

b. How much information is conveyed about a student's academic standing by specifying whether or not he belongs to a fraternity?

c. If a student's academic standing, car-owning status, and fraternity status are transmitted by three successive binary digits, how much information is conveyed by each digit?

1.4 Establish the following:

a. $$H(Y, Z \mid X) \le H(Y \mid X) + H(Z \mid X)$$

with equality if and only if $p(y_j, z_k \mid x_i) = p(y_j \mid x_i)p(z_k \mid x_i)$ for all i, j, k.

b. $$H(Y, Z \mid X) = H(Y \mid X) + H(Z \mid X, Y).$$

c. $$H(Z \mid X, Y) \le H(Z \mid X)$$

with equality if and only if $p(y_j, z_k \mid x_i) = p(y_j \mid x_i)p(z_k \mid x_i)$ for all i, j, k.

Note that these results hold if the random variables are replaced by random vectors, that is,

$$X = (X_1, \ldots, X_n), \qquad Y = (Y_1, \ldots, Y_m), \qquad Z = (Z_1, \ldots, Z_r).$$

The condition $p(y_j, z_k \mid x_i) = p(y_j \mid x_i)p(z_k \mid x_i)$ becomes

$$p(y_1, \ldots, y_m, z_1, \ldots, z_r \mid x_1, \ldots, x_n) \\ = p(y_1, \ldots, y_m \mid x_1, \ldots, x_n)p(z_1, \ldots, z_r \mid x_1, \ldots, x_n)$$

for all $x_1, \ldots, x_n, y_1, \ldots, y_m, z_1, \ldots, z_r$. This condition is sometimes expressed by saying that *Y and Z are conditionally independent given X.*

1.5 Use Lemma 1.4.1 to prove the *inequality of the arithmetic and geometric means*:

Let $x_1, \ldots, x_n$ be arbitrary positive numbers; let $a_1, \ldots, a_n$ be positive numbers whose sum is unity. Then

$$x_1^{a_1}x_2^{a_2} \cdots x_n^{a_n} \le \sum_{i=1}^{n} a_i x_i$$

with equality if and only if all x_i are equal.

Note that the inequality still holds if some of the a_i are allowed to be zero (keeping $\sum_{i=1}^{n} a_i = 1$) since if $a = 0$ we may multiply the left-hand side of the inequality by x^a and add ax to the right-hand side without affecting the result. However, the condition for equality becomes:

All x_i corresponding to positive a_i are equal.

1.6 A change that tends in the following sense to equalize a set of probabilities $p_1, \ldots, p_M$ always results in an increase in uncertainty:

Suppose $p_1 > p_2$. Define

$$p_1' = p_1 - \Delta p$$
$$p_2' = p_2 + \Delta p$$
$$p_i' = p_i, \; i = 3, \ldots, M$$

where $\Delta p > 0$ and $p_1 - \Delta p \geq p_2 + \Delta p$. Show that $H(p_1', \ldots, p_M') > H(p_1, \ldots, p_M)$.

1.7 (Feinstein 1958). Let $A = [a_{ij}]$ be a doubly stochastic matrix, that is, $a_{ij} \geq 0$ for all i, j; $\sum_{j=1}^{M} a_{ij} = 1, i = 1, \ldots, M$; $\sum_{i=1}^{M} a_{ij} = 1, j = 1, \ldots, M$. Given a set of probabilities $p_1, \ldots, p_M$, define a new set of probabilities $p_1', \ldots, p_M'$ by

$$p_i' = \sum_{j=1}^{M} a_{ij} p_j, \qquad i = 1, 2, \ldots, M.$$

Show that $H(p_1', \ldots, p_M') \geq H(p_1, \ldots, p_M)$ with equality if and only if $(p_1', \ldots, p_M')$ is a rearrangement of $(p_1, \ldots, p_M)$. Show also that Problem 1.6 is a special case of this result.

1.8 Given a discrete random variable X with values $x_1, \ldots, x_M$, define a random variable Y by $Y = g(X)$ where g is an arbitrary function. Show that $H(Y) \leq H(X)$. Under what conditions on the function g will there be equality?

1.9 Let X and Y be random variables with numerical values $x_1, \ldots, x_M$; $y_1, \ldots, y_L$ respectively. Let $Z = X + Y$.

a. Show that $H(Z \mid X) = H(Y \mid X)$: hence if X and Y are independent, $H(Z \mid X) = H(Y)$ so that $H(Y) \leq H(Z)$, and similarly $H(X) \leq H(Z)$.

b. Give an example in which $H(X) > H(Z)$, $H(Y) > H(Z)$.

1.10 Prove the *generalized grouping axiom*.

$$\begin{aligned} H(p_1, \ldots, p_{r_1}; p_{r_1+1}, \ldots, p_{r_2}; \ldots; p_{r_{k-1}+1}, \ldots, p_{r_k}) \\ = H(p_1 + \cdots + p_{r_1}, p_{r_1+1} + \cdots + p_{r_2}, \ldots, p_{r_{k-1}+1} + \cdots + p_{r_k}) \\ + \sum_{i=1}^{k} (p_{r_{i-1}+1} + \cdots + p_{r_i}) H\left(\frac{p_{r_{i-1}+1}}{\sum_{j=r_{i-1}+1}^{r_i} p_j}, \ldots, \frac{p_{r_i}}{\sum_{j=r_{i-1}+1}^{r_i} p_j} \right) \end{aligned}$$

1.11 Show that if $h(p)$, $0 < p \leq 1$, is a continuous function such that $\sum_{i=1}^{M} p_i h(p_i) = -C \sum_{i=1}^{M} p_i \log p_i$ for all M and all $p_1, \ldots, p_M$ such that $p_i > 0$, $\sum_{i=1}^{M} p_i = 1$, then $h(p) = -C \log p$.

1.12 Given a function $h(p)$, $0 < p \leq 1$, satisfying

a. $h(p_1 p_2) = h(p_1) + h(p_2), \qquad 0 < p_1 \leq 1, \quad 0 < p_2 \leq 1.$

b. $h(p)$ is a monotonically decreasing and continuous function of p, $0 < p \leq 1$.

Show that the only function satisfying the given conditions is $h(p) = -C \log_b p$ where $C > 0$, $b > 1$.

CHAPTER TWO

Noiseless Coding

2.1. *Introduction*

Our first application of the notion of uncertainty introduced in Chapter 1 will be to the problem of efficient coding of messages to be sent over a "noiseless" channel, that is, a channel allowing perfect transmission from input to output. Thus we do not consider the problem of error correction; our only concern is to maximize the number of messages that can be sent over the channel in a given time. To be specific, assume that the messages to be transmitted are generated by a random variable X whose values are $x_1, \ldots, x_M$. A noiseless channel may be thought of intuitively as a device that accepts an input from a specified set of "code characters" $a_1, \ldots, a_D$ and reproduces the input symbol at the output with no possibility of error. (The formal definition of an information channel will be deferred until Chapter 3; it will not be needed here.) If the symbols x_i are to be communicated properly, each x_i must be represented by a sequence of symbols chosen from the set $\{a_1, \ldots, a_D\}$. Thus, we assign a sequence of code characters to each x_i; such a sequence is called a "code word." Since the problem of error correction does not arise, efficient communication would involve transmitting a given message in the shortest possible time. If the rate at which the symbols a_j can be sent through the channel is fixed, the requirement of efficiency suggests that we make the code words as short as possible. In calculating the long-run efficiency of communication, the *average* length of a code word is of interest; it is this quantity which we choose to minimize.

To summarize, the ingredients of the noiseless coding problem are:

1. A *random variable X*, taking on the values $x_1, \ldots, x_M$ with probabilities $p_1, \ldots, p_M$ respectively. X is to be observed independently over and over again, thus generating a sequence whose components belong to the set $\{x_1, \ldots, x_M\}$; such a sequence is called a *message*.

2. A set $\{a_1, \ldots, a_D\}$ called the set of *code characters* or the *code alphabet;* each symbol x_i is to be assigned a finite sequence of code characters called the *code word* associated with x_i (for example, x_1 might correspond to a_1a_2, and x_2 to $a_3a_7a_3a_8$). The collection of all code words is called a *code*. The code words are assumed to be distinct.

3. The *objective* of noiseless coding is to minimize the average code-word length. If the code word associated with x_i is of length n_i, $i = 1, 2, \ldots, M$, we will try to find codes that minimize $\sum_{i=1}^{M} p_i n_i$.

2.2. *The problem of unique decipherability*

It becomes clear very quickly that some restriction must be placed on the assignment of code words. For example, consider the following binary code:

x_1	0
x_2	010
x_3	01
x_4	10

The binary sequence 010 could correspond to any one of the three messages x_2, x_3x_1, or x_1x_4. Thus the sequence 010 cannot be decoded accurately. We would like to rule out ambiguities of this type; hence the following definition.

A code is *uniquely decipherable* if every finite sequence of code characters corresponds to at most one message.

One way to insure unique decipherability is to require that no code word be a "prefix" of another code word. If A_1 and A_2 are finite (nonempty) sequences of code characters then the *juxtaposition* of A_1 and A_2, written A_1A_2, is the sequence formed by writing A_1 followed by A_2. We say that the sequence A is a *prefix* of the sequence B if B may be written as AC for some sequence C.

A code having the property that no code word is a prefix of another code word is said to be *instantaneous*. The code below is an example of an instantaneous code.

x_1	0
x_2	100
x_3	101
x_4	11

Notice that the sequence 11111 does not correspond to any message; such a sequence will never appear and thus can be disregarded. Before turning to the problem of characterizing uniquely decipherable codes, we note that *every instantaneous code is uniquely decipherable, but not conversely*. For given a finite sequence of code characters of an instantaneous code, proceed from the left until a code word W is formed. (If there is no such word, the unique decipherability condition is vacuously satisfied.) Since W is not the prefix of another code word, W must correspond to the first symbol of the message. Now continue until another code word is formed. The process may be repeated until the end of the message.

For example, in the instantaneous code {0, 100, 101, 11} above, the sequence 101110100101 is decoded as $x_3x_4x_1x_2x_3$.

Now consider the code

$$\begin{array}{ll} x_1 & 0 \\ x_2 & 01 \end{array}$$

This code is not instantaneous since 0 is a prefix of 01. The code is uniquely decipherable, however, since any sequence of code characters may be decoded by noting the position of the ones in the sequence. For example, the sequence 0010000101001 is decoded as $x_1x_2\,x_1x_1x_1\,x_2x_2x_1x_2$. The word "instantaneous" refers to the fact that a sequence of code characters may be decoded step by step. If, proceeding from the left, W is the first word formed, we known immediately that W is the first word of the message. In a uniquely decipherable code which is not instantaneous, we may have to wait a long time before we know the identity of the first word. For example, if in the code

$$\begin{array}{ll} x_1 & 0 \\ x_2 & \underbrace{00\cdots001}_{n} \end{array}$$

we received the sequence $\underbrace{00\cdots001}_{n+1}$ we would have to wait until the end of the sequence to find out that the corresponding message starts with x_1.

We now present a testing procedure that can always be used to determine whether or not a code is uniquely decipherable. To see how the procedure works, consider the code of Fig. 2.2.1, which is not instantaneous but could conceivably be uniquely decipherable. We construct a sequence of sets $S_0, S_1, S_2, \ldots$, as follows. Let S_0 be the original set of code words. To form S_1, we look at all pairs of code words in S_0. If a code word W_i is a prefix of another code word W_j, that is, $W_j = W_iA$, we place the suffix A in S_1. In the above code, the word a is a prefix of the word abb, so that bb is one of the members of S_1. In general, to form S_n, $n > 1$, we compare S_0 and S_{n-1}. If a code word $W \in S_0$ is a prefix of a sequence $A = WB \in S_{n-1}$, the suffix B is placed in S_n, and if a sequence $A' \in S_{n-1}$ is a prefix of a code word $W' = A'B' \in S_0$, we place the suffix $B' \in S_n$. The sets S_n, $n = 0, 1, \ldots$, for the code of Fig. 2.2.1 are shown in Fig. 2.2.2.

x_1	a
x_2	c
x_3	ad
x_4	abb
x_5	bad
x_6	deb
x_7	$bbcde$

Fig. 2.2.1. A code.

We shall prove

Theorem 2.2.1. A code is uniquely deciperable if and only if none of the sets $S_1, S_2, S_3, \ldots$ contains a code word, that is, a member of S_0.

In Fig. 2.2.1, the code word *ad* belongs to the set S_5; hence, according to Theorem 2.2.1, the code is not uniquely decipherable. In fact, the sequence *abbcdebad* is ambiguous, having the two possible interpretations *a*, *bbcde*, *bad* and *abb*, *c*, *deb*, *ad*. A systematic method of constructing ambiguous sequences will be given as part of the proof of Theorem 2.2.1.

S_0	S_1	S_2	S_3	S_4	S_5	S_6	S_7
a	*d*	*eb*	*de*	*b*	*ad*	*d*	*eb*
c	*bb*	*cde*			*bcde*		
ad							
abb							
bad				S_n empty $(n > 7)$			
deb							
bbcde							

Fig. 2.2.2. Test for unique decipherability.

Proof. First suppose that the code is not uniquely decipherable, so that there is a sequence of code characters which is ambiguous, that is, corresponds to more than one possible message. Pick an ambiguous sequence G with the smallest possible number of symbols. Then G may be written in at least two distinct ways:

$$G = W_1 W_2 \cdots W_n = W_1' W_2' \cdots W_m'$$

where the W_i and W_j' are code words (assume $n \geq 2$, $m \geq 2$; otherwise the conclusion is immediate).

Now define the *index* of the word W_i (respectively W_j') in G as the number of letters in $W_1 W_2 \cdots W_{i-1}$ (respectively $W_1' \cdots W_{j-1}'$), $i = 2, \ldots, n$, $j = 2, \ldots, m$. The minimality of the number of letters of G implies that the indices of $W_2, \ldots, W_n, W_2', \ldots, W_m'$ are distinct. If W_1 has fewer letters than W_1', define the index of W_1 to be -1 and that of W_1' to be 0; reverse this procedure if W_1' has fewer letters than W_1. (Note that W_1' cannot equal W_1 for if so, $W_2' \cdots W_m' = W_2 \cdots W_n$, contradicting the minimality of the number of letters of G.) Let $U_1, U_2, \ldots, U_{n+m}$ be the words of G, arranged in order of increasing index. If $j < i$ and index $U_i >$ index U_j, but index $U_{i+1} <$ index $U_j +$ the number of letters in U_j, we say that U_i is *embedded* in U_j. We claim that for each $i = 3, \ldots, n + m$, either U_i is embedded in some U_j, $j < i$, or the subsequence A_i of G which begins with the first letter of U_i and ends with the letter immediately preceding the first letter of U_{i+1}, is in one of the sets S_n, $n \geq 1$. (The sequence A_{n+m} is defined to be U_{n+m} itself.) The claim is true for $i = 3$ by inspection. The various possibilities are indicated in Fig. 2.2.3a and b. If the claim has been verified for $i \leq r$, consider U_{r+1}. If U_{r+1} is not embedded in some U_j, $j < r + 1$, we have:

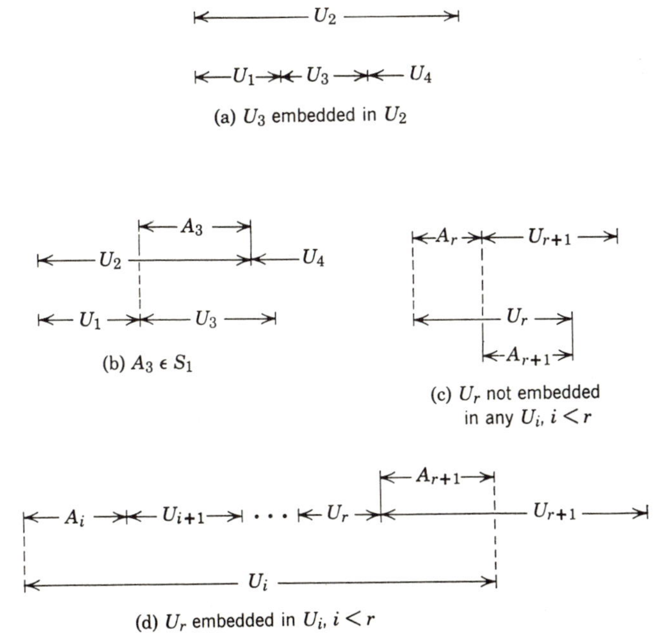

Fig. 2.2.3. Proof of Theorem 2.2.1.

CASE 1. U_r is not embedded in any U_i, $i < r$ (see Fig. 2.2.3c). By induction hypothesis, $A_r \in$ some S_n, $n \geq 1$. But then $A_{r+1} \in S_{n+1}$.

CASE 2. U_r is embedded in some U_i, $i < r$ (see Fig. 2.2.3d). Then $U_i = A_i U_{i+1} \cdots U_r A_{r+1}$. By induction hypothesis, $A_i \in$ some S_n, $n \geq 1$. By definition of the sets S_n, $U_{i+1} \cdots U_r A_{r+1} \in S_{n+1}$, $U_{i+2} \cdots U_r A_{r+1} \in S_{n+2}, \ldots, U_r A_{r+1} \in S_{n+r-i}$, $A_{r+1} \in S_{n+r-i+1}$.

Now U_{n+m} cannot be embedded in any U_i; hence $A_{n+m} = U_{n+m} \in$ some S_n, $n \geq 1$, and the first half of the theorem is proved.

Conversely, suppose that one of the sets S_k, $k \geq 1$, contains a code word. Let n be the smallest integer ≥ 1 such that S_n contains a code word W. If we retrace the steps by which W arrived in S_n, we obtain a sequence

$$A_0, W_0, A_1, W_1, \ldots, A_n, W_n$$

such that $A_0, W_0, W_1, \ldots, W_n$ are code words, $A_1, \ldots, A_n$ are sequences of code characters such that $A_i \in S_i$, $i = 1, \ldots, n$, $A_n = W_n$, $W_0 = A_0 A_1$, and for each $i = 1, 2, \ldots, n-1$ either $A_i = W_i A_{i+1}$ or $W_i = A_i A_{i+1}$. For example, for the code of Fig. 2.2.2, we obtain

$A_0 = a$	$A_1 = bb \in S_1$	$A_2 = cde \in S_2$	$A_3 = de \in S_3$	$A_4 = b \in S_4$	$A_5 = ad \in S_5$
$W_0 = abb$	$W_1 = bbcde$	$W_2 = c$	$W_3 = deb$	$W_4 = bad$	$W_5 = ad$

We now give a systematic way of constructing an ambiguous sequence. We construct two sequences, one starting with A_0W_1 and the other with W_0. The sequences are formed in accordance with the following rules.

Having placed W_i at the end of one of the sequences:

CASE 1. $A_i = W_iA_{i+1}$. Add W_{i+1} at the end of the sequence containing W_i.

CASE 2. $W_i = A_iA_{i+1}$. Add W_{i+1} at the end of the sequence not containing W_i. Continue until W_n is reached.

We shall illustrate this procedure for the code of Fig. 2.2.2.

$$A_0W_1 = abbcde, \quad W_0 = abb$$

$$W_1 = A_1A_2 \text{ so form } W_0W_2 = abbc.$$

(Notice that the sequence A_0W_1 is longer than the sequence W_0, hence W_2 is added to the shorter sequence.)

$$A_2 = W_2A_3 \text{ so form } W_0W_2W_3 = abbcdeb.$$

(After the addition of W_2, the sequence beginning with W_0 is still the shorter, so W_3 is added to that sequence.)

$$W_3 = A_3A_4 \text{ so form } A_0W_1W_4 = abbcdebad.$$

(After the addition of W_3, the sequence beginning with W_0 exceeds the sequence beginning with A_0 in length, and thus W_4 is added to the latter sequence.)

$$W_4 = A_4A_5 \text{ so form } W_0W_2W_3W_5 = abbcdebad.$$

The sequence $abbcdebad = A_0W_1W_4 = W_0W_2W_3W_5$ is ambiguous.

We now show that the procedure outlined above always yields an ambiguous sequence. We may establish by induction that after the word W_i is assigned ($i = 1, \dots, n-1$), one of the sequences is a prefix of the other. By inspection, this is true after W_1 is assigned. The inductive step is accomplished by a tedious case-by-case analysis. For example, suppose that W_i is assigned to the first sequence and that after W_i is written down, sequence 2 is a prefix of sequence 1. In addition, suppose that W_{i-1} was assigned to sequence 2 (see Fig. 2.2.4). Then necessarily $W_{i-1} = A_{i-1}A_i$ and $W_i = A_iA_{i+1}$, therefore W_{i+1} is assigned to sequence 2. If $W_{i+1} = A_{i+1}A_{i+2}$ then A_{i+1} is a prefix of W_{i+1}. If $A_{i+1} = W_{i+1}A_{i+2}$ then W_{i+1} is a prefix of A_{i+1}. In either case, one of the sequences is still a prefix of the other. The other cases may be handled in a similar fashion. Now after W_{n-1} is assigned we have either $W_{n-1} = A_{n-1}W_n$ or $A_{n-1} = W_{n-1}W_n$. Another case-by-case analysis shows that after W_n is assigned, the two sequences are identical. Since $W_0 \neq A_0$, we have produced an ambiguous sequence.

We remark that since the sequences in the sets S_i, $i \geq 0$, cannot be longer than the longest code word, only finitely many of the S_i can be distinct. Thus eventually the S_n must exhibit a periodicity; that is, there must be integers N and k such that $S_i = S_{i+k}$ for $i \geq N$ (it may happen that S_i is empty for $i \geq N$; in fact a code is instantaneous if and only if S_i is

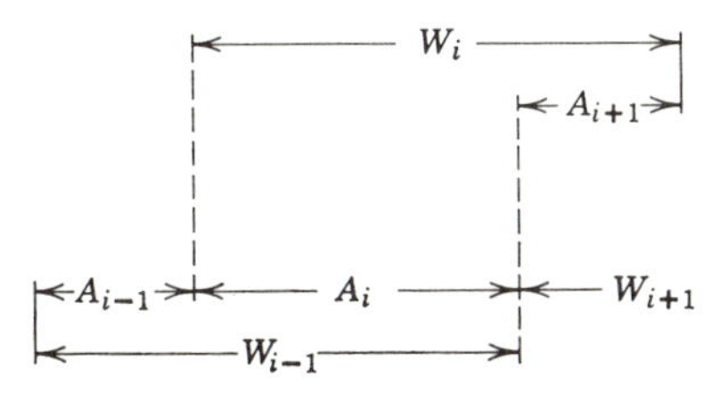

Fig. 2.2.4. Proof of Theorem 2.2.1.

empty for $i \geq 1$). Thus the testing procedure to determine unique decipherability must terminate in a finite number of steps, and an upper bound on the number of steps required may be readily calculated for a given code.

2.3. *Necessary and sufficient conditions for the existence of instantaneous codes*

As we have seen in the previous section, an instantaneous code is quite easy to decode compared to the general uniquely decipherable code. Moreover, we shall prove in Section 2.6 that for the purpose of solving the noiseless coding problem, we may without loss of generality restrict our attention to instantaneous codes. Thus it would be desirable to examine the properties of such codes. We start by posing the following problem. Given a set of symbols $x_1, x_2, \ldots, x_M$, a code alphabet $a_1, a_2, \ldots, a_D$, and a set of positive integers $n_1, n_2, \ldots, n_M$, is it possible to construct an instantaneous code such that n_k is the length of the code word corresponding to x_k? For example, if $M = 3$, $D = 2$, $n_1 = 1$, $n_2 = 2$, $n_3 = 3$, a possible code is $\{0, 10, 110\}$. If $M = 3$, $D = 2$, $n_1 = n_2 = n_3 = 1$, no uniquely decipherable code, instantaneous or otherwise, will meet the specifications. The complete solution to the problem is provided by the following theorem.

Theorem 2.3.1. An instantaneous code with word lengths $n_1, n_2, \ldots, n_M$ exists if and only if $\sum_{i=1}^{M} D^{-n_i} \leq 1$ (D = size of the code alphabet).

Proof. To prove the theorem we make use of the idea of a *tree of order D and size k*, which is simply a system of points and lines such that

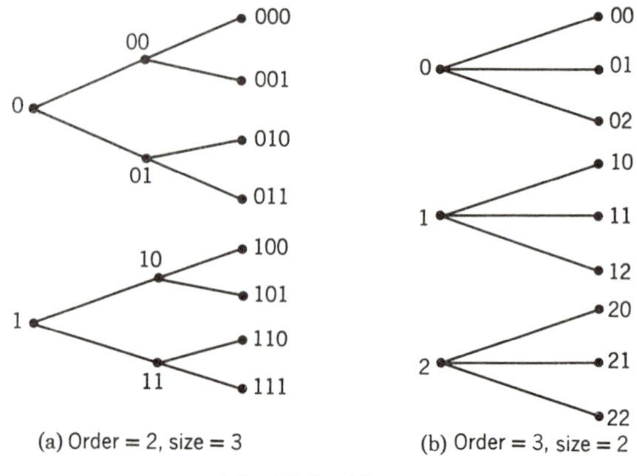

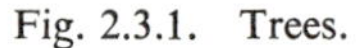
Fig. 2.3.1. Trees.

each sequence s of length $\leq k$ formed from the symbols $\{0, 1, \ldots, D-1\}$ is represented by a distinct point V_s, and such that if the sequence s' is formed by adding a single digit to s, then V_s and V_s' are connected by a line. Typical trees are shown in Fig. 2.3.1.

Now suppose we are given a base D instantaneous code with word lengths $n_1, n_2, \ldots, n_M$. We may assume without loss of generality that the code alphabet is $\{0, 1, \ldots, D-1\}$. Assume for simplicity that $n_1 \leq n_2 \leq \cdots \leq n_M$. Each code word may be identified with a point on the tree of order D and size n_M. For example, the binary code $\{0, 10, 111\}$ may be represented as shown in Fig. 2.3.2. Since no code word can be a prefix of another code word, once a point P on the tree is reserved for a particular code word, no other code word can correspond to any point on the branches emanating from P. In Fig. 2.3.2, the dashed lines indicate the forbidden parts of the tree. By construction of the tree, a code word of

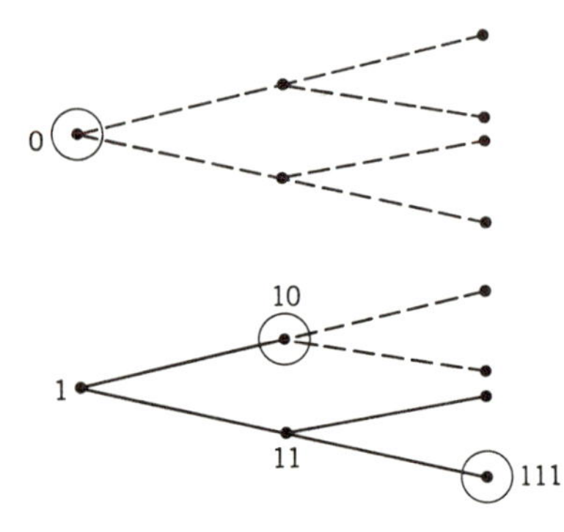

Fig. 2.3.2. Tree representation of a binary code.

length n_k excludes $D^{n_M-n_k}$ "terminal points," that is, points corresponding to sequences of length n_M. Thus the total number of terminal points excluded by the given code is $\sum_{i=1}^M D^{n_M-n_i}$. Since a tree of order D and size n_M has D^{n_M} terminal points, the total number of terminal points excluded cannot exceed D^{n_M}. Thus, $\sum_{i=1}^M D^{n_M-n_i} \le D^{n_M}$, or $\sum_{i=1}^M D^{-n_i} \le 1$.

Conversely, suppose we are given positive integers $n_1, n_2, \ldots, n_M$ satisfying $\sum_{i=1}^M D^{-n_i} \le 1$. Again assume $n_1 \le n_2 \le \cdots \le n_M$. To construct the required instantaneous code, select any point on the tree of order D and size n_M corresponding to a sequence of length n_1. This excludes $D^{n_M-n_1}$ terminal points. Since $\sum_{i=1}^M D^{-n_i} \le 1$, or equivalently $\sum_{i=1}^M D^{n_M-n_i} \le D^{n_M}$, we have $D^{n_M-n_1} < D^{n_M}$, and at least one terminal point remains. Hence a point corresponding to a sequence of length n_2 remains, and we may choose this point to correspond to the code word of length n_2. The total number of points excluded is now $D^{n_M-n_1} + D^{n_M-n_2}$, and again it follows from the hypothesis that this number is less than D^{n_M}, and hence we may choose a point corresponding to a sequence of length n_3. The process may be continued until all code words are assigned.

We remark that the condition $\sum_{i=1}^M D^{-n_i} \le 1$ is equivalent to

$$\sum_{j=1}^r \omega_j D^{-j} \le 1,$$

where ω_j is the number of code words of length j, and r is the maximum code-word length. This follows since the sum $\sum_{i=1}^M D^{-n_i}$ may be written as

$$\underbrace{D^{-1} + D^{-1} + \cdots + D^{-1}}_{\omega_1 \text{ times}} + \underbrace{D^{-2} + \cdots + D^{-2}}_{\omega_2} + \cdots + \underbrace{D^{-r} + \cdots + D^{-r}}_{\omega_r}.$$

2.4. *Extension of the condition $\sum_{i=1}^M D^{-n_i} \le 1$ to uniquely decipherable codes*

We have shown that the condition $\sum_{i=1}^M D^{-n_i} \le 1$ is necessary and sufficient for the existence of a base D instantaneous code with word lengths $n_1, n_2, \ldots, n_M$. In this section, we prove a stronger result, namely, that the same condition is necessary and sufficient for the existence of a *uniquely decipherable* code. The sufficiency part is immediate, since every instantaneous code is uniquely decipherable. Thus it remains to prove necessity.

Theorem 2.4.1. If a uniquely decipherable code has word lengths $n_1, n_2, \ldots, n_M$, then $\sum_{i=1}^M D^{-n_i} \le 1$.

Proof. For convenience we use $\sum_{j=1}^{r} \omega_j D^{-j}$ instead of $\sum_{i=1}^{M} D^{-n_i}$. Now

$$\left(\sum_{j=1}^{r} \omega_j D^{-j}\right)^n = (\omega_1 D^{-1} + \cdots + \omega_r D^{-r})^n. \tag{2.4.1}$$

Each term in the expansion of (2.4.1) is of the form $\omega_{i_1} D^{-i_1} \omega_{i_2} D^{-i_2} \cdots \omega_{i_n} D^{-i_n}$ where $1 \leq i_k \leq r$ for each k, and hence $n \leq i_1 + \cdots + i_n \leq nr$. Thus

$$\left(\sum_{j=1}^{r} \omega_j D^{-j}\right)^n = \sum_{k=n}^{nr} N_k D^{-k} \tag{2.4.2}$$

where

$$N_k = \sum_{i_1 + \ldots + i_n = k} \omega_{i_1} \omega_{i_2} \cdots \omega_{i_n}. \tag{2.4.3}$$

We claim that N_k is the total number of messages using n (not necessarily distinct) symbols x_i whose coded representation is of length k. To see this, we note that the number of sequences of code characters that can be formed using code words of length $i_1, \ldots, i_n$, *in that order*, where $i_1 + \ldots + i_n = k$ (that is, the number of sequences that start with a code word of length i_1, follow with a code word of length $i_2, \ldots,$ and end with a code word of length i_n) is $\omega_{i_1} \ldots \omega_{i_n}$. By unique decipherability, every sequence of code characters corresponds to at most one message. Thus N_k cannot exceed the total number of coded sequences of length k, that is, $N_k \leq D^k$. By (2.4.2),

$$\left(\sum_{j=1}^{r} \omega_j D^{-j}\right)^n \leq \sum_{k=n}^{nr} 1 = nr - n + 1 \leq nr.$$

Taking nth roots, we have

$$\sum_{j=1}^{r} \omega_j D^{-j} \leq n^{1/n} r^{1/n}. \tag{2.4.4}$$

Since (2.4.4) holds for any n, we may let $n \to \infty$ to obtain $\sum_{j=1}^{r} \omega_j D^{-j} \leq 1$.

2.5. *The noiseless coding theorem*

Let us return to the noiseless coding problem. A random variable X takes on values $x_1, \ldots, x_M$ with probabilities $p_1, \ldots, p_M$ respectively. Code words $W_1, \ldots, W_M$, of length $n_1, \ldots, n_M$ respectively, are assigned to the symbols $x_1, \ldots, x_M$. The code words are combinations of characters taken from a code alphabet $a_1, \ldots, a_D$. We wish to construct a uniquely decipherable code which minimizes the average code-word length $\bar{n} = \sum_{i=1}^{M} p_i n_i$. We shall approach the problem in three steps. First we establish an absolute lower bound on $\bar{n}$; then we find out how close we can come to the lower bound; then we attempt to synthesize the "best" code. The lower bound is set by the following theorem.

Theorem 2.5.1. (Noiseless Coding Theorem). If $\bar{n} = \sum_{i=1}^{M} p_i n_i$ is the average code-word length of a uniquely decipherable code for the random variable X, then $\bar{n} \geq H(X)/\log D$ with equality if and only if $p_i = D^{-n_i}$ for $i = 1, 2, \ldots, M$. Note that $H(X)/\log D$ is the uncertainty of X computed using logs to the base D, that is,

$$\frac{H(X)}{\log D} = -\sum_{i=1}^{M} p_i \frac{\log_2 p_i}{\log_2 D} = -\sum_{i=1}^{M} p_i \log_D p_i.$$

Proof. The condition $\bar{n} \geq H(X)/\log D$ is equivalent to

$$\log D \sum_{i=1}^{M} p_i n_i \geq -\sum_{i=1}^{M} p_i \log p_i.$$

Since $p_i n_i \log D = p_i \log D^{n_i} = -p_i \log D^{-n_i}$, the above condition may be written as

$$-\sum_{i=1}^{M} p_i \log D^{-n_i} \geq -\sum_{i=1}^{M} p_i \log p_i.$$

This looks suspiciously like Lemma 1.4.1, except that the terms D^{-n_i} do not necessarily add to unity. This difficulty may be circumvented by defining $q_i = D^{-n_i}/\sum_{j=1}^{M} D^{-n_j}$. Then the q_i's add to unity, and Lemma 1.4.1 yields

$$-\sum_{i=1}^{M} p_i \log p_i \leq -\sum_{i=1}^{M} p_i \log \frac{D^{-n_i}}{\sum_{j=1}^{M} D^{-n_j}} \tag{2.5.1}$$

with equality if and only if $p_i = D^{-n_i}/\sum_{j=1}^{M} D^{-n_j}$ for all i. Hence by (2.5.1),

$$H(X) \leq -\sum_{i=1}^{M} p_i \log D^{-n_i} + \left(\sum_{i=1}^{M} p_i\right) \log \left(\sum_{j=1}^{M} D^{-n_j}\right),$$

or

$$H(X) \leq \bar{n} \log D + \log \left(\sum_{j=1}^{M} D^{-n_j}\right) \tag{2.5.2}$$

$$\text{with equality if and only if} \quad p_i = \frac{D^{-n_i}}{\sum_{j=1}^{M} D^{-n_j}} \quad \text{for all } i.$$

But by unique decipherability (Theorem 2.4.1) $\sum_{j=1}^{M} D^{-n_j} \leq 1$; hence

$$\log \left(\sum_{j=1}^{M} D^{-n_j}\right) \leq 0,$$

and $H(X) \leq \bar{n} \log D$.

Now if $p_i = D^{-n_i}$ for all i, then

$$H(X) = -\sum_{i=1}^{M} p_i \log p_i = \sum_{i=1}^{M} p_i n_i \log D = \bar{n} \log D.$$

It therefore remains to prove that if $H(X) = \bar{n} \log D$, then $p_i = D^{-n_i}$ for all i. By (2.5.2), $H(X) = \bar{n} \log D$ implies that

$$\log \left(\sum_{j=1}^{M} D^{-n_j} \right) \geq 0;$$

by unique decipherability

$$\log \left(\sum_{j=1}^{M} D^{-n_j} \right) \leq 0,$$

and thus

$$\log \left(\sum_{j=1}^{M} D^{-n_j} \right) = 0$$

or $\sum_{j=1}^{M} D^{-n_j} = 1$. It now follows from (2.5.2) that $p_i = D^{-n_i}$ for all i. This completes the proof.

We shall call a code that achieves the lower bound set by the noiseless coding theorem *absolutely optimal*. An example of an absolutely optimal binary code is the following.

X	Probabilities	Code Words	
x_1	1/2	0	
x_2	1/4	10	
x_3	1/8	110	$H(X) = \bar{n} = \frac{7}{4}$.
x_4	1/8	111	

In general we cannot hope to construct an absolutely optimal code for a given set of probabilities $p_1, p_2, \ldots, p_M$, since if we choose n_i to satisfy $p_i = D^{-n_i}$, then $n_i = (-\log p_i)/(\log D)$ may not be an integer. However, we can do the next best thing and select the integer n_i such that

$$\frac{-\log p_i}{\log D} \leq n_i < \frac{-\log p_i}{\log D} + 1 \qquad (i = 1, 2, \ldots, M). \tag{2.5.3}$$

We claim that an instantaneous code can be constructed with word lengths $n_1, n_2, \ldots, n_M$. To prove this, we must show that $\sum_{i=1}^{M} D^{-n_i} \leq 1$. From the left-hand inequality of (2.5.3), it follows that $\log p_i \geq -n_i \log D$, or $p_i \geq D^{-n_i}$. Thus $\sum_{i=1}^{M} D^{-n_i} \leq \sum_{i=1}^{M} p_i = 1$. To estimate the average code-word length, we multiply (2.5.3) by p_i and sum over i, to obtain

$$-\sum_{i=1}^{M} p_i \frac{\log p_i}{\log D} \leq \sum_{i=1}^{M} p_i n_i < -\sum_{i=1}^{M} p_i \frac{\log p_i}{\log D} + \sum_{i=1}^{M} p_i \, .$$

Thus we have proved the following theorem.

Theorem 2.5.2. Given a random variable X with uncertainty $H(X)$, there exists a base D instantaneous code for X whose average code-word length $\bar{n}$ satisfies

$$\frac{H(X)}{\log D} \leq \bar{n} < \frac{H(X)}{\log D} + 1.$$

Consequently the average code-word length may be brought within one digit of the lower bound set by the noiseless coding theorem. In fact, we can always approach the lower bound as closely as desired if we are allowed to use "block coding." Suppose that instead of assigning a code word to each symbol x_i, we take a series of s independent observations of X and assign a code word to the resulting group of s symbols. In other words, we construct a code for the random vector $Y = (X_1, X_2, \ldots, X_s)$, where the X_i are independent and each X_i has the same distribution as X. If X (and hence each X_i) assumes M values, then Y assumes M^s values. Block coding will in general decrease the average code-word length *per value of* X as illustrated by the following example.

X	p	Code Word	$Y = (X_1, X_2)$	p	Code Word
x_1	3/4	0	x_1x_1	9/16	0
x_2	1/4	1	x_1x_2	3/16	10
			x_2x_1	3/16	110
			x_2x_2	1/16	111

$\bar{n} = 1$

$$\begin{aligned}\bar{n} &= \tfrac{9}{16} + \tfrac{3}{16}(2) + \tfrac{1}{4}(3) \\ &= \tfrac{27}{16} \text{ code characters/2 values of } X \\ &= \tfrac{27}{32} \text{ code characters/value of } X\end{aligned}$$

By Theorem 2.5.2, we may construct an instantaneous code for Y whose average code-word length $\bar{n}_s$ satisfies

$$\frac{H(Y)}{\log D} \leq \bar{n}_s < \frac{H(Y)}{\log D} + 1 \text{ code characters/value of } Y.$$

Since $X_1, \ldots, X_s$ are independent and identically distributed, $H(Y) = H(X_1, \ldots, X_s) = H(X_1) + \cdots + H(X_s) = sH(X)$. Thus

$$\frac{sH(X)}{\log D} \leq \bar{n}_s < \frac{sH(X)}{\log D} + 1,$$

or

$$\frac{H(X)}{\log D} \leq \frac{\bar{n}_s}{s} < \frac{H(X)}{\log D} + \frac{1}{s}. \tag{2.5.4}$$

But $\bar{n}_s/s$ is the average number of code characters per value of X; it follows from (2.5.4) that $\bar{n}_s/s$ may be made arbitrarily close to $H(X)/\log D$ by choosing s large enough. Thus $H(X)/\log D$ may be interpreted as the minimum average number of base D digits required to encode one observation of X. If we consider the case $D = 2$, we make the observation that any instantaneous binary code corresonds to a sequence of "yes or no" questions, as illustrated in Fig. 2.5.1. In the example, an answer "yes" corresponds to a zero in a code word; an answer "no" to a one. Thus,

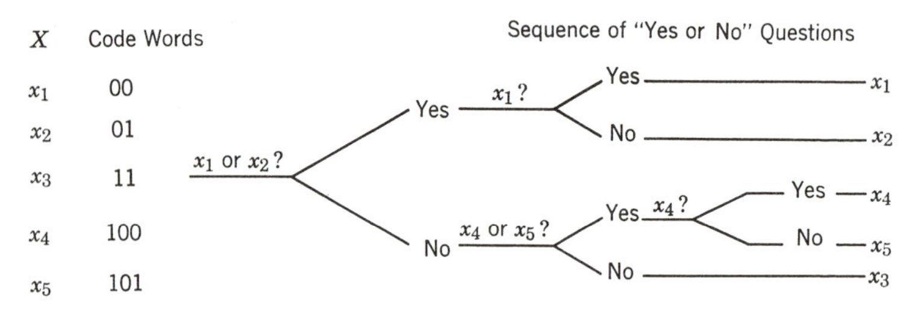

Fig. 2.5.1. A binary instantaneous code and its corresponding sequence of "yes or no" questions.

as we indicated in Section 1.3, the uncertainty $H(X)$ may be interpreted as the minimum average number of "yes or no" questions required to determine the result of one observation of X.

2.6. *Construction of optimal codes*

The only part of the noiseless coding problem that remains is the construction of the code which minimizes the average code-word length for a given set of probabilities $p_1, p_2, \ldots, p_M$. We first show that in the search for optimal codes, we may restrict our attention to instantaneous codes. More precisely, we prove:

Lemma 2.6.1. Suppose a code C is optimal within the class of instantaneous codes for the given probabilities $p_1, p_2, \ldots, p_M$; in other words, no other instantaneous code for $p_1, \ldots, p_M$ has a smaller average code-word length than C. Then C is optimal within the entire class of uniquely decipherable codes.

Proof. Suppose a uniquely decipherable code C' has a smaller average code-word length than C. Let $n_1', n_2', \ldots, n_M'$ be the code-word lengths of C'. By Theorem 2.4.1, $\sum_{i=1}^{M} D^{-n_i'} \leq 1$. But by Theorem 2.3.1 there exists an instantaneous code C'' with word lengths $n_1', \ldots, n_M'$. Hence the average word length of C'' is the same as the average word length of C', contradicting the fact that C is the best instantaneous code.

Before turning to the synthesis of optimal instantaneous codes, we imagine that we have constructed an optimal code and determine what properties such a code must have. In other words, we shall derive *necessary* conditions on an optimal instantaneous code. To simplify the presentation we consider only the binary case. For an extension to nonbinary code alphabets see Problem 2.5. The results are summarized as follows.

Lemma 2.6.2. Given a binary instantaneous code C with word lengths $n_1, n_2, \ldots, n_M$ associated with a set of symbols with probabilities $p_1, p_2, \ldots, p_M$. For convenience, assume that the symbols are arranged in order of decreasing probability ($p_1 \geq p_2 \geq \cdots \geq p_M$) and that a group of symbols with the same probability is arranged in order of increasing code-word length. (If $p_i = p_{i+1} = \cdots = p_{i+r}$, then $n_i \leq n_{i+1} \leq \cdots \leq n_{i+r}$.) Then if C is optimal within the class of instantaneous codes, C must have the following properties:

a. Higher probability symbols have shorter code words, that is, $p_j > p_k$ implies $n_j \leq n_k$.

b. The two least probable symbols have code words of equal length, that is, $n_{M-1} = n_M$.

c. Among the code words of length n_M there must be at least two words that agree in all digits except the last. For example, the following code cannot be optimal since code words 4 and 5 do not agree in the first three places.

x_1	0
x_2	100
x_3	101
x_4	1101
x_5	1110

Proof. To prove (a), we note that if $p_j > p_k$ and $n_j > n_k$, it is possible to construct a better code C' by interchanging code words j and k. The difference between the average code-word length of C' and of C is

$$(\bar{n})' - \bar{n} = p_j n_k + p_k n_j - (p_j n_j + p_k n_k) = (p_j - p_k)(n_k - n_j) < 0;$$

hence C' is better than C.

To prove (b) we note first that $n_{M-1} \leq n_M$, for if $p_{M-1} > p_M$ then $n_{M-1} \leq n_M$ by (a); if $p_{M-1} = p_M$, then $n_{M-1} \leq n_M$ by our assumption about arrangement of code words associated with equally probable symbols. Now, if $n_M > n_{M-1}$, we may drop the last digit of the Mth code word to obtain a code that is still instantaneous and better than the original code.

Finally, (c) is proved by observing that if no two code words of maximal length agree in all places but the last, then we may drop the last digit of all such code words to obtain a better code.

We now describe a method, due to Huffman, for the construction of optimal (instantaneous) codes. If we have an array of symbols $x_1, \ldots, x_M$ with probabilities $p_1, \ldots, p_M$ (again assume $p_1 \geq p_2 \geq \cdots \geq p_M$), then we combine the last two symbols x_{M-1} and x_M into an equivalent symbol $x_{M,M-1}$ with probability $p_M + p_{M-1}$. Now suppose that somehow we could construct an optimal code C_2 for the new set of symbols (see Table 2.6.1). We now construct a code C_1 for the original set of symbols $x_1, \ldots, x_M$. The code words associated with $x_1, x_2, \ldots, x_{M-2}$ are exactly the same as the corresponding code words of C_2. The code words associated with x_{M-1} and x_M are formed by adding a zero and a one, respectively, to the code word $W_{M,M-1}$ associated with the symbol $x_{M,M-1}$ in C_2. We now claim that C_1 *is an optimal code for the set of probabilities* $p_1, p_2, \ldots, p_M$.

To establish the claim, suppose that C_1 is not optimal. Let C_1' be an optimal instantaneous code for $x_1, x_2, \ldots, x_M$. Then C_1' will have code words $W_1', W_2', \ldots, W_M'$ with word lengths $n_1', n_2', \ldots, n_M'$ respectively. By Lemma 2.6.2b, $n'_{M-1} = n_M'$. By Lemma 2.6.2c, at least two words of length n_M' agree in all digits but the last. We may assume without loss of generality that W'_{M-1} and W_M' are two of these words. (If necessary we may interchange words of the same length without affecting the average word length.) At this point we again combine the symbols x_{M-1} and x_M and construct a code C_2' by taking as the code word for $x_{M,M-1}$ the word W_M' (or W'_{M-1}) with the last digit removed (see Table 2.6.2). We now establish that C_2' has a smaller average word length than C_2, contradicting the optimality of C_2. The average word length of C_2' is

$$\begin{aligned}(\bar{n}_2)' &= p_1 n_1' + \cdots + p_{M-2}n'_{M-2} + (p_{M-1} + p_M)(n'_{M-1} - 1) \\ &= p_1 n_1' + \cdots + p_{M-2}n'_{M-2} + p_{M-1}n'_{M-1} \\ &\qquad\qquad + p_M n_M' - (p_{M-1} + p_M).\end{aligned}$$

Since C_1' has a smaller average word length than C_1,

$$\begin{aligned}p_1 n_1' + \cdots + p_{M-2}n'_{M-2} + p_{M-1}n'_{M-1} + p_M n'_M &< p_1 n_1 + \cdots \\ &\quad + p_{M-2}n_{M-2} + p_{M-1}n_{M-1} + p_M n_M.\end{aligned}$$

Using the fact that $n_{M-1} = n_M$ (by construction of C_1) we obtain

$$(\bar{n}_2)' < p_1 n_1 + \cdots + p_{M-2}n_{M-2} + (p_{M-1} + p_M)n_{M-1} - (p_{M-1} + p_M)$$

or

$$(\bar{n}_2)' < p_1 n_1 + \cdots + p_{M-2}n_{M-2} + (p_{M-1} + p_M)(n_{M-1} - 1).$$

Table 2.6.1. Construction of the Huffman code

X	Probabilities	Code words of C_1	Word lengths	X	Probabilities	Code words of C_2	Word lengths
x_1	p_1	W_1	n_1	x_1	p_1	W_1	n_1
x_2	p_2	W_2	n_2	x_2	p_2	W_2	n_2
$\vdots$				$\vdots$	$\vdots$	$\vdots$	
				$x_{M,M-1}$	p_M+p_{M-1}	$W_{M,M-1}$	$n_{M,M-1}$
				$\vdots$	$\vdots$	$\vdots$	
x_{M-2}	p_{M-2}	W_{M-2}	n_{M-2}				
x_{M-1}	p_{M-1}	$[W_{M,M-1}\ 0]$	n_{M-1}	x_{M-2}	p_{M-2}	W_{M-2}	n_{M-2}
x_M	p_M	$[W_{M,M-1}\ 1]$	n_M				

Table 2.6.2. Proof of optimality of the Huffman procedure

X	Probabilities	Code words of C_1'	Word lengths	X	Probabilities	Code words of C_2'	Word lengths
x_1	p_1	W_1'	n_1'	x_1	p_1	W_1'	n_1'
x_2	p_2	W_2'	n_2'	x_2	p_2	W_2'	n_2'
$\vdots$				$\vdots$			
				$x_{M,M-1}$	p_M+p_{M-1}	U'	$n'_M-1 = n'_{M-1}-1$
				$\vdots$			
x_{M-2}	p_{M-2}	W'_{M-2}	n'_{M-2}				
x_{M-1}	p_{M-1}	W'_{M-1}	n'_{M-1}	x_{M-2}	p_{M-2}	W'_{M-2}	n'_{M-2}
x_M	p_M	W_M'	$n_M'=n'_{M-1}$				

$U' = W_M'$ with last digit removed.

Since $n_{M-1} - 1 = n_{M,M-1}$, the expression on the right is the average word length of C_2, and thus the proof is complete.

The above result tells us exactly how to construct optimal codes. An example is carried through in Fig. 2.6.1. The idea is simply to combine the two least probable symbols until only two symbols remain. An optimal binary code for two symbols must consist of the two code words 0 and 1; we can then work backwards, constructing optimal codes for each array until we arrive at the original set of symbols.

2.7. *Notes and remarks*

The material of this chapter has become fairly standard, and is treated in Feinstein (1958), Abramson (1963), and Fano (1961). Theorem 2.2.1 is due to Sardinas and Patterson (1950, 1953), whose proof exploited the fact that the collection of finite sequences of code characters is a semigroup under juxtaposition. The somewhat cumbersome direct argument in the present text is the author's. The phrase "instantaneous code" was

x_1	.3	x_1	.3	x_1	.3	$x_{2,3}$	.45	$x_{1,456}$	.55 (0)
x_2	.25	x_2	.25	$x_{4,56}$	.25	x_1	.3 (0)	$x_{2,3}$	.45 (1)
x_3	.2	x_3	.2	x_2	.25 (0)	$x_{4,56}$	.25 (1)		
x_4	.1	$x_{5,6}$	.15 (0)	x_3	.2 (1)				
x_5	.1 (0)	x_4	.1 (1)						
x_6	.05 (1)								

Optimal Codes

$x_{1,456}$	0	$x_{2,3}$	1	x_1	00	x_1	00	x_1	00
$x_{2,3}$	1	x_1	00	$x_{4,56}$	01	x_2	10	x_2	10
		$x_{4,56}$	01	x_2	10	x_3	11	x_3	11
				x_3	11	$x_{5,6}$	010	x_4	011
						x_4	011	x_5	0100
								x_6	0101

Fig. 2.6.1. Huffman code.

coined by Abramson (1963). Theorem 2.3.1 is due to Kraft (1949); Theorem 2.4.1 was proved by McMillan (1956); the proof in the text is due to Karush (1961). Theorem 2.5.1 is due to Shannon (1948); the proof follows Feinstein (1958). The construction of Section 2.6 is due to Huffman (1952).

PROBLEMS

2.1 Determine whether or not each of the following codes is uniquely decipherable. If a code is not uniquely decipherable, construct an ambiguous sequence.

a.

x_1	010
x_2	0001
x_3	0110
x_4	1100
x_5	00011
x_6	00110
x_7	11110
x_8	101011

b.

x_1	*abc*
x_2	*abcd*
x_3	*e*
x_4	*dba*
x_5	*bace*
x_6	*ceac*
x_7	*ceab*
x_8	*eabd*

2.2 a. For the binary code below, let $N(k)$ be the number of messages that can be formed using exactly k code characters. For example, $N(1) = 1$ (that is, x_1), $N(2) = 3$ (x_1x_1, x_2, x_3), $N(3) = 5$ ($x_1x_1x_1, x_1x_2, x_1x_3, x_2x_1, x_3x_1$).

Find a general expression for $N(k)$ ($k = 1, 2, \ldots$).

x_1	0
x_2	10
x_3	11

b. Repeat part (a) for the code below.

x_1	0
x_2	10
x_3	110
x_4	111

2.3 Construct a Huffman code for the symbols below. Compare the average code-word length with the uncertainty $H(X)$.

Symbols	Probabilities	Symbols	Probabilities
x_1	.2	x_7	.059
x_2	.18	x_8	.04
x_3	.1	x_9	.04
x_4	.1	x_{10}	.04
x_5	.1	x_{11}	.04
x_6	.061	x_{12}	.03
		x_{13}	.01

2.4 Show that the average word length of a Huffman binary code satisfies

$$\bar{n} < H(X) + 1$$

2.5 Explain how to construct a base D instantaneous code that minimizes the average code-word length and apply the results to construct an optimal ternary code for the set of symbols below.

Symbols	Probabilities	Symbols	Probabilities
x_1	.3	x_5	.1
x_2	.2	x_6	.08
x_3	.15	x_7	.05
x_4	.1	x_8	.02

CHAPTER THREE

The Discrete Memoryless Channel

3.1. Models for communication channels

In this chapter we shall begin the analysis of communication channels whose inputs are subject to random disturbances in transmission. In the usual communication situation, an object of some kind (for example, a letter of the alphabet, a pulse or other waveform) is selected from a specified class of inputs. The channel is a device that acts on the input to produce an output belonging to another specified class. The random nature of the channel may in many cases be described by giving a probability distribution over the set of possible outputs. The distribution will in general depend on the particular input chosen for transmission and in addition may depend on the internal structure of the channel at the time the input is applied. We shall try to arrive at a definition of an information channel that reflects these considerations.

We are going to specialize to the so-called "discrete case," that is, the situation in which the information to be transmitted consists of a sequence of symbols, each symbol belonging to a finite alphabet. More general models will be discussed in Chapter 8. If we apply a sequence $\alpha_1, \ldots, \alpha_n$ at the input of a channel, then at the output, perhaps after an appropriate delay, we will receive a sequence $\beta_1, \ldots, \beta_n$. It is reasonable to describe the action of the channel by giving a probability distribution over the output sequences $\beta_1, \ldots, \beta_n$ for each input sequence $\alpha_1, \ldots, \alpha_n$; the family of distributions should also reflect the fact that the "internal state" of the channel at the time the input is applied will affect the transmission of information. Physically, we expect that many channels have "memory"; that is, the distribution of the output symbol β_n may depend on previous inputs and outputs. We do not, however, expect "anticipatory" behavior; in our model the distribution of β_n should not depend on future inputs or outputs. Thus in giving the distribution of $\beta_1, \ldots, \beta_n$, we need not consider inputs beyond $\alpha_1, \ldots, \alpha_n$. We are led to the following definition.

Definition. Given finite sets Γ and Γ', to be called respectively the *input alphabet* and *output alphabet*, and an arbitrary set S called the *set of states*, a *discrete channel* is a system of probability functions

$$p_n(\beta_1, \ldots, \beta_n \mid \alpha_1, \ldots, \alpha_n;\ s)$$
$$\alpha_1, \ldots, \alpha_n \in \Gamma$$
$$\beta_1, \ldots, \beta_n \in \Gamma'$$
$$s \in S$$
$$n = 1, 2, \ldots,$$

that is, a system of functions satisfying

1. $p_n(\beta_1, \ldots, \beta_n \mid \alpha_1, \ldots, \alpha_n;\ s) \geq 0$ for all $n, \alpha_1, \ldots, \alpha_n,$ $\beta_1, \ldots, \beta_n, s.$

2. $\sum_{\beta_1, \ldots, \beta_n} p_n(\beta_1, \ldots, \beta_n \mid \alpha_1, \ldots, \alpha_n;\ s) = 1$ for all $n,$ $\alpha_1, \ldots, \alpha_n, s.$

Physically we interpret $p_n(\beta_1, \ldots, \beta_n \mid \alpha_1, \ldots, \alpha_n;\ s)$ as the probability that the sequence $\beta_1, \ldots, \beta_n$ will appear at the output if the input sequence $\alpha_1, \ldots, \alpha_n$ is applied and the initial state of the channel, that is, the state just prior to the appearance of α_1, is s. The state of the channel may change as the components of the input sequence are applied; our model assumes that knowledge of the initial state and the input sequence determines the distribution of the output sequence.

In this chapter we consider only a very special case of the discrete channel. However, the techniques of analysis which will be developed are basic to the analysis of more general channels. Specifically, we are going to consider the discrete channel without memory; such a channel is characterized by the requirements that successive symbols be acted on independently and that the functions p_n do not depend on the state s. Formally we have the following:

Definition. A discrete channel is *memoryless* if

1. the functions $p_n(\beta_1, \ldots, \beta_n \mid \alpha_1, \ldots, \alpha_n;\ s)$ do not depend on s, hence may be written $p_n(\beta_1, \ldots, \beta_n \mid \alpha_1, \ldots, \alpha_n)$, and

2. $p_n(\beta_1, \ldots, \beta_n \mid \alpha_1, \ldots, \alpha_n) = p_1(\beta_1 \mid \alpha_1)\, p_1(\beta_2 \mid \alpha_2) \cdots p_1(\beta_n \mid \alpha_n)$ for all $\alpha_1, \ldots, \alpha_n \in \Gamma$, $\beta_1, \ldots, \beta_n \in \Gamma'$, $n = 1, 2, \ldots.$

The second condition may be replaced by the conjunction of two other conditions as follows:

Lemma 3.1.1. Given probability functions satisfying condition (1), define

$$p_n(\beta_1, \ldots, \beta_{n-k} \mid \alpha_1, \ldots, \alpha_n) = \sum_{\beta_{n-k+1}, \ldots, \beta_n} p_n(\beta_1, \ldots, \beta_n \mid \alpha_1, \ldots, \alpha_n),$$
$$1 \leq k \leq n-1.$$

(This quantity is interpreted as the probability that the first $n - k$ output symbols will be $\beta_1, \ldots, \beta_{n-k}$ when the input sequence $\alpha_1, \ldots, \alpha_n$ is applied.) Also define

$$p_n(\beta_n \mid \alpha_1, \ldots, \alpha_n;\ \beta_1, \ldots, \beta_{n-1}) = \frac{p_n(\beta_1, \ldots, \beta_{n-1}, \beta_n \mid \alpha_1, \ldots, \alpha_n)}{p_n(\beta_1, \ldots, \beta_{n-1} \mid \alpha_1, \ldots, \alpha_n)};$$

this expression is interpreted as the conditional probability that the nth output symbol will be β_n, given that the input sequence $\alpha_1, \ldots, \alpha_n$ is applied and the first $n - 1$ output symbols are $\beta_1, \ldots, \beta_{n-1}$. (Note that the subscript n on the functions above corresponds to the length of the input sequence.)

The functions satisfy condition (2) if and only if for all $n = 1, 2, \ldots,$ both of the following conditions are satisfied.

a. $p_n(\beta_n \mid \alpha_1, \ldots, \alpha_n;\ \beta_1, \ldots, \beta_{n-1}) = p_1(\beta_n \mid \alpha_n)$ for all $\alpha_1, \ldots, \alpha_n \in \Gamma, \quad \beta_1, \ldots, \beta_n \in \Gamma'$

b. $p_n(\beta_1, \ldots, \beta_{n-k} \mid \alpha_1, \ldots, \alpha_n) = p_{n-k}(\beta_1, \ldots, \beta_{n-k} \mid \alpha_1, \ldots, \alpha_{n-k})$ for all $\alpha_1, \ldots, \alpha_n \in \Gamma, \quad \beta_1, \ldots, \beta_{n-k} \in \Gamma', \quad 1 \leq k \leq n - 1.$

Condition (a) puts into evidence the memoryless feature of the channel, and condition (b) indicates the nonanticipatory behavior.

Proof. Suppose that condition (2) is satisfied. Then

$$\begin{aligned}
p_n(\beta_n \mid \alpha_1, \ldots, \alpha_n;\ \beta_1, \ldots, \beta_{n-1}) &= \frac{p_n(\beta_1, \ldots, \beta_{n-1}, \beta_n \mid \alpha_1, \ldots, \alpha_n)}{p_n(\beta_1, \ldots, \beta_{n-1} \mid \alpha_1, \ldots, \alpha_n)} \\
&= \frac{\prod_{k=1}^{n} p_1(\beta_k \mid \alpha_k)}{\sum_{\beta_n} p_n(\beta_1, \ldots, \beta_n \mid \alpha_1, \ldots, \alpha_n)} \\
&= \frac{\prod_{k=1}^{n} p_1(\beta_k \mid \alpha_k)}{\sum_{\beta_n} \prod_{k=1}^{n} p_1(\beta_k \mid \alpha_k)} \\
&= \frac{\prod_{k=1}^{n} p_1(\beta_k \mid \alpha_k)}{\prod_{k=1}^{n-1} p_1(\beta_k \mid \alpha_k) \sum_{\beta_n} p_1(\beta_n \mid \alpha_n)} \\
&= p_1(\beta_n \mid \alpha_n), \qquad \text{proving (a).}
\end{aligned}$$

To prove (b), note that the above argument shows that

$$p_n(\beta_1, \ldots, \beta_{n-1} \mid \alpha_1, \ldots, \alpha_n) = \prod_{k=1}^{n-1} p_1(\beta_k \mid \alpha_k)$$
$$= p_{n-1}(\beta_1, \ldots, \beta_{n-1} \mid \alpha_1, \ldots, \alpha_{n-1}).$$

An induction argument now establishes (b).

Conversely, if (a) and (b) are satisfied then

$$p_n(\beta_1, \ldots, \beta_n \mid \alpha_1, \ldots, \alpha_n)$$
$$= p_n(\beta_1, \ldots, \beta_{n-1} \mid \alpha_1, \ldots, \alpha_n) p_n(\beta_n \mid \alpha_1, \ldots, \alpha_n;\ \beta_1, \ldots, \beta_{n-1}).$$

By (b), the first term on the right is $p_{n-1}(\beta_1, \ldots, \beta_{n-1} \mid \alpha_1, \ldots, \alpha_{n-1})$; by (a) the second term is $p_1(\beta_n \mid \alpha_n)$. Proceeding inductively, we establish (2).

Note that we could have specified $k = 1$ in the statement of condition (b) and the lemma would still hold.

Thus a discrete memoryless channel is characterized by a matrix whose elements are $a_{\alpha\beta} = p_1(\beta \mid \alpha)$, $\alpha \in \Gamma$, $\beta \in \Gamma'$. ($a_{\alpha\beta}$ is the element in row α and column β.) The matrix $[p_1(\beta \mid \alpha)]$ is called the *channel matrix;* henceforth we shall drop the subscript and write $p(\beta \mid \alpha)$ for $p_1(\beta \mid \alpha)$. In this chapter, the word "channel" will mean "discrete memoryless channel" unless otherwise specified.

3.2. The information processed by a channel; channel capacity; classification of channels

Consider a discrete memoryless channel with input alphabet* $x_1, \ldots, x_M$, output alphabet $y_1, \ldots, y_L$ and channel matrix $[a_{ij}]$, $a_{ij} = p(y_j \mid x_i)$, $i = 1, \ldots, M$, $j = 1, \ldots, L$. If an input symbol is chosen at random, that is, if X is a random variable taking on the values $x_1, \ldots, x_M$ with probabilities $p(x_1), \ldots, p(x_M)$ respectively, then the channel output also becomes a random variable. The joint distribution of the input X and the output Y is given by $P\{X = x_i, Y = y_j\} = p(x_i)p(y_j \mid x_i)$, $i = 1, 2, \ldots, M$, $j = 1, 2, \ldots, L$, and the distribution of Y is given by

$$P\{Y = y_j\} = \sum_{i=1}^{M} p(x_i)p(y_j \mid x_i), \qquad j = 1, 2, \ldots, L.$$

---

* In Chapter 2 we distinguished between the symbols $x_1, x_2, \ldots, x_M$, which were the building blocks for a message, and the code alphabet $a_1, a_2, \ldots, a_D$. Coding is necessary since the inputs accepted by the channel may have no relation to the symbols $x_1, x_2, \ldots x_M$. In this chapter we assume for convenience that $x_1, x_2, \ldots, x_M$ is the code alphabet. We do this because we would like to reserve the symbol $H(X)$ for the uncertainty at the input to a channel. If it becomes necessary to distinguish uncoded information from coded information, then we shall denote the original (uncoded) symbols by m_1, m_2, etc.; thus the m_i will be the components of a message. Each message will be assigned a code word, that is, a sequence of symbols from the alphabet $x_1, x_2, \ldots, x_M$.

Thus the specification of an input distribution induces in a natural way an output distribution and a joint distribution on input and output. We may therefore calculate the *input uncertainty* $H(X)$, the *output uncertainty* $H(Y)$, and the *joint uncertainty* of input and output $H(X, Y)$ as well as the conditional uncertainties $H(Y \mid X)$ and $H(X \mid Y)$. It is natural to define the *information processed by the channel* as

$$I(X \mid Y) = H(X) - H(X \mid Y). \tag{3.2.1}$$

By the discussion in Section 1.5, we have

$$I(X \mid Y) = H(Y) - H(Y \mid X) = I(Y \mid X) = H(X) + H(Y) - H(X, Y).$$

It is important to notice that the information processed by a channel depends on the input distribution $p(x)$. We may vary the input distribution until the information reaches a maximum; the maximum information is called the *channel capacity*. Specifically, we define the channel capacity as

$$C = \max_{p(x)} I(X \mid Y). \tag{3.2.2}$$

(There is a true maximum rather than just a least upper bound; see Problem 3.12.)

The significance of the channel capacity is not at all clear from what we have done so far. However, we shall prove later in this chapter that (loosely speaking) it is possible to transmit information through a channel at any rate less than the channel capacity with an arbitrarily small probability of error; completely reliable transmission is not possible if the information processed is greater than the channel capacity. The calculation of the capacity of a channel is a difficult problem in general; the solution in certain special cases will be discussed in the next section.

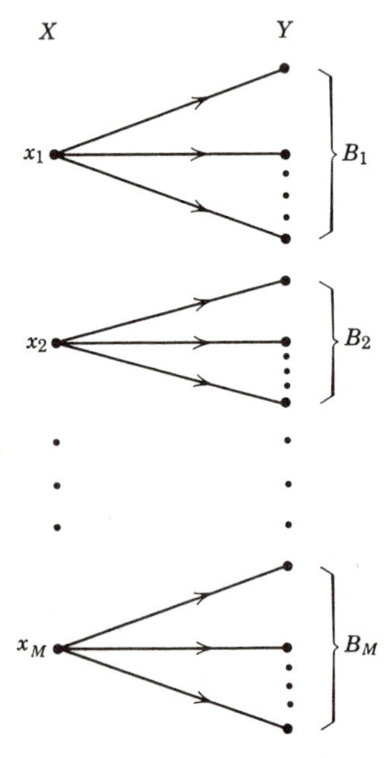

Fig. 3.2.1. Lossless channel.

It is convenient at this point to introduce certain classes of channels which are easy to analyze.

a. A channel is *lossless* if $H(X \mid Y) = 0$ for all input distributions. In other words, a lossless channel is characterized by the fact that the input is determined by the output and hence that no transmission errors can occur. Equivalently, the values of Y may be partitioned into disjoint sets

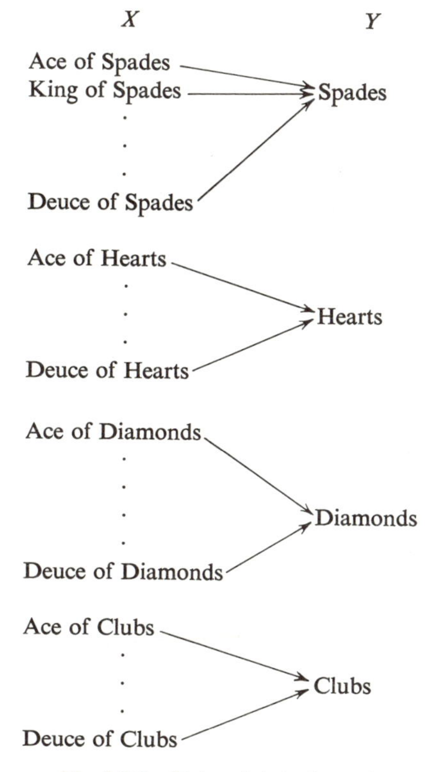

Fig. 3.2.2. Deterministic channel.

$B_1, B_2, \ldots, B_M$ such that $P\{Y \epsilon B_i \mid X = x_i\} = 1$ $(i = 1, \ldots, M)$. The structure of a lossless channel is indicated in Fig. 3.2.1.

b. A channel is *deterministic* if $p(y_j \mid x_i) = 1$ or 0 for all i, j; that is, if Y is determined by X, or equivalently $H(Y \mid X) = 0$ for all input distributions. An example of a deterministic channel is one whose input X is the identity of a playing card picked from an ordinary 52-card pack, and whose output Y is the suit of the card (see Fig. 3.2.2). If the card is picked at random so that all values of X (and hence of Y) are equally likely, then the information processed is $H(Y) - H(Y \mid X) = H(Y) = \log 4$.

c. A channel is *noiseless* if it is lossless and deterministic (see Fig. 3.2.3). Note that the definition agrees with the intuitive notion of a noiseless channel used in Chapter 2.

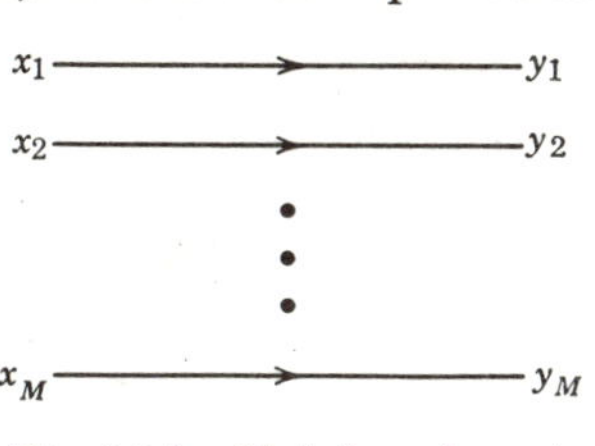

Fig. 3.2.3. Noiseless channel.

d. A channel is *useless* (or *zero-capacity*) if $I(X \mid Y) = 0$ for all input distributions.

Equivalently, a useless channel may be characterized by the condition that $H(X \mid Y) = H(X)$ for all $p(x)$, or alternately X and Y are independent for all $p(x)$. Since independence of X and Y means that $p(y_j \mid x_i) = p(y_j)$ for all i, j, a channel is useless if and only if its channel matrix has identical rows. A lossless channel and a useless channel represent extremes of possible channel behavior. The output symbol of a lossless channel uniquely specifies the input symbol, so that perfect transmission of information is possible. A useless channel completely scrambles all input messages. Since $p(x_i \mid y_j) = p(x_i)$ for all i, j, the conditional distribution of X after Y is received is the same as the original distribution of X. Roughly speaking, knowledge of the output tells us nothing about the input; for the purpose of determining the input we may as well ignore the output completely. An example of a useless channel is shown in Fig. 3.2.4.

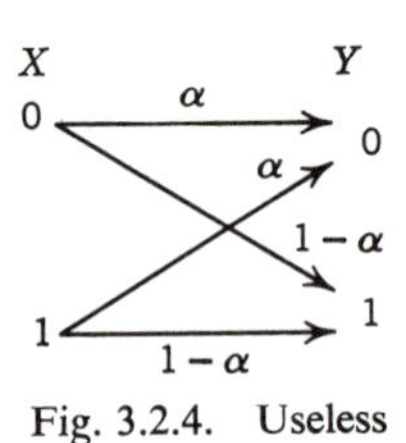

Fig. 3.2.4. Useless channel.

e. A channel is *symmetric* if each row of the channel matrix $[p(y_j \mid x_i)]$ contains the same set of numbers $p_1', p_2', \ldots, p_L'$ and each column of $[p(y_j \mid x_i)]$ contains the same set of numbers $q_1', q_2', \ldots, q_M'$. For example, the matrices

$$\begin{array}{c} \\ x_1 \\ x_2 \end{array}\!\!\begin{array}{c} \begin{array}{cccc} y_1 & y_2 & y_3 & y_4 \end{array} \\ \begin{bmatrix} \frac{1}{3} & \frac{1}{3} & \frac{1}{6} & \frac{1}{6} \\ \frac{1}{6} & \frac{1}{6} & \frac{1}{3} & \frac{1}{3} \end{bmatrix} \end{array} \quad \text{and} \quad \begin{array}{c} \\ x_1 \\ x_2 \\ x_3 \end{array}\!\!\begin{array}{c} \begin{array}{ccc} y_1 & y_2 & y_3 \end{array} \\ \begin{bmatrix} \frac{1}{2} & \frac{1}{3} & \frac{1}{6} \\ \frac{1}{6} & \frac{1}{2} & \frac{1}{3} \\ \frac{1}{3} & \frac{1}{6} & \frac{1}{2} \end{bmatrix} \end{array}$$

represent symmetric channels. The rows of the channel matrix are identical except for permutations, and similarly for the columns.

It is an immediate consequence of the definition of a symmetric channel that for such a channel, $H(Y \mid X)$ is independent of the input distribution $p(x)$ and depends only on the channel probabilities $p(y_j \mid x_i)$. To show this, we note that if $X = x_i$, the probabilities associated with the output symbols $y_1, \ldots, y_L$ are (not necessarily in this order) $p_1', p_2', \ldots, p_L'$. Hence

$$H(Y \mid X = x_i) = -\sum_{j=1}^{L} p_j' \log p_j', \qquad i = 1, 2, \ldots, M.$$

Therefore

$$\begin{aligned} H(Y \mid X) &= \sum_{i=1}^{M} p(x_i) H(Y \mid X = x_i) \\ &= -\sum_{j=1}^{L} p_j' \log p_j', \end{aligned} \tag{3.2.3}$$

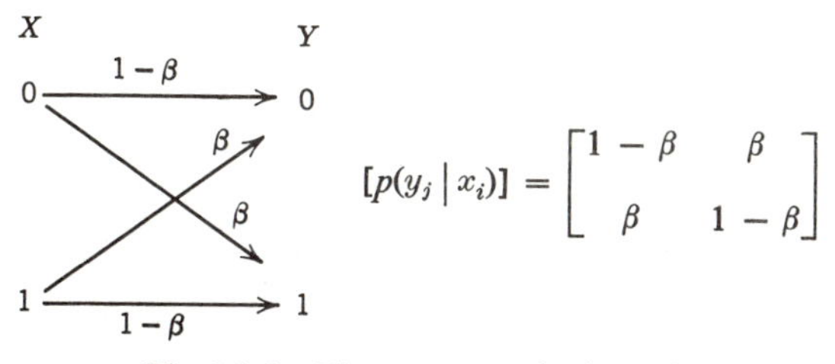

Fig. 3.2.5. Binary symmetric channel.

for any input distribution $p(x)$. The most celebrated example of a symmetric channel is the "binary symmetric channel," shown in Fig. 3.2.5.

3.3. *Calculation of channel capacity*

Before attacking the general problem of calculating the capacity of an arbitrary discrete memoryless channel, we consider a special case in which a closed-form expression for the capacity can be found, namely, the case of a *symmetric channel.*

Consider a symmetric channel with input alphabet $x_1, \ldots, x_M$, output alphabet $y_1, \ldots, y_L$, and channel matrix with row probabilities $p_1', \ldots, p_L'$ and column probabilities $q_1', \ldots, q_M'$. Since $H(Y \mid X)$ does not depend on the input distribution, the problem of maximizing the information $I(X \mid Y) = H(Y) - H(Y \mid X)$ reduces to the problem of maximizing the output uncertainty $H(Y)$. Now we know from Theorem 1.4.2 that $H(Y) \leq \log L$ with equality if and only if all values of Y are equally likely. Thus if we can find an input distribution under which all values of Y have the same probability, that input distribution would maximize $I(X \mid Y)$. We claim that the uniform input distribution will do the trick; in other words, if all input symbols of a symmetric channel are equally likely, then all output symbols are equally likely as well. To prove this, we let $p(x_i) = 1/M$ for $i = 1, 2, \ldots, M$ and write

$$p(y_j) = \sum_{i=1}^{M} p(x_i, y_j) = \sum_{i=1}^{M} p(x_i)p(y_j \mid x_i) = \frac{1}{M}\sum_{i=1}^{M} p(y_j \mid x_i).$$

However, the term $\sum_{i=1}^{M} p(y_j \mid x_i)$ is the sum of the entries in the jth column of the channel matrix. Since the channel is symmetric, we have $\sum_{i=1}^{M} p(y_j \mid x_i) = \sum_{k=1}^{M} q_k'$, independent of j. Thus $p(y_j)$ does not depend on j, or equivalently all values of Y have the same probability. The maximum $H(Y) = \log L$ is therefore attainable, and we conclude, using (3.2.3), that the capacity of a symmetric channel is

$$C_{\text{sym}} = \log L + \sum_{j=1}^{L} p_j' \log p_j'. \tag{3.3.1}$$

The capacity of the binary symmetric channel of Fig. 3.2.5 is $C_{\text{BSC}} = \log 2 + \beta \log \beta + (1 - \beta) \log (1 - \beta)$, or

$$C_{\text{BSC}} = 1 - H(\beta, 1 - \beta) \tag{3.3.2}$$

(see Fig. 3.3.1).

Before trying to compute the capacity in the general case, we need some further properties of the information $I(X \mid Y)$. Suppose that X is the input to a discrete memoryless channel and Y the corresponding

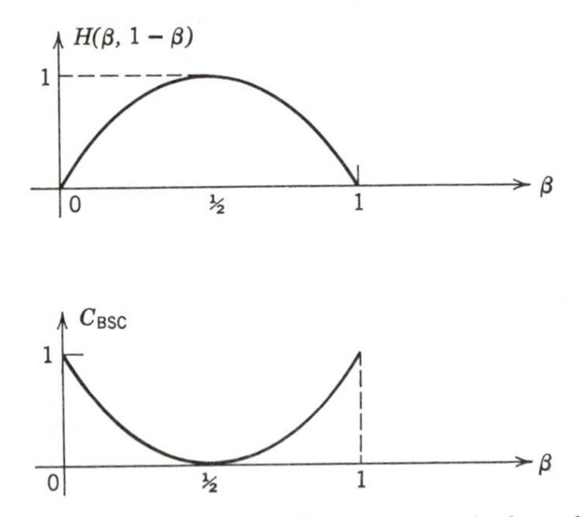

Fig. 3.3.1. Capacity of the binary symmetric channel.

output. Let $I_k(X \mid Y)$ $(k = 1, 2, \ldots, r)$ be the information processed when the input distribution is given by $P\{X = x\} = p_k(x)(x = x_1, \ldots, x_M)$. In other words, we consider the operation of the channel under various input distributions $p_1(x), \ldots, p_r(x)$. We have the following theorem.

Theorem 3.3.1. The information processed by a channel is a convex function of the input probabilities. Specifically, if $a_1, \ldots, a_r$ are non-negative numbers whose sum is unity, and we define an input distribution $p_0(x) = \sum_{i=1}^{r} a_i p_i(x)$ [a convex linear combination of the $p_i(x)$], then the information $I_0(X \mid Y)$ corresponding to $p_0(x)$ satisfies

$$I_0(X \mid Y) \geq \sum_{k=1}^{r} a_k I_k(X \mid Y).$$

In particular if we take a lossless channel then $I_k(X \mid Y) = H_k(X) = -\sum_x p_k(x) \log p_k(x)$, and we have

$$H_0(X) \geq \sum_{k=1}^{r} a_k H_k(X).$$

Hence the uncertainty is also a convex function.

Proof. The subscript k will indicate that the quantity in question is to be evaluated when the input distribution is $p_k(x)$.

Let

$$\Delta I = I_0(X \mid Y) - \sum_{k=1}^{r} a_k I_k(X \mid Y).$$

Then

$$\Delta I = H_0(Y) - H_0(Y \mid X) - \sum_{k=1}^{r} a_k[H_k(Y) - H_k(Y \mid X)].$$

But

$$H_0(Y \mid X) = -\sum_{i=1}^{M}\sum_{j=1}^{L} p_0(x_i, y_j) \log p_0(y_j \mid x_i).$$

Since $p_0(y_j \mid x_i)$ is just the channel probability $p(y_j \mid x_i)$ we have

$$\begin{aligned} H_0(Y \mid X) &= \sum_{k=1}^{r} a_k\Big[-\sum_{i=1}^{M}\sum_{j=1}^{L} p_k(x_i, y_j) \log p(y_j \mid x_i)\Big] \\ &= \sum_{k=1}^{r} a_k H_k(Y \mid X). \end{aligned}$$

Consequently

$$\begin{aligned} \Delta I &= H_0(Y) - \sum_{k=1}^{r} a_k H_k(Y) \\ &= -\sum_{j=1}^{L} p_0(y_j) \log p_0(y_j) + \sum_{k=1}^{r} a_k\left[\sum_{j=1}^{L} p_k(y_j) \log p_k(y_j)\right] \\ &= \sum_{k=1}^{r} a_k\left[-\sum_{j=1}^{L} p_k(y_j) \log p_0(y_j) + \sum_{j=1}^{L} p_k(y_j) \log p_k(y_j)\right]. \end{aligned}$$

Since $a_k \geq 0$, Lemma 1.4.1 yields $\Delta I \geq 0$, completing the argument. Note that the hypothesis $\sum_{k=1}^{r} a_k = 1$ is needed to insure that $p_0(x)$ is a legitimate probability function, that is, $\sum_{i=1}^{M} p_0(x_i) = 1$.

Theorem 3.3.1 yields valuable information about the problem of maximizing $I(X \mid Y)$. If $P\{X = x_i\} = p_i$, $i = 1, 2, \ldots, M$, the resulting information $I(X \mid Y)$ may be regarded as a function $I(p_1, \ldots, p_M)$ of the M variables $p_1, \ldots, p_M$. In the problem of evaluating channel capacity, we are trying to find the particular $p_1, \ldots, p_M$ that maximize I. Theorem 3.3.1 assures us that any solution found by the methods of the differential calculus yields an absolute maximum rather than a relative maximum or minimum or saddle point. In particular we have:

Lemma 3.3.2. Let $g(\mathbf{p})$ be a real-valued function of the vector variable $\mathbf{p} = (p_1, \ldots, p_M)$, defined for $p_i \geq 0$, $i = 1, \ldots, M$. Suppose that g is convex on the set

$$S = \left\{\mathbf{p}\colon\ p_i \geq 0,\ i = 1.\ldots, M, \sum_{i=1}^{M} p_i = 1\right\},$$

that is, $g((1-a)\mathbf{p}+a\mathbf{p}') \geq (1-a)g(\mathbf{p})+ag(\mathbf{p}')$ for all $\mathbf{p}, \mathbf{p}' \in S$ and all $a \in [0, 1]$. [Note that this implies by induction that

$$g\left(\sum_{k=1}^{r} a_k \mathbf{p}_k\right) \geq \sum_{k=1}^{r} a_k g(\mathbf{p}_k)$$

if $a_k \geq 0, \sum_{k=1}^{r} a_k = 1, \mathbf{p}_k = (p_{k1}, \ldots, p_{kM}) \in S, k = 1, 2, \ldots, r$.] Suppose also that g is continuously differentiable for $p_i > 0, i = 1, 2, \ldots, M$. Then if $p_i^* > 0, i = 1, 2, \ldots, M$, and

$$\frac{\partial g(p_1, \ldots, p_M)}{\partial p_i} = 0$$

for all i at the point $\mathbf{p}^* = (p_1^*, \ldots, p_M^*) \in S$, then an absolute maximum of $g(\mathbf{p})$ *on* S, that is, an absolute maximum of the function g restricted to S, is attained at $\mathbf{p} = \mathbf{p}^*$.

Proof. Suppose $g(\mathbf{p}') > g(\mathbf{p}^*)$ for some $\mathbf{p}' \in S$. The line segment joining $\mathbf{p}^*$ to $\mathbf{p}'$ is the set of points $(1-a)\mathbf{p}^* + a\mathbf{p}', 0 \leq a \leq 1$. Note that the line segment lies entirely within S. If we examine the values of g along the segment we find by convexity that

$$\frac{g[(1-a)\mathbf{p}^* + a\mathbf{p}'] - g(\mathbf{p}^*)}{a} \geq \frac{(1-a)g(\mathbf{p}^*) + ag(\mathbf{p}') - g(\mathbf{p}^*)}{a}$$
$$= g(\mathbf{p}') - g(\mathbf{p}^*) > 0 \tag{3.3.3}$$

as long as $0 < a \leq 1$. The hypothesis that $\dfrac{\partial g(\mathbf{p})}{\partial p_i} = 0$ for all i at $\mathbf{p} = \mathbf{p}^*$ implies that the gradient of g is zero at $\mathbf{p}^*$, and hence that the directional derivative of g is zero for all possible directions. In particular, the difference quotient (3.3.3) must approach zero as $a \to 0$, a contradiction. We conclude that for all $\mathbf{p}$ in S, $g(\mathbf{p}) \leq g(\mathbf{p}^*)$.

We can now exhibit a closed-form expression for channel capacity which is valid in certain cases.

Theorem 3.3.3. Suppose that the channel matrix Π of a discrete memoryless channel is square and nonsingular. Let q_{ij} be the element in row i and column j of Π^{-1}, $i, j = 1, \ldots, M$. Suppose that for each $k = 1, 2, \ldots M$,

$$d_k = \sum_{j=1}^{M} q_{jk} \exp_2 \left[-\sum_{i=1}^{M} q_{ji} H(Y \mid X = x_i) \right] > 0. \tag{3.3.4}$$

The channel capacity is given by

$$C = \log \sum_{j=1}^{M} \exp_2 \left[-\sum_{i=1}^{M} q_{ji} H(Y \mid X = x_i) \right] \tag{3.3.5}$$

and a distribution that achieves capacity is given by

$$p(x_k) = 2^{-C}d_k, \qquad k = 1, 2, \ldots, M.$$

Proof. The function that we are trying to maximize is of the form

$$I(X \mid Y) = -\sum_{j=1}^{M} p(y_j) \log p(y_j) + \sum_{i,j=1}^{M} p(x_i, y_j) \log p(y_j \mid x_i), \quad (3.3.6)$$

where

$$p(y_j) = \sum_{i=1}^{M} p(x_i)\, p(y_j \mid x_i); \qquad p(x_i, y_j) = p(x_i)\, p(y_j \mid x_i).$$

The expression (3.3.6) may be regarded as being defined for all nonnegative real values of the variables $p_i = p(x_i)$. Of course, the expression is equal to the information processed by the channel only for $0 \leq p_i \leq 1$, $i = 1, \ldots, M$, $\sum_{i=1}^{M} p_i = 1$. Thus $I(X \mid Y)$ is a continuously differentiable function of the variables $p(x_1), \ldots, p(x_M)$ when all $p(x_i)$ are greater than 0. Note that the quantities $p(y_j \mid x_i)$ are fixed for a given channel. Let us try to maximize (3.3.6) subject to the condition $\sum_{i=1}^{M} p(x_i) = 1$, and let us hope that the solution to the maximization problem does not involve negative values for the $p(x_i)$. (The techniques of the differential calculus do not cover constraints of the form $p(x_i) \geq 0$.) Using the method of Lagrange multipliers, we try to maximize $I(X \mid Y) + \lambda \sum_{i=1}^{M} p(x_i)$ by setting the partial derivatives with respect to each of the $p(x_i)$ equal to zero.

Since $H(Y \mid X) = \sum_{i=1}^{M} p(x_i) H(Y \mid X = x_i)$, we have

$$\frac{\partial H(Y \mid X)}{\partial p(x_k)} = H(Y \mid X = x_k).$$

For the remainder of the proof we write $H(Y \mid x_k)$ for $H(Y \mid X = x_k)$. We are also going to assume that the logarithms are to the base e for convenience in differentiation; we will switch back to base 2 at the end. Now

$$H(Y) = -\sum_{j=1}^{M} p(y_j) \log p(y_j) \quad \text{where} \quad p(y_j) = \sum_{i=1}^{M} p(x_i) p(y_j \mid x_i).$$

Thus

$$\begin{aligned} \frac{\partial H(Y)}{\partial p(x_k)} &= \sum_{j=1}^{M} \frac{\partial H(Y)}{\partial p(y_j)} \frac{\partial p(y_j)}{\partial p(x_k)} \\ &= -\sum_{j=1}^{M} [1 + \log p(y_j)] p(y_j \mid x_k) \\ &= -1 - \sum_{j=1}^{M} p(y_j \mid x_k) \log p(y_j). \end{aligned}$$

Thus the equations

$$\frac{\partial}{\partial p(x_k)}\left[I(X \mid Y) + \lambda \sum_{i=1}^{M} p(x_i)\right] = 0$$

become

$$-1 - \sum_{j=1}^{M} p(y_j \mid x_k) \log p(y_j) - H(Y \mid x_k) + \lambda = 0, \qquad k = 1, 2, \ldots, M, \tag{3.3.7}$$

where

$$p(y_j) = \sum_{i=1}^{M} p(x_i)p(y_j \mid x_i), \qquad j = 1, 2, \ldots, M$$

and

$$\sum_{i=1}^{M} p(x_i) = 1.$$

In matrix form, (3.3.7) becomes

$$\Pi \begin{bmatrix} 1 - \lambda + \log p(y_1) \\ \cdot \\ \cdot \\ \cdot \\ 1 - \lambda + \log p(y_M) \end{bmatrix} = \begin{bmatrix} -H(Y \mid x_1) \\ \cdot \\ \cdot \\ \cdot \\ -H(Y \mid x_M) \end{bmatrix}.$$

[Note that $1 - \lambda = \sum_{j=1}^{M} p(y_j \mid x_k)(1 - \lambda)$.] Since Π is nonsingular we have

$$1 - \lambda + \log p(y_j) = -\sum_{i=1}^{M} q_{ji} H(Y \mid x_i)$$

or

$$p(y_j) \exp(1 - \lambda) = \exp\left[-\sum_{i=1}^{M} q_{ji} H(Y \mid x_i)\right]. \tag{3.3.8}$$

Summing (3.3.8) over j and then taking logarithms, we have [observing that $\sum_{i=1}^{M} p(x_i) = 1$ implies $\sum_{j=1}^{M} p(y_j) = 1$ and vice versa]

$$1 - \lambda = \log \sum_{j=1}^{M} \exp\left[-\sum_{i=1}^{M} q_{ji} H(Y \mid x_i)\right]. \tag{3.3.9}$$

To find the numbers $p(x_i)$ we note that $p(y_j) = \sum_{i=1}^{M} p(x_i)p(y_j \mid x_i)$, or in matrix form,

$$[p(y_1) \cdots p(y_M)] = [p(x_1) \cdots p(x_M)]\Pi.$$

Postmultiplying both sides of this equation by Π^{-1} we have

$$p(x_k) = \sum_{j=1}^{M} p(y_j) q_{jk}. \tag{3.3.10}$$

From (3.3.8) and (3.3.10) we obtain

$$p(x_k) = \exp(\lambda - 1)\sum_{j=1}^{M} q_{jk} \exp\left[-\sum_{i=1}^{M} q_{ji} H(Y \mid x_i)\right]. \qquad (3.3.11)$$

By hypothesis, the right side of (3.3.11) is strictly positive for each k. Thus (3.3.11) and (3.3.9) give us numbers $p(x_1), \ldots, p(x_M), \lambda$ with all $p(x_k) > 0$, $\sum_{k=1}^{M} p(x_k) = 1$, such that

$$\frac{\partial}{\partial p(x_k)}\left[I(X \mid Y) + \lambda \sum_{i=1}^{M} p(x_i)\right] = 0, \qquad k = 1, 2, \ldots, M.$$

By Theorem 3.3.1, $I(X \mid Y) + \lambda \sum_{k=1}^{M} p(x_k)$ is the sum of a convex and a linear (hence convex) function and is therefore convex on the set of nonnegative numbers whose sum is unity. By Lemma 3.3.2, we have found, for the given λ, an absolute maximum of the function $I(X \mid Y) + \lambda \sum_{k=1}^{M} p(x_k)$ over the domain $p(x_i) \geq 0$, $\sum_{i=1}^{M} p(x_i) = 1$.

We claim that the solution yields an absolute maximum for the information processed. For if the channel capacity C is greater than the number $I(X \mid Y)$ found above, then let $p^*(x_1), \ldots, p^*(x_M)$ be an input distribution which achieves capacity. Let $p'(x_1), \ldots, p'(x_M)$ be the input distribution found in the Lagrange multiplier calculation, and let $I(X \mid Y) = I'$ be the resulting information. Then

$$I' + \lambda < C + \lambda = C + \lambda \sum_{i=1}^{M} p^*(x_i).$$

But since $p'(x_i)$, $i = 1, \ldots, M$, yields an absolute maximum,

$$C + \lambda \sum_{i=1}^{M} p^*(x_i) \leq I' + \lambda \sum_{i=1}^{M} p'(x_i) = I' + \lambda,$$

a contradiction.

It remains to find the channel capacity explicitly. If we multiply (3.3.7) by $p(x_k)$ and sum over k, we obtain

$$H(Y) - H(Y \mid X) = 1 - \lambda$$

Hence $1 - \lambda$ is the channel capacity C.

From (3.3.9) we obtain

$$\exp(C) = \sum_{j=1}^{M} \exp\left[-\sum_{i=1}^{M} q_{ji} H(Y \mid x_i)\right]. \qquad (3.3.12)$$

The theorem now follows from (3.3.12) and (3.3.11). (The transition from base e back to base 2 is easily accomplished if we observe that uncertainty in binary units is $\log_2 e$ times the corresponding uncertainty in natural units.)

If the channel matrix Π is square and nonsingular, but the assumption (3.3.4) is not satisfied, then the Lagrange maximization will yield negative or zero values for at least one of the $p(x_i)$, indicating that the maximizing distribution is located on the boundary. It then becomes necessary to set some of the $p(x_i)$ equal to zero and try to maximize $I(X \mid Y)$ as a function of the remaining variables. However, if we set one or more of the $p(x_i)$ equal to zero, we are essentially reducing the number of channel inputs without, in general, reducing the number of possible outputs. The reduced channel matrix is no longer square so that the formulas of Theorem 3.3.3 do not apply. The general problem of computing channel capacity is a problem in numerical analysis, best treated by convex programming methods. (See the notes and remarks at the end of the chapter.)

3.4. *Decoding schemes; the ideal observer*

We now consider for the first time the problem of reliable transmission of messages through a noisy communication channel. In order to achieve reliability, we must be able to determine the input message with a high degree of accuracy after seeing the received sequence of symbols. We are looking for the "best" way of finding the correct input, in other words, for the best "decoding scheme." To formulate the problem more precisely, suppose that a channel has an input alphabet $x_1, x_2, \ldots, x_M$, an output alphabet $y_1, y_2, \ldots, y_L$ and a channel matrix $[p(y_j \mid x_i)]$. For simplicity we consider first the special case in which a single symbol, chosen at random according to a known input distribution $p(x)$, is transmitted through the channel. A *decoder* or *decision scheme* is an assignment to every output symbol y_j of an input symbol $x_j{}^*$ from the alphabet $x_1, \ldots, x_M$. The interpretation is of course that if y_j is received, it will be decoded as $x_j{}^*$. The decoder may be thought of as a deterministic channel with input alphabet $y_1, \ldots, y_L$ and output alphabet $x_1, \ldots, x_M$. If Z is the output of the decoder, then we may express Z as a function of Y, say $Z = g(Y)$. (See Fig. 3.4.1.) Equivalently, we may think of the decoder as partitioning the values of Y into disjoint subsets $B_1, \ldots, B_M$, such that every y in B_i is decoded as x_i.

X		Y		$Z = g(Y)$
x_1		y_1		x_1
x_2	channel	y_2	decoder	x_2
$\cdot$				
$\cdot$				
$\cdot$				
x_M		y_L		x_M

Fig. 3.4.1. Representation of a decoder.

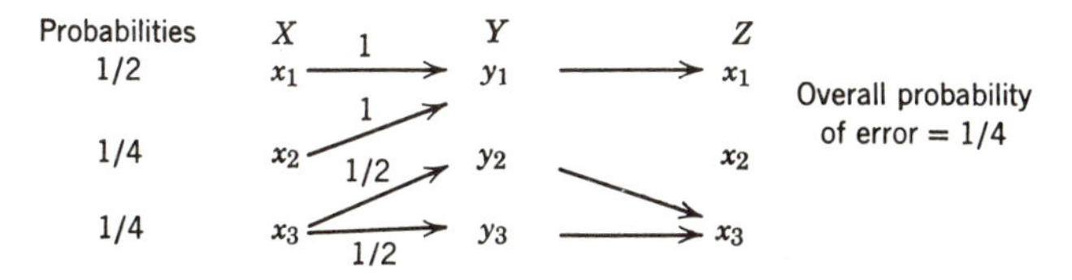

Fig. 3.4.2. Example of channel and decoder.

As an example, consider the channel and decoder combination of Fig. 3.4.2. The symbol y_1 is decoded as x_1 and the symbols y_2 and y_3 as x_3. The probability of error in this case is just the probability that x_2 is chosen, since x_1 and x_3 are always decoded perfectly.

We now propose the following problem: For a given input distribution $p(x)$, construct the decision scheme that minimizes the overall probability of error. Such a decision scheme is called the *ideal observer*. To find the required decoder, assume that each output symbol y_j is associated with an input symbol x_j^* ($j = 1, 2, \dots, L$) (see Fig. 3.4.3). Let $p(e)$ be the overall probability of error and $p(e')$ the overall probability of correct transmission. Given that y_j is received, the probability of correct transmission is the probability that the actual input is x_j^*. Thus we may write the probability of correct transmission as

$$p(e') = \sum_{j=1}^{L} p(y_j)p(e' \mid y_j) = \sum_{j=1}^{L} p(y_j)P\{X = x_j^* \mid y_j\}. \tag{3.4.1}$$

The probability $p(y_j)$ is the same for any decision scheme since $p(y_j)$ depends only on the input distribution and the channel matrix. For each symbol y_j we are free to choose the corresponding x_j^*. It follows from (3.4.1) that if we choose x_j^* as that value of X which maximizes $p(x \mid y_j)$, we have maximized $p(e' \mid y_j)$ for each j, and therefore we have maximized the probability of correct transmission. To summarize, the *ideal-observer*

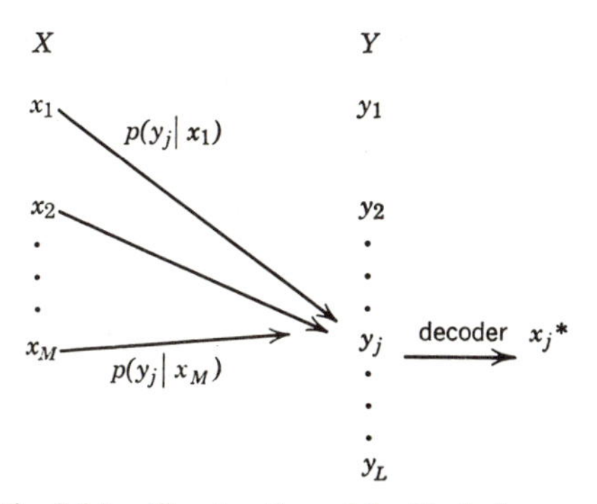

Fig. 3.4.3. Construction of the ideal observer.

decision scheme associates with each output symbol y_j the input symbol x that maximizes $p(x \mid y_j)$. (If more than one input symbol yields a maximum, any one of the maximizing inputs may be chosen; the probability of error will not be affected.)

In a similar fashion we may consider a situation in which a sequence $\mathbf{x} = (\alpha_1, \ldots, \alpha_n)$ chosen in accordance with a distribution $p(\mathbf{x}) = p(\alpha_1, \ldots, \alpha_n)$ is transmitted through the channel. The probability that an output sequence $(\beta_1, \ldots, \beta_n)$ is produced is then given by

$$\begin{aligned} p(\beta_1, \ldots, \beta_n) &= \sum_{\alpha_1, \ldots, \alpha_n} p(\alpha_1, \ldots, \alpha_n) p(\beta_1, \ldots, \beta_n \mid \alpha_1, \ldots, \alpha_n) \\ &= \sum_{\alpha_1, \ldots, \alpha_n} p(\alpha_1, \ldots, \alpha_n) p(\beta_1 \mid \alpha_1) p(\beta_2 \mid \alpha_2) \ldots p(\beta_n \mid \alpha_n). \end{aligned}$$

(Note that $\alpha_1, \ldots, \alpha_n$ need not be chosen independently.)

We may define a decoder or decision scheme as a function that assigns to each output sequence $(\beta_1, \ldots, \beta_n)$ an input sequence $(\alpha_1, \ldots, \alpha_n)$. The ideal observer, that is, the decision scheme minimizing the overall probability of error for the given input distribution, is, just as before, the decoder that selects, for each $(\beta_1, \ldots, \beta_n)$, the input sequence

$$(\alpha_1, \ldots, \alpha_n)$$

which maximizes the conditional probability

$$p(\alpha_1, \ldots, \alpha_n \mid \beta_1, \ldots, \beta_n) = \frac{p(\alpha_1, \ldots, \alpha_n) \prod_{k=1}^{n} p(\beta_k \mid \alpha_k)}{p(\beta_1, \ldots, \beta_n)}$$

An important special case occurs when all inputs are equally likely. If, say, $x_1, \ldots, x_M$ have the same probability, then

$$p(x_i \mid y) = \frac{p(x_i) p(y \mid x_i)}{p(y)} = \frac{1}{M p(y)} p(y \mid x_i)$$

Hence for a fixed y, maximizing $p(x_i \mid y)$ is equivalent to maximizing $p(y \mid x_i)$. Thus *when all inputs are equally likely, the ideal observer selects the input x_i for which $p(y \mid x_i)$ is a maximum.* The resulting decoder is sometimes referred to as a *maximum-likelihood* decision scheme.

The ideal observer suffers from several disadvantages. It is defined only for a particular input distribution; if the input probabilities change, the decision scheme will in general also change. In addition, it may happen that certain inputs are never received correctly. For example, in the channel of Fig. 3.4.2 an error is always made whenever x_2 is sent. It would be more desirable to have a decision scheme with a *uniform error bound.* A decoding scheme with uniform error bound ε is a decoder for which the probability of error given that x_i is sent is less than ε for all i. For such a decision scheme the overall error probability is less than ε for any input distribution. On the other hand, no general method is known for actually constructing decision schemes with uniform error bounds, while the ideal observer is quite easily obtained as soon as the input distribution is known. Also, as we shall see in Chapter 4, the ideal observer for the binary symmetric channel has a significant interpretation which gives considerable insight into the theory of error correcting codes.

Finally, we shall see in the proof of the fundamental theorem (Section 3.5) that if we are studying the behavior of channels under long sequences of inputs, then in a sense (to be made precise in the next section) we may without loss of generality restrict attention to the ideal observer, calculated with all inputs equally likely.

3.5. The fundamental theorem

We now try to formulate precisely the idea that it is possible to transmit information through a noisy channel at any rate less than channel capacity with an arbitrarily small probability of error. Suppose we are given a source of information that produces a sequence of binary digits at the fixed rate of R digits per second. Then in n seconds the source will produce nR digits; hence the total number of messages the source can produce in a period of n seconds is 2^{nR}. In any n-second interval, the source will produce one of the 2^{nR} possible sequences. (In the informal discussion we ignore the fact that nR may not be an integer; in general the number of possible messages will be at most $[2^{nR}]$ = the largest integer $\leq 2^{nR}$. Note also that for the purpose of this discussion we need not consider nonbinary sources; if a source produces S base D digits per second, then we can find an equivalent binary rate R by setting $D^{nS} = 2^{nR}$: $R = S \log D$.)

Suppose that the information produced by the source is to be transmitted through a discrete memoryless channel. Assume for the sake of definiteness that it is possible to send symbols through the channel at any rate ≤ 1 symbol per second. The way we are going to achieve reliable transmission is the following. Suppose that we wait for a period of n seconds and observe the message produced by the source. We then assign a "code word," that is, a sequence of symbols from the input alphabet of the channel, to each message. Thus instead of coding individual digits produced by the source we are coding an entire block of digits. We transmit the code word and, after receiving the output sequence, make a hopefully accurate decision as to the identity of the input sequence. Now if we are to maintain the source transmission rate R, we can take no longer than n seconds to process the information produced by the source in an n-second interval. Since the channel rate is at most 1 symbol per second, the code word assigned to the source message must not contain more than

n symbols. Thus in order to maintain the source rate R we need a vocabulary of 2^{nR} sequences of length $\leq n$, whose components belong to the channel input alphabet. The basic idea of the fundamental theorem is that given $\varepsilon > 0$, if we choose n sufficiently large we can find 2^{nR} code words of length n along with a corresponding decision scheme with a probability of error uniformly $<\varepsilon$, that is, $<\varepsilon$ no matter which code word is transmitted. Thus we can maintain the transmission rate R and at the same time reduce the probability of error toward zero.

A price must be paid in order to achieve the results of the fundamental theorem. First of all, since we must wait n seconds before encoding the source message, there is a possibly long delay between production and transmission of a message. (Further delay may be introduced in the encoding and decoding operations.) In addition, the coding and decoding schemes needed to meet the specifications of the fundamental theorem are in general exceedingly complex and difficult to construct and to implement.

An example may help to clarify the previous ideas. Suppose that $R = \frac{2}{5}$ and $n = 5$. Then in 5 seconds the source will produce one of $2^{nR} = 4$ possible messages, say m_1, m_2, m_3, and m_4. If we are using a binary symmetric channel, then we should assign a binary sequence of length ≤ 5 to each message. For example, we might make the following assignment.

m_1 00000
m_2 01101
m_3 11010
m_4 10111

Now we could assign distinct code words to the messages m_1, m_2, m_3, m_4 by using binary sequences of length 2:

m_1 00
m_2 01
m_3 10
m_4 11

In using sequences of length 5, however, we gain error correcting ability. For example, in the second assignment above an error in a single digit in any code word will cause the word to be confused with another word. In the first assignment, there is a decision scheme to correct all single errors in transmission. If for a given received sequence $\mathbf{v}$ we select the code word $\mathbf{w}$ that differs from $\mathbf{v}$ in the fewest number of places, then noting that for all $i \neq j$, the words $\mathbf{w}_i$ and $\mathbf{w}_j$ disagree in at least 3 digits, we see that if the received sequence $\mathbf{v}$ differs from the transmitted sequence $\mathbf{w}$ in at most 1 digit, the decoder will make a correct decision.

We shall have much more to say about the properties of codes in Chapter 4. We hope that the intuitive idea behind the fundamental theorem is clear; if further discussion is still needed, perhaps a rereading of Section 1.1 will help.

We turn now to the statement and proof of the fundamental theorem. First, a few definitions. Given a discrete memoryless channel, an *input n-sequence* (respectively *output n-sequence*) for the channel is a sequence of n symbols, each belonging to the channel input (respectively output) alphabet. A *code* (s, n) is a set of s input n-sequences $\mathbf{x}^{(1)}, \ldots, \mathbf{x}^{(s)}$, called *code words*, together with a corresponding decision scheme, that is, a function which assigns to each output n-sequence one of the sequences $\mathbf{x}^{(1)}, \ldots, \mathbf{x}^{(s)}$, or equivalently a partition of the set of all output n-sequences into disjoint sets $B_1, \ldots, B_s$, called the *decoding sets*, with the interpretation that an output in B_i will be decoded as $\mathbf{x}^{(i)}$. Note that we do not require that the $\mathbf{x}^{(i)}$ be distinct. However, if $\mathbf{x}^{(i)} = \mathbf{x}^{(j)}$ for $i \neq j$, and the transmission of $\mathbf{x}^{(i)}$ results in an output sequence belonging to B_j, we shall regard this as a decoding error, since according to our previous discussion the indices i and j will correspond to distinct messages produced by the source.

We shall use the symbol $\mathbf{X} = (X_1, \ldots, X_n)$ to denote a randomly chosen input n-sequence, and $\mathbf{Y} = (Y_1, \ldots, Y_n)$ for the corresponding output n-sequence.

Once the sequences in a code and the decoding sets are specified, various error probabilities can be computed. For example, the probability of error given that $\mathbf{x}^{(i)}$ is transmitted is

$$p(e \mid \mathbf{x}^{(i)}) = P\{\mathbf{Y} \notin B_i \mid \mathbf{X} = \mathbf{x}^{(i)}\} = \sum_{\mathbf{y} \notin B_i} p(\mathbf{y} \mid \mathbf{x}^{(i)}).$$

Suppose that a code word is chosen at random in accordance with a distribution $\mathfrak{D}$ which assigns to the code word $\mathbf{x}^{(i)}$ the probability $p(\mathbf{x}^{(i)})$, $i = 1, 2, \ldots, s$. Suppose that the code word chosen is transmitted through the channel and a decision is made at the output in accordance with the given decision scheme. The resulting probability of error is given by

$$p_{\mathfrak{D}}(e) = \sum_{i=1}^{s} p(\mathbf{x}^{(i)}) p(e \mid \mathbf{x}^{(i)}).$$

The *average probability of error* of a code is defined by

$$\overline{p(e)} = s^{-1} \sum_{i=1}^{s} p(e \mid \mathbf{x}^{(i)}).$$

Thus $\overline{p(e)}$ is the error probability $p_{\mathfrak{D}}(e)$ associated with the uniform distribution, that is, the distribution assigning the same probability to all code words.

The *maximum probability of error* of a code is defined by $p_m(e) = \max_{1 \leq i \leq s} p(e \mid \mathbf{x}^{(i)})$. Thus if $p_m(e) \leq \varepsilon$ then, no matter which code word is transmitted, the probability of error is $\leq \varepsilon$; in other words, we have a decoding scheme with uniform error bound ε. Finally, a *code* (s, n, λ) is a code consisting of s input n-sequences, whose maximum probability of error is $\leq \lambda$.† We may now formally state the basic result.

Theorem 3.5.1. (Fundamental Theorem of Information Theory). Given a discrete memoryless channel with capacity $C > 0$ and a positive number $R < C$, there exists a sequence of codes $\mathcal{A}_1, \mathcal{A}_2, \ldots$ such that $\mathcal{A}_n$ is a code $([2^{nR}], n, \lambda_n)$ and $\lambda_n \to 0$ as $n \to \infty$.

Thus it is possible, by choosing n sufficiently large, to reduce the maximum probability of error to a figure as low as desired while at the same time maintaining the transmission rate R.

We shall try to approach the proof gradually. Shannon's original argument was informal and went something like this. Suppose that an input n-sequence $\mathbf{X} = (X_1, \ldots, X_n)$ is chosen at random, with the components X_i being chosen independently. Suppose further that $p(\alpha)$, $\alpha \in \Gamma =$ channel input alphabet, is an input distribution that achieves channel capacity and that each X_i is chosen in accordance with this distribution, that is, $P\{X_i = \alpha\} = p(\alpha)$, $\alpha \in \Gamma$. Let $\mathbf{X}$ be transmitted through the channel and let $\mathbf{Y}$ be the corresponding received sequence. Write $H(X)$ for the common uncertainty $H(X_i)$, $i = 1, 2, \ldots, n$, and similarly define $H(Y)$, $H(X \mid Y)$, $H(Y \mid X)$, and $H(X, Y)$.

Now from Section 1.3 we know that for large n there are approximately $2^{nH(X)}$ "typical" input n-sequences, each with probability roughly $2^{-nH(X)}$; similarly there are $2^{nH(Y)}$ typical output n-sequences and $2^{nH(X,Y)}$ "typical pairs" of input and output n-sequences. A typical pair may be generated by first selecting a typical output n-sequence $\mathbf{y}$, and then selecting a typical input n-sequence $\mathbf{x}$ such that $(\mathbf{x}, \mathbf{y})$ is a typical pair. Since the number of typical output sequences is $2^{nH(Y)}$ and the number of typical pairs is $2^{nH(X,Y)}$, we may conclude that for each typical output sequence $\mathbf{y}$, there are $2^{n[H(X,Y)-H(Y)]} = 2^{nH(X|Y)}$ input sequences $\mathbf{x}$ such that $(\mathbf{x}, \mathbf{y})$ is a typical pair. In other words, if we receive a typical sequence $\mathbf{y}$ and are faced with the problem of determining the corresponding input sequence, there are roughly $2^{nH(X|Y)}$ (*not* $2^{nH(X)}$) possibilities, each with approximately the same probability. (See Fig. 3.5.1.)

Now suppose that a code is formed by selecting 2^{nR} $(R < C)$ typical input n-sequences successively, with the sequences chosen at random,

† This terminology was introduced by Wolfowitz (1961).

independently, and with the $2^{nH(X)}$ possible input sequences being equally likely at each choice. If a code word $\mathbf{x}^{(i)}$ is transmitted through the channel and the typical sequence $\mathbf{y}$ is received, an error in decoding is possible only if at least one code word $\mathbf{x}^{(j)}$, $j \neq i$, belongs to the set S of inputs associated with $\mathbf{y}$.† But

$$P\{\text{at least one } \mathbf{x}^{(j)}, j \neq i, \text{ belongs to } S\} \leq \sum_{\substack{j=1 \\ j \neq i}}^{2^{nR}} P\{\mathbf{x}^{(j)} \in S\}$$

$$\leq 2^{nR} \frac{2^{nH(X|Y)}}{2^{nH(X)}} = \frac{2^{nR}}{2^{nC}} \to 0 \text{ as } n \to \infty.$$

The above discussion suggests the possibility of constructing codes that maintain any transmission rate $R < C$ with an arbitrarily small probability

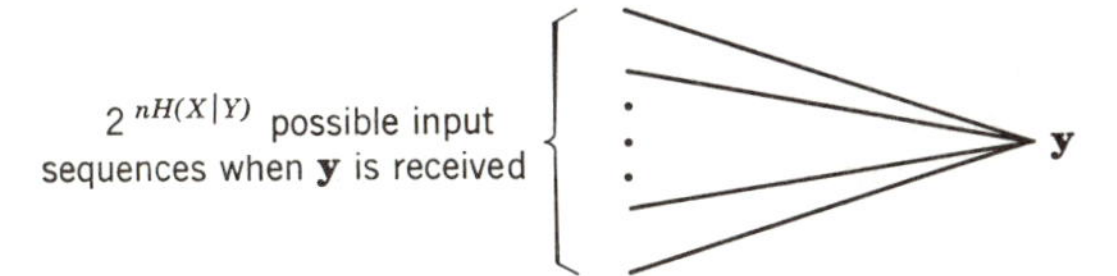

Fig. 3.5.1. Intuitive justification of the fundamental theorem.

of error. If a code is selected by choosing 2^{nR} input n-sequences independently and at random, with the components of each code word chosen independently according to the distribution that achieves capacity (equivalently, if each code word is selected by choosing a typical input n-sequence at random, with all typical sequences having the same probability), then "on the average," accurate decoding can be accomplished. Furthermore, if $(\mathbf{x}, \mathbf{y})$ is a typical pair, then

$$\frac{p(\mathbf{y} \mid \mathbf{x})}{p(\mathbf{y})} = \frac{p(\mathbf{x}, \mathbf{y})}{p(\mathbf{x})p(\mathbf{y})} \approx \frac{2^{-nH(X,Y)}}{2^{-nH(X)}2^{-nH(Y)}} = 2^{nC}.$$

Let b be a positive number less than C. It follows that for "most" pairs $(\mathbf{x}, \mathbf{y})$,

$$\frac{1}{n} \log \frac{p(\mathbf{y} \mid \mathbf{x})}{p(\mathbf{y})} > b;$$

hence if a typical input sequence $\mathbf{x}$ is transmitted, it is very likely that the output sequence $\mathbf{y}$ will belong to the set

$$A_{\mathbf{x}} = \left\{\mathbf{y} \colon \frac{1}{n} \log \frac{p(\mathbf{y} \mid \mathbf{x})}{p(\mathbf{y})} > b\right\}.$$

Thus if $\mathbf{x}$ is a code word and we impose the requirement that the decoding set associated with $\mathbf{x}$ be a subset of $A_{\mathbf{x}}$, the effect on the probability of

†We assume the ideal observer decision scheme with all code words equally likely.

error will be small. This observation gives some information about the structure of the decoding sets for the desired codes, and will be useful in following the reasoning to be used in the proof of the fundamental theorem.

Although the above discussion is not in the form of a mathematical proof, Shannon's remarkable intuitive argument is the ancestor of all known proofs of the fundamental theorem.

We shall give two proofs of the fundamental theorem. The first proof will be based on an (unfortunately impractical) algorithm for constructing the desired codes. We shall require that the decoding set associated with a given code word $\mathbf{x}$ be a subset of the set $A_{\mathbf{x}}$ described above; this device will allow us to estimate the relationship between the number of words in a code and the probability of error. The second proof will be a precise translation of the idea that a code selected randomly will under certain conditions exhibit, on the average, desirable properties.

We first proceed to the "algorithmic" proof. The key part of the argument is the following result.

Lemma 3.5.2. (Fundamental Lemma). Given an arbitrary discrete memoryless channel, let n be a fixed positive integer, and $p(\mathbf{x})$ an arbitrary probability function defined on the set of all input n-sequences. If $\mathbf{x}$ is an input n-sequence and $\mathbf{y}$ an output n-sequence, let

$$p(\mathbf{y} \mid \mathbf{x}) = P\{\mathbf{Y} = \mathbf{y} \mid \mathbf{X} = \mathbf{x}\},$$

$$p(\mathbf{y}) = P\{\mathbf{Y} = \mathbf{y}\} = \sum_{\mathbf{x}} p(\mathbf{x}, \mathbf{y}) \quad \text{where} \quad p(\mathbf{x}, \mathbf{y}) = p(\mathbf{x})p(\mathbf{y} \mid \mathbf{x})$$

For any real number a, let

$$A = \left\{(\mathbf{x}, \mathbf{y}) \colon \log \frac{p(\mathbf{y} \mid \mathbf{x})}{p(\mathbf{y})} > a\right\}.$$

[Note A is a set of $(2n)$-tuples, with the first n components in the input alphabet and the last n components in the output alphabet; an intuitive justification for the appearance of such a set has been given above.] Then for each positive integer s there exists a code (s, n, λ) such that

$$\lambda \leq s \cdot 2^{-a} + P\{(\mathbf{X}, \mathbf{Y}) \notin A\} \tag{3.5.1}$$

where $P\{(\mathbf{X}, \mathbf{Y}) \notin A\}$ is computed under the distribution $p(\mathbf{x})$.

The essential content of the fundamental lemma is that given any probability function $p(\mathbf{x})$, then by means of an algorithm to be described below, an algorithm whose steps depend in a significant way on the particular function $p(\mathbf{x})$, we can construct a code whose maximum probability of error is bounded by the expression (3.5.1). In order to use this result to prove the fundamental theorem, we will take $s = [2^{nR}]$, $R < C$, and then choose the number a and the distribution $p(\mathbf{x})$ so that the bound on the maximum probability of error will approach 0 as n approaches infinity.

Proof. We shall construct the code by means of an iterative procedure as diagrammed schematically in Fig. 3.5.2. For any input n-sequence $\mathbf{x}$,

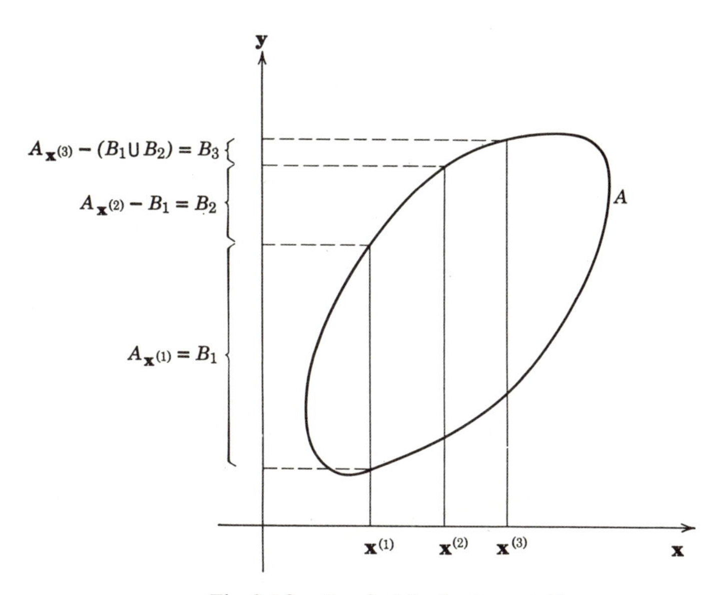

Fig. 3.5.2. Proof of the fundamental lemma.

let $A_{\mathbf{x}}$ be the "section of A at $\mathbf{x}$," that is, $A_{\mathbf{x}} = \{\mathbf{y} : (\mathbf{x}, \mathbf{y}) \in A\}$. Let $\varepsilon = s \cdot 2^{-a} + P\{(\mathbf{X}, \mathbf{Y}) \notin A\}$. We are attempting to construct a code (s, n, λ) with $\lambda \leq \varepsilon$. If $\varepsilon \geq 1$, there is nothing to prove since in that case any code (s, n, λ) has $\lambda \leq \varepsilon$. Thus assume $0 < \varepsilon < 1$.

First pick if possible any input n-sequence $\mathbf{x}^{(1)}$ such that

$$P\{\mathbf{Y} \in A_{\mathbf{x}^{(1)}} \mid \mathbf{X} = \mathbf{x}^{(1)}\} \geq 1 - \varepsilon;$$

take the decoding set B_1 to be $A_{\mathbf{x}^{(1)}}$. (Thus if $\mathbf{x}^{(1)}$ is transmitted, the probability of error cannot exceed ε.)

Then select any input n-sequence $\mathbf{x}^{(2)}$ such that

$$P\{\mathbf{Y} \in A_{\mathbf{x}^{(2)}} - B_1 \mid \mathbf{X} = \mathbf{x}^{(2)}\} \geq 1 - \varepsilon,$$

where $A_{\mathbf{x}^{(2)}} - B_1$ is the set of elements that belong to $A_{\mathbf{x}^{(2)}}$ but not to B_1. Take $B_2 = A_{\mathbf{x}^{(2)}} - B_1$. (Note that we cannot choose $B_2 = A_{\mathbf{x}^{(2)}}$ since the decoding sets must be disjoint.) In general, having chosen $\mathbf{x}^{(1)}, \ldots, \mathbf{x}^{(k-1)}$ and $B_1, \ldots, B_{k-1}$, pick any input n-sequence $\mathbf{x}^{(k)}$ such that

$$P\left\{\mathbf{Y} \in A_{\mathbf{x}^{(k)}} - \bigcup_{i=1}^{k-1} B_i \,\middle|\, \mathbf{X} = \mathbf{x}^{(k)}\right\} \geq 1 - \varepsilon;$$

take

$$B_k = A_{\mathbf{x}^{(k)}} - \bigcup_{i=1}^{k-1} B_i.$$

Since there are only finitely many input sequences, the process must terminate, say, after t steps (conceivably $t = 0$). If we show that t is necessarily $\geq s$, the sequence $\mathbf{x}^{(1)}, \ldots, \mathbf{x}^{(s)}$ and corresponding decoding sets $B_1, \ldots, B_s$ form, by construction, a code (s, n, λ) with

$$\lambda \leq \max_{1 \leq i \leq s} P\{\mathbf{Y} \notin B_i \mid \mathbf{X} = \mathbf{x}^{(i)}\} \leq \varepsilon.$$

To show $t \geq s$, we will prove that $\varepsilon \leq t \cdot 2^{-a} + P\{(\mathbf{X}, \mathbf{Y}) \notin A\}$; the result $t \geq s$ then follows from the definition of ε.

Let us examine $P\{(\mathbf{X}, \mathbf{Y}) \in A\}$. We have

$$P\{(\mathbf{X}, \mathbf{Y}) \in A\} = \sum_{(\mathbf{x},\mathbf{y}) \in A} p(\mathbf{x}, \mathbf{y}) = \sum_{\mathbf{x}} p(\mathbf{x}) \left[\sum_{\mathbf{y} \in A_{\mathbf{x}}} p(\mathbf{y} \mid \mathbf{x})\right].$$

Let $B = \bigcup_{j=1}^{t} B_j$. (If $t = 0$ take B to be the empty set $\emptyset$. The argument below will go through essentially unchanged.) Now

$$P\{(\mathbf{X}, \mathbf{Y}) \in A\} = \sum_{\mathbf{x}} p(\mathbf{x}) \left[\sum_{\mathbf{y} \in B \cap A_{\mathbf{x}}} p(\mathbf{y} \mid \mathbf{x})\right] + \sum_{\mathbf{x}} p(\mathbf{x}) \left[\sum_{\mathbf{y} \in \bar{B} \cap A_{\mathbf{x}}} p(\mathbf{y} \mid \mathbf{x})\right] \tag{3.5.2}$$

($\bar{B}$ denotes the complement of B). Since $B \cap A_{\mathbf{x}} \subset B$, the first term of (3.5.2) cannot exceed

$$\sum_{\mathbf{x}} p(\mathbf{x}) \sum_{\mathbf{y} \in B} p(\mathbf{y} \mid \mathbf{x}) = P\{\mathbf{Y} \in B\}.$$

But since $B_i \subset A_{\mathbf{x}^{(i)}}$,

$$P\{\mathbf{Y} \in B\} = \sum_{i=1}^{t} P\{\mathbf{Y} \in B_i\} \leq \sum_{i=1}^{t} P\{\mathbf{Y} \in A_{\mathbf{x}^{(i)}}\}.$$

Now if $\mathbf{y} \in A_{\mathbf{x}^{(i)}}$, then $(\mathbf{x}^{(i)}, \mathbf{y}) \in A$, and hence $\log [p(\mathbf{y} \mid \mathbf{x}^{(i)})/p(\mathbf{y})] > a$, or

$$p(\mathbf{y}) < p(\mathbf{y} \mid \mathbf{x}^{(i)}) 2^{-a}.$$

Consequently

$$P\{\mathbf{Y} \in A_{\mathbf{x}^{(i)}}\} = \sum_{\mathbf{y} \in A_{\mathbf{x}^{(i)}}} p(\mathbf{y}) \leq 2^{-a} \sum_{\mathbf{y} \in A_{\mathbf{x}^{(i)}}} p(\mathbf{y} \mid \mathbf{x}^{(i)}) \leq 2^{-a}.$$

Therefore the first term of (3.5.2) is $\leq t \cdot 2^{-a}$.

In order to estimate the second term of (3.5.2), we note first that

$$\bar{B} \cap A_{\mathbf{x}} = \emptyset \quad \text{when} \quad \mathbf{x} \text{ is a code word } \mathbf{x}^{(k)}. \tag{3.5.3}$$

To verify this, we will show that if a sequence $\mathbf{y}$ belongs to $A_{\mathbf{x}^{(k)}}$, then $\mathbf{y}$ must also belong to B. Now we know that $B_k = A_{\mathbf{x}^{(k)}} - \bigcup_{i=1}^{k-1} B_i$; thus if $\mathbf{y} \notin \bigcup_{i=1}^{k-1} B_i$ then $\mathbf{y} \in B_k \subset B$. On the other hand if $\mathbf{y} \in \bigcup_{i=1}^{k-1} B_i$, then since B is the union of all the decoding sets, $\mathbf{y} \in B$. This establishes (3.5.3). We now claim that

$$\sum_{\mathbf{y} \in \bar{B} \cap A_{\mathbf{x}}} p(\mathbf{y} \mid \mathbf{x}) < 1 - \varepsilon \quad \text{for} \quad \text{all input } n\text{-sequences } \mathbf{x}. \tag{3.5.4}$$

By (3.5.3), the statement (3.5.4) holds for $\mathbf{x} = \mathbf{x}^{(1)}, \ldots, \mathbf{x}^{(t)}$, so assume $\mathbf{x}$ not equal to any of the $\mathbf{x}^{(i)}$. If $P\{\mathbf{Y} \in \bar{B} \cap A_{\mathbf{x}} \mid \mathbf{X} = \mathbf{x}\} \geq 1 - \varepsilon$ then since $\bar{B} \cap A_{\mathbf{x}} = A_{\mathbf{x}} - \bigcup_{i=1}^{t} B_i$, we could enlarge the code by adding the input n-sequence $\mathbf{x}^{(t+1)} = \mathbf{x}$ and the decoding set $B_{t+1} = \bar{B} \cap A_{\mathbf{x}}$. This contradicts the assumption that the process terminates after t steps. This proves (3.5.4) and shows that the second term of (3.5.2) is $\leq 1 - \varepsilon$. Thus $P\{(\mathbf{X}, \mathbf{Y}) \in A\} \leq t \cdot 2^{-a} + 1 - \varepsilon$, or $\varepsilon \leq t \cdot 2^{-a} + P\{(\mathbf{X}, \mathbf{Y}) \notin A\}$; the proof is complete.

Notice that *the fact that the channel is memoryless is not used in the proof.* We used only the existence of a system of probability functions $p(\mathbf{y} \mid \mathbf{x})$, where $\mathbf{x}$ is an input n-sequence and $\mathbf{y}$ an output n-sequence. Thus the lemma applies to any discrete channel whose probability functions do not depend on the state, or equally well to an arbitrary discrete channel whose initial state is chosen at random in accordance with a known distribution. This observation will be useful in Chapter 7.

Proof of Theorem 3.5.1. Let n be any positive integer. Let $p_0(\alpha)$ be an input distribution that achieves channel capacity. In Lemma 3.5.2, take $s = [2^{nR}]$, $a = n[(R + C)/2] < nC$, and determine the distribution $p(\mathbf{x})$ by choosing $X_1, \ldots, X_n$ independently, each X_i having the distribution $p_0(\alpha)$. In other words, if $\mathbf{x} = (\alpha_1, \ldots, \alpha_n)$, let

$$p(\mathbf{x}) = p(\alpha_1, \ldots, \alpha_n) = \prod_{i=1}^{n} p_0(\alpha_i).$$

By Lemma 3.5.2, there is a code $([2^{nR}], n, \lambda_n)$ with

$$\lambda_n \leq [2^{nR}]2^{-\frac{1}{2}n(R+C)} + P\{(\mathbf{X}, \mathbf{Y}) \notin A\}. \tag{3.5.5}$$

Since $R < C$, the first term of (3.5.5) approaches zero as n approaches infinity. Now

$$P\{(\mathbf{X}, \mathbf{Y}) \notin A\} = P\left\{(\mathbf{x}, \mathbf{y}) : \log \frac{p(\mathbf{y} \mid \mathbf{x})}{p(\mathbf{y})} \leq a\right\}.$$

In order to estimate this expression, we first observe that since $X_1, \ldots, X_n$ are independent, so are $Y_1, \ldots, Y_n$. (The converse is not true; see Problem 3.13.) To see this, note that

$$\begin{aligned}
P\{Y_1 = \beta_1, \ldots, Y_n = \beta_n\} &= \sum_{\alpha_1, \ldots, \alpha_n} P\{X_1 = \alpha_1, \ldots, X_n = \alpha_n\} \\
&\qquad \times P\{Y_1 = \beta_1, \ldots, Y_n = \beta_n \mid X_1 = \alpha_1, \ldots, X_n = \alpha_n\} \\
&= \sum_{\alpha_1, \ldots, \alpha_n} p_0(\alpha_1) \cdots p_0(\alpha_n) p(\beta_1 \mid \alpha_1) \cdots p(\beta_n \mid \alpha_n) \\
&= \prod_{i=1}^{n} \left[\sum_{\alpha_i} p_0(\alpha_i) p(\beta_i \mid \alpha_i) \right] = \prod_{i=1}^{n} P\{Y_i = \beta_i\}.
\end{aligned}$$

Therefore

$$\begin{aligned}
P\{(\mathbf{X}, \mathbf{Y}) \notin A\} &= P\left\{(\alpha_1, \ldots, \alpha_n, \beta_1, \ldots, \beta_n) : \log \left[\prod_{i=1}^{n} \frac{p(\beta_i \mid \alpha_i)}{p(\beta_i)} \right] \leq a \right\} \\
&= P\left\{(\alpha_1, \ldots, \alpha_n, \beta_1, \ldots, \beta_n) : \frac{1}{n} \sum_{i=1}^{n} \log \frac{p(\beta_i \mid \alpha_i)}{p(\beta_i)} \leq \frac{R + C}{2} \right\}.
\end{aligned}$$

Now for each $k = 1, 2, \ldots, n$, define a random variable $U(X_k, Y_k)$ as follows: If $X_k = \alpha_k$, $Y_k = \beta_k$, let

$$U(X_k, Y_k) = \log \frac{p(\beta_k \mid \alpha_k)}{p(\beta_k)}.$$

Then $U(X_1, Y_1), \ldots, U(X_n, Y_n)$ are independent, identically distributed random variables with expected value

$$\sum_{\alpha_k, \beta_k} p(\alpha_k, \beta_k) \log \frac{p(\beta_k \mid \alpha_k)}{p(\beta_k)} = H(Y_k) - H(Y_k \mid X_k) = C.$$

Finally,

$$P\{(\mathbf{X}, \mathbf{Y}) \notin A\} = P\left\{\frac{1}{n} \sum_{i=1}^{n} U(X_i, Y_i) \leq \frac{R + C}{2}\right\}.$$

But by the weak law of large numbers, $(1/n) \sum_{i=1}^{n} U(X_i, Y_i)$, the arithmetic average of n independent, identically distributed random variables with expected value C, converges in probability to C. Since $\frac{1}{2}(R + C) < C$, $P\{(\mathbf{X}, \mathbf{Y}) \notin A\} \to 0$ as $n \to \infty$, and the proof is complete. Note that in order for the above argument to succeed, we must choose the number a large enough so that the first term of (3.5.1) will approach zero, but small enough so that the second term of (3.5.1) will approach zero.

We are now going to give the second proof of the fundamental theorem.

As we have indicated previously, the idea is that if a code is selected by choosing the code words at random in a certain way, then with high probability a "good" code is obtained. Before we proceed, we note that for the purpose of proving the fundamental theorem, we may assume without loss of generality that we are using the ideal observer with all code words equally likely. More precisely we have the following:

Lemma 3.5.3. Given a discrete memoryless channel and a positive number R_0, suppose that for each $R < R_0$ we can find a sequence of codes $([2^{nR}], n)$ whose *average* probability of error approaches zero as $n \to \infty$. Then it follows that for each $R' < R_0$ we can find a sequence of codes $([2^{nR'}], n, \lambda_n)$ such that $\lambda_n \to 0$ as $n \to \infty$.

[Recall that the average probability of error $\overline{p(e)}$ of a code with words $\mathbf{x}^{(1)}, \ldots, \mathbf{x}^{(s)}$ and a given decision scheme $\mathcal{S}$ is the probability of error associated with a random experiment; this involves selecting a code word at random with all words equally likely, transmitting the word through the channel, and then making a decision at the output according to the scheme $\mathcal{S}$. In particular we may take $\mathcal{S}$ to be the ideal-observer decision scheme (calculated for the uniform input distribution). Thus Lemma 3.5.3 implies that if we can construct codes to maintain any transmission rate below capacity and at the same time have an overall probability of error, calculated for the ideal-observer decision scheme with all words equally likely, approaching zero with increasing code-word length, then it is possible to construct a (possibly) different set of codes which maintain any rate below capacity and whose *maximum* probability of error approaches zero.]

Proof. Let the code words of the code $([2^{nR}], n)$ be $\mathbf{x}^{(1)}, \ldots, \mathbf{x}^{(s)}$, $s = [2^{nR}]$. If the average probability of error is $\overline{p_n(e)}$, then

$$\sum_{i=1}^{s} p_n(e \mid \mathbf{x}^{(i)}) = s\,\overline{p_n(e)}. \tag{3.5.6}$$

We claim that $p_n(e \mid \mathbf{x}^{(i)}) \leq 2\,\overline{p_n(e)}$ for at least half of the words $\mathbf{x}^{(1)}, \ldots, \mathbf{x}^{(s)}$. For if $p_n(e \mid \mathbf{x}^{(i)}) > 2\,\overline{p_n(e)}$ for (say) $i = 1, 2, \ldots, r$ where $r \geq s/2$, then

$$\sum_{i=1}^{s} p_n(e \mid \mathbf{x}^{(i)}) \geq \sum_{i=1}^{r} p_n(e \mid \mathbf{x}^{(i)}) > \frac{s}{2}\,2\,\overline{p_n(e)} = s\,\overline{p_n(e)},$$

contradicting (3.5.6). (All we are saying here is that no more than half of a set of nonnegative numbers can exceed twice the average of the set.)

Thus we can find code words $\mathbf{x}^{(i_1)}, \ldots, \mathbf{x}^{(i_r)}$, $r \geq s/2$, such that $p_n(e \mid \mathbf{x}^{(i_k)}) \leq 2\,\overline{p_n(e)} \to 0$ as $n \to \infty$. Thus by using the words $\mathbf{x}^{(i_1)}, \ldots, \mathbf{x}^{(i_r)}$ and the corresponding decoding sets $B_{i_1}, \ldots, B_{i_r}$ to form a subcode of the original code, we can form a sequence of codes (r_n, n, λ_n) where $r_n \geq \frac{1}{2}[2^{nR}]$ and $\lambda_n \leq 2\overline{p_n(e)} \to 0$ as $n \to \infty$. Now

$$\tfrac{1}{2}[2^{nR}] \geq [\tfrac{1}{2} \cdot 2^{nR}] = [2^{n(R-1/n)}] = [2^{nR_n'}]$$

where $R_n' \to R$ as $n \to \infty$. Hence if we are given $R' < R_0$, we need only choose R such that $R' < R < R_0$ and construct the corresponding sequence of subcodes (r_n, n, λ_n). For sufficiently large n, $r_n \geq [2^{nR'}]$; this proves the lemma.

We may now give the second proof of Theorem 3.5.1.

Proof of the Fundamental Theorem by "Random Coding." As in the first proof of the theorem, let $p_0(\alpha)$ be an input distribution that achieves channel capacity, and let $p(\mathbf{x})$ be the distribution on input n-sequences determined by choosing $X_1, \ldots, X_n$ independently, each X_i having the distribution $p_0(\alpha)$. We construct a random experiment as follows.

1. Choose code words (of length n) $\mathbf{x}^{(1)}, \ldots, \mathbf{x}^{(s)}$ independently ($s = [2^{nR}]$); for each $i = 1, 2, \ldots, s$, let $P\{\mathbf{x}^{(i)} = \mathbf{x}\} = p(\mathbf{x})$. In other words, choose the components of each $\mathbf{x}^{(i)}$ independently in accordance with the capacity-achieving distribution. For example, if the channel is binary and the distribution $p(0) = 2/3$, $p(1) = 1/3$ achieves capacity, and $s = 3$, $n = 4$, then the probability of obtaining a code with the words $\mathbf{x}^{(1)} = 1101$, $\mathbf{x}^{(2)} = 0010$, $\mathbf{x}^{(3)} = 0101$, is

$$p(\mathbf{x}^{(1)})p(\mathbf{x}^{(2)})p(\mathbf{x}^{(3)}) = \frac{2}{3}\left(\frac{1}{3}\right)^3 \cdot \left(\frac{2}{3}\right)^3 \frac{1}{3} \cdot \left(\frac{2}{3}\right)^2 \left(\frac{1}{3}\right)^2.$$

2. Calculate the average probability of error $\bar{\varepsilon}(\mathbf{x}^{(1)}, \ldots, \mathbf{x}^{(s)})$ for the code selected in step 1, using the decision scheme determined by the ideal observer with all words equally likely. The notation emphasizes that $\bar{\varepsilon}$ is a random variable which depends on $\mathbf{x}^{(1)}, \ldots, \mathbf{x}^{(s)}$.

We are going to show that $E[\bar{\varepsilon}]$ approaches zero as n approaches infinity. Since there must be at least one code with $\bar{\varepsilon} \leq E[\bar{\varepsilon}]$, it will then follow that we can construct a sequence of codes whose average probability of error goes to zero. Lemma 3.5.3 then yields the desired result.

By definition of the average probability of error, $\bar{\varepsilon} = (1/s)\sum_{i=1}^{s} \varepsilon_i$, where $\varepsilon_i = \varepsilon_i(\mathbf{x}^{(1)}, \ldots, \mathbf{x}^{(s)})$ is the probability of error when $\mathbf{x}^{(i)}$ is transmitted for the code chosen in step 1. Hence the expected value of $\bar{\varepsilon}$ is

$$E[\bar{\varepsilon}] = \frac{1}{s}\sum_{i=1}^{s} E[\varepsilon_i]. \tag{3.5.7}$$

We may write $E[\varepsilon_i]$ as follows:

$$E[\varepsilon_i] = \sum_{\mathbf{x}} P\{\mathbf{x}^{(i)} = \mathbf{x}\}E[\varepsilon_i \mid \mathbf{x}^{(i)} = \mathbf{x}]. \tag{3.5.8}$$

Now if B_i is the decoding set corresponding to the code word $\mathbf{x}^{(i)}$, then

$$\varepsilon_i = \sum_{\mathbf{y} \notin B_i} p(\mathbf{y} \mid \mathbf{x}^{(i)}).$$

Equivalently,

$$\varepsilon_i = \sum_{\mathbf{y}} p(\mathbf{y} \mid \mathbf{x}^{(i)}) \gamma_i$$

where

$$\begin{aligned} \gamma_i = \gamma_i(\mathbf{x}^{(1)}, \ldots, \mathbf{x}^{(s)}; \mathbf{y}) &= 0 \quad \text{if} \quad \mathbf{y} \in B_i \\ &= 1 \quad \text{if} \quad \mathbf{y} \notin B_i \end{aligned}$$

(note that the $B_i = B_i(\mathbf{x}^{(1)}, \ldots, \mathbf{x}^{(s)})$ are "random sets" which depend on the code chosen in step 1). Consequently

$$\begin{aligned} E[\varepsilon_i \mid \mathbf{x}^{(i)} = \mathbf{x}] &= \sum_{\mathbf{y}} p(\mathbf{y} \mid \mathbf{x}) E[\gamma_i \mid \mathbf{x}^{(i)} = \mathbf{x}] \\ &= \sum_{\mathbf{y}} p(\mathbf{y} \mid \mathbf{x}) P\{\gamma_i = 1 \mid \mathbf{x}^{(i)} = \mathbf{x}\} \\ &= \sum_{\mathbf{y}} p(\mathbf{y} \mid \mathbf{x}) P\{\mathbf{y} \notin B_i \mid \mathbf{x}^{(i)} = \mathbf{x}\} \\ &= \sum_{\mathbf{y}} p(\mathbf{y} \mid \mathbf{x}) p^*(e \mid \mathbf{x}^{(i)}, \mathbf{x}, \mathbf{y}) \end{aligned} \tag{3.5.9}$$

where $p^*(e \mid \mathbf{x}^{(i)}, \mathbf{x}, \mathbf{y}) = P\{\mathbf{y} \notin B_i \mid \mathbf{x}^{(i)} = \mathbf{x}\}$ may be interpreted as the probability of a decoding error given that

a. $\mathbf{x}^{(i)}$ is transmitted,

b. $\mathbf{x}^{(i)} = \mathbf{x}$,

c. $\mathbf{y}$ is received.

From (3.5.8) and (3.5.9) we have

$$E[\varepsilon_i] = \sum_{\mathbf{x},\mathbf{y}} p(\mathbf{x}, \mathbf{y}) p^*(e \mid \mathbf{x}^{(i)}, \mathbf{x}, \mathbf{y}) \tag{3.5.10}$$

To estimate $p^*(e \mid \mathbf{x}^{(i)}, \mathbf{x}, \mathbf{y})$, we note that if $\mathbf{x}^{(i)} = \mathbf{x}$ is transmitted and $\mathbf{y}$ is received, then by definition of the ideal observer with equally likely inputs, an error in decoding will be made only if there is an input n-sequence $\mathbf{x}'$ such that $p(\mathbf{y} \mid \mathbf{x}') \geq p(\mathbf{y} \mid \mathbf{x})$, and at least one code word $\mathbf{x}^{(j)}, j \neq i$, is chosen to be $\mathbf{x}'$ (see Fig. 3.5.3).

Now let $G(\mathbf{x}, \mathbf{y})$ be the set of all input n-sequences $\mathbf{x}'$ such that $p(\mathbf{y} \mid \mathbf{x}') \geq p(\mathbf{y} \mid \mathbf{x})$; thus $G(\mathbf{x}, \mathbf{y})$ is the set of all input n-sequences that could conceivably appear at the output of the decoder if $\mathbf{x}$ is transmitted and $\mathbf{y}$ is received. If $\mathbf{x}^{(i)} = \mathbf{x}$ is transmitted and $\mathbf{y}$ is received, the occurrence of an error implies

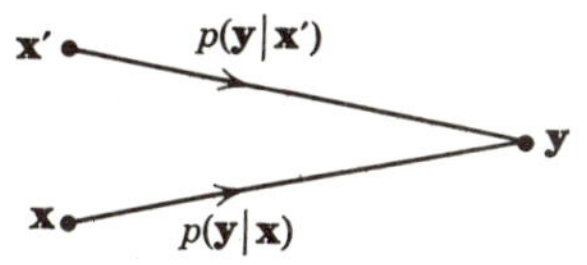

Fig. 3.5.3. Proof of the fundamental theorem by random coding.

that at least one $\mathbf{x}^{(j)}, j \neq i$, belongs to $G(\mathbf{x}, \mathbf{y})$. Therefore

$$P\{\mathbf{y} \notin B_i \mid \mathbf{x}^{(i)} = \mathbf{x}\} \leq P\{\mathbf{x}^{(j)} \in G(\mathbf{x}, \mathbf{y}) \quad \text{for at least one } j \neq i \mid \mathbf{x}^{(i)} = \mathbf{x}\}$$
$$\leq \sum_{j \neq i} P\{\mathbf{x}^{(j)} \in G(\mathbf{x}, \mathbf{y}) \mid \mathbf{x}^{(i)} = \mathbf{x}\}.$$

Since the $\mathbf{x}^{(i)}$ are independent,

$$p^*(e \mid \mathbf{x}^{(i)}, \mathbf{x}, \mathbf{y}) \leq \sum_{j \neq i} P\{\mathbf{x}^{(j)} \in G(\mathbf{x}, \mathbf{y})\}.$$

(Observe that only pairwise independence is required here, that is, all we need is that $\mathbf{x}^{(i)}$ and $\mathbf{x}^{(j)}$ be independent for $i \neq j$.) But since each $\mathbf{x}^{(j)}$ has the same distribution,

$$P\{\mathbf{x}^{(j)} \in G(\mathbf{x}, \mathbf{y})\} = \sum_{\mathbf{x}' \in G(\mathbf{x}, \mathbf{y})} p(\mathbf{x}'), \qquad j = 1, 2, \ldots, s. \tag{3.5.11}$$

If we define $Q(\mathbf{x}, \mathbf{y})$ to be the probability (3.5.11), we have

$$p^*(e \mid \mathbf{x}^{(i)}, \mathbf{x}, \mathbf{y}) \leq (s-1)Q(\mathbf{x}, \mathbf{y}) \leq sQ(\mathbf{x}, \mathbf{y}). \tag{3.5.12}$$

Now consider the set A of Lemma 3.5.2 with $a = \frac{1}{2}n(R + C)$, that is,

$$A = \left\{(\mathbf{x}, \mathbf{y}) : \frac{1}{n} \log \frac{p(\mathbf{y} \mid \mathbf{x})}{p(\mathbf{y})} > \tfrac{1}{2}(R + C)\right\}$$
$$= \left\{(\mathbf{x}, \mathbf{y}) : \frac{1}{n} \log \frac{p(\mathbf{x} \mid \mathbf{y})}{p(\mathbf{x})} > \tfrac{1}{2}(R + C)\right\}.$$

We are going to estimate $Q(\mathbf{x}, \mathbf{y})$ for $(\mathbf{x}, \mathbf{y}) \in A$. If $(\mathbf{x}, \mathbf{y}) \in A$ then $\log [p(\mathbf{x} \mid \mathbf{y})/p(\mathbf{x})] > \frac{1}{2}n(R + C)$, or $p(\mathbf{x}) < 2^{-\frac{1}{2}n(R+C)} p(\mathbf{x} \mid \mathbf{y})$. Hence $(\mathbf{x}, \mathbf{y}) \in A$ implies [note $(\mathbf{x}, \mathbf{y}) \in A$, $\mathbf{x}' \in G(\mathbf{x}, \mathbf{y}) \Rightarrow (\mathbf{x}', \mathbf{y}) \in A$]

$$Q(\mathbf{x}, \mathbf{y}) = \sum_{\mathbf{x}' \in G(\mathbf{x}, \mathbf{y})} p(\mathbf{x}') \leq 2^{-\frac{1}{2}n(R+C)} \sum_{\mathbf{x}' \in G(\mathbf{x}, \mathbf{y})} p(\mathbf{x}' \mid \mathbf{y}) \leq 2^{-\frac{1}{2}n(R+C)}.$$

It follows from (3.5.12) that

$$p^*(e \mid \mathbf{x}^{(i)}, \mathbf{x}, \mathbf{y}) \leq s \cdot 2^{-\frac{1}{2}n(R+C)} \quad \text{for} \quad (\mathbf{x}, \mathbf{y}) \in A. \tag{3.5.13}$$

Now from (3.5.10) we have

$$E[\varepsilon_i] = \sum_{\mathbf{x}, \mathbf{y}} p(\mathbf{x}, \mathbf{y}) p^*(e \mid \mathbf{x}^{(i)}, \mathbf{x}, \mathbf{y})$$
$$= \sum_{(\mathbf{x}, \mathbf{y}) \in A} p(\mathbf{x}, \mathbf{y}) p^*(e \mid \mathbf{x}^{(i)}, \mathbf{x}, \mathbf{y}) + \sum_{(\mathbf{x}, \mathbf{y}) \notin A} p(\mathbf{x}, \mathbf{y}) p^*(e \mid \mathbf{x}^{(i)}, \mathbf{x}, \mathbf{y}).$$

By (3.5.13) and the fact that $p^*(e \mid \mathbf{x}^{(i)}, \mathbf{x}, \mathbf{y})$ is always ≤ 1, we have

$$E[\varepsilon_i] \leq s \cdot 2^{-\frac{1}{2}n(R+C)} + P\{\mathbf{X}, \mathbf{Y}) \notin A\}. \tag{3.5.14}$$

Since $s \leq 2^{nR}$, the first term of (3.5.14) approaches zero as n approaches infinity. The second term approaches zero by the argument used in the previous proof of the fundamental theorem [see text after (3.5.5)]. It follows that $E[\varepsilon_i] \le b_n$, $i = 1, 2, \ldots, s$, where $\lim_{n\to\infty} b_n = 0$. By (3.5.7), $E[\bar{\varepsilon}] \le b_n \to 0$, and the proof is complete.

Note that the proofs of the fundamental theorem do not give us a convenient algorithm for constructing the required codes. In fact, all efforts to construct such an algorithm have so far failed. All that can be said is that if a code is selected at random, on the average it will be good. We can sharpen this statement as follows. Suppose we desire a code with average probability of error $<\delta$. In the random-coding proof, choose n large enough so that $E[\bar{\varepsilon}] < \delta/K$, where K is a fixed positive number. It then follows that the probability of obtaining a code whose average probability of error $\bar{\varepsilon}$ is $\ge \delta$ is less than $1/K$. For suppose $P\{\bar{\varepsilon} \ge \delta\} \ge 1/K$. Then we would have

$$\begin{aligned} E[\bar{\varepsilon}] &= \sum_{\mathbf{x}^{(1)},\ldots,\mathbf{x}^{(s)}} p(\mathbf{x}^{(1)}, \ldots, \mathbf{x}^{(s)})\bar{\varepsilon}(\mathbf{x}^{(1)}, \ldots, \mathbf{x}^{(s)}) \\ &\ge \sum_{\{(\mathbf{x}^{(1)},\ldots,\mathbf{x}^{(s)}):\bar{\varepsilon}(\mathbf{x}^{(1)},\ldots,\mathbf{x}^{(s)})\ge\delta\}} p(\mathbf{x}^{(1)}, \ldots, \mathbf{x}^{(s)})\bar{\varepsilon}(\mathbf{x}^{(1)}, \ldots, \mathbf{x}^{(s)}) \\ &\ge \delta P\{\bar{\varepsilon} \ge \delta\} \ge \delta / K, \qquad \text{a contradiction.} \end{aligned}$$

3.6. *Exponential error bounds*

The fundamental theorem asserts the existence of codes $([2^{nR}], n, \lambda_n)$ with $\lim_{n\to\infty} \lambda_n = 0$. In this section we study the rate at which λ_n approaches zero and show that $\lambda_n \to 0$ exponentially with n. The estimation of λ_n is based on the following lemma.

Lemma 3.6.1. If Y is any nonnegative (discrete) random variable and d is any positive real number, then $P\{Y \ge d\} \le E(Y)/d$.

Proof.

$$\begin{aligned} E(Y) &= \sum_y yp(y) = \sum_{0\le y<d} yp(y) + \sum_{y\ge d} yp(y) \\ &\ge \sum_{y\ge d} yp(y) \ge d\sum_{y\ge d} p(y) = dP\{Y \ge d\}. \end{aligned}$$

Now taking $Y = e^{rX}$ where X is any random variable and r is any positive real number, we obtain the following lemma.

Lemma 3.6.2. If X is an arbitrary (discrete) random variable, b any real number, and r any positive real number, then

$$P\{X \ge b\} \le e^{-rb}E[e^{rX}].$$

For by Lemma 3.6.1, $P\{e^{rX} \ge e^{rb}\} \le E[e^{rX}]/e^{rb}$.

We digress for a moment to consider a property of characteristic functions.

Lemma 3.6.3. Let X be a discrete random variable having only finitely many possible values. Assume X is nondegenerate, that is, not identically constant. Let $\varphi(s) = E[e^{sX}]$, $\mu(s) = \log_e \varphi(s)$. Assume s is real; then

a. $\mu''(s) > 0$ for all s.

b. $\mu(s) - s\mu'(s) < 0$ for $s < 0$.

(The primes indicate differentiation.)

Proof. Suppose X takes the values $x_1, x_2, \ldots, x_k$. Then

$$\mu'(s) = \frac{\varphi'(s)}{\varphi(s)} = \frac{\dfrac{d}{ds}\sum_{i=1}^{k} p(x_i)e^{sx_i}}{\sum_{i=1}^{k} p(x_i)e^{sx_i}} = \frac{\sum_{i=1}^{k} x_i p(x_i)e^{sx_i}}{\sum_{i=1}^{k} p(x_i)e^{sx_i}}$$

and

$$\mu''(s) = \frac{\left[\sum_i p(x_i)e^{sx_i}\right]\left[\sum_j x_j^2 p(x_j)e^{sx_j}\right] - \left[\sum_i x_i p(x_i)e^{sx_i}\right]^2}{\left[\sum_{i=1}^{k} p(x_i)e^{sx_i}\right]^2}$$

$$= \frac{E[e^{sX}]E[X^2e^{sX}] - [E(Xe^{sX})]^2}{[E(e^{sX})]^2}. \tag{3.6.1}$$

Now the *Schwarz inequality* states that for arbitrary random variables Y_1 and Y_2, $[E(Y_1Y_2)]^2 \le E(Y_1^2)E(Y_2^2)$ with equality if and only if $aY_1 + bY_2 \equiv 0$ for some constants a and b, not both zero. (To prove the Schwarz inequality, note that

$$E[(\lambda Y_1 + Y_2)^2] = \lambda^2E(Y_1^2) + 2\lambda E(Y_1Y_2) + E(Y_2^2).$$

If $Y_1 \equiv 0$, the inequality is immediate, so assume $Y_1 \not\equiv 0$. Then

$$E[(\lambda Y_1 + Y_2)^2]$$

is a quadratic polynomial $f(\lambda)$ which is nonnegative for all λ; hence the quadratic equation $f(\lambda) = 0$ has either two complex roots or one real repeated root. Thus the discriminant $[2E(Y_1Y_2)]^2 - 4E(Y_1^2)E(Y_2^2)$ is ≤ 0 and the result follows. Now if $Y_1 \equiv 0$, equality holds and also

$aY_1 + bY_2 \equiv 0$ provided $b = 0$. Thus assume $Y_1 \not\equiv 0$. Then equality holds if and only if the discriminant of $f(\lambda)$ is zero, that is, if and only if $E[(\lambda Y_1 + Y_2)^2] = 0$ for some λ, or equivalently if and only if $\lambda Y_1 + Y_2 \equiv 0$ for some λ.)

In the Schwarz inequality, take $Y_1 = e^{sX/2}$, $Y_2 = Xe^{sX/2}$; we then find from (3.6.1) that $\mu''(s) \geq 0$, with equality if and only if $ae^{sX/2} + bXe^{sX/2} \equiv 0$, that is, $a + bX \equiv 0$, for some a and b, not both zero. Since X is nondegenerate by hypothesis, equality cannot occur, and therefore $\mu''(s) > 0$ for all s, proving (a).

To establish (b), let $g(s) = \mu(s) - s\mu'(s)$. Then

$$\frac{dg(s)}{ds} = -s\,\frac{d^2\mu(s)}{ds^2} > 0 \quad \text{for} \quad s < 0$$

by (a). Thus $g(s)$ is montone increasing for $s < 0$. Since $g(0) = \mu(0) = \log \varphi(0) = 0$, (b) is proved.

We now establish the exponential bound on the probability of error.

Theorem 3.6.4. Under the hypothesis of Theorem 3.5.1, there exist positive constants K and α, depending on the channel and on R, and a sequence of codes $([2^{nR}], n, \lambda_n)$, such that $\lambda_n \leq Ke^{-\alpha n}$.

Proof. In view of (3.5.5) [or (3.5.14)] we need only show that in the proof of Theorem 3.5.1, $P\{(\mathbf{X}, \mathbf{Y}) \notin A\} \to 0$ exponentially as $n \to \infty$. As in the proof of Theorem 3.5.1, write

$$P\{(\mathbf{X}, \mathbf{Y}) \notin A\} = P\left\{\sum_{i=1}^{n} U(X_i, Y_i) \leq n\left(\frac{R + C}{2}\right)\right\}$$

where the random variables $U(X_i, Y_i)$ are independent and identically distributed with expected value C.

Let $U_i = U(X_i, Y_i)$, and $Z = \sum_{i=1}^{n} U_i$. Then by Lemma 3.6.2,

$$P\{Z \leq d\} = P\{-Z \geq -d\} \leq e^{rd}E[e^{-rZ}], \qquad r > 0, \quad d \text{ any real number},$$

or

$$P\{Z \leq d\} \leq e^{-rd}E[e^{rZ}], \qquad r < 0.$$

But

$$E\,[e^{rZ}] = E\left[\prod_{i=1}^{n} e^{rU_i}\right] = \prod_{i=1}^{n} E\,[e^{rU_i}] \qquad \text{by independence of the } U_i$$

$$= (E[e^{rU_1}])^n.$$

Let $\varphi(r) = E[e^{rU_1}]$, $\mu(r) = \log_e \varphi(r)$ as in Lemma 3.6.3. Then

$$P\{Z \leq d\} \leq e^{-rd}[\varphi(r)]^n = e^{-rd}e^{n\mu(r)}, \qquad r < 0. \tag{3.6.2}$$

In (3.6.2) we are free to choose d and r, subject to $r < 0$. To obtain the "tightest" bound for a given d we might try to differentiate the exponent with respect to r and set the result equal to zero. If we set $d = n\mu'(r)$ in (3.6.2) we obtain

$$P\left\{\frac{Z}{n} \leq \mu'(r)\right\} \leq e^{n[\mu(r) - r\mu'(r)]} \tag{3.6.3}$$

By Lemma 3.6.3b, $\mu(r) - r\mu'(r) < 0$ for $r < 0$. In addition, $\mu'(r)$ is a continuous function (which is also monotone increasing for all r by Lemma 3.6.3a) such that $\mu'(0) = \varphi'(0)/\varphi(0) = \varphi'(0) = E(U_1) = C$. It follows that there is an $r_0 < 0$ such that $\mu'(r_0) \geq (C + R)/2$. Using (3.6.3) we have

$$P\left\{\frac{Z}{n} \leq \frac{R + C}{2}\right\} \leq P\left\{\frac{Z}{n} \leq \mu'(r_0)\right\} \leq e^{n[\mu(r_0) - r_0\mu'(r_0)]}$$

where $\mu(r_0) - r_0\mu'(r_0) < 0$. This establishes the exponential bound.

3.7. *The weak converse to the fundamental theorem*

In this section we prove a result that may be regarded as a converse to the fundamental theorem. Intuitively, what we prove is that it is not possible to maintain a transmission rate $R > C$ while at the same time reducing the probability of error to zero.

First we relate the probability of error of a code to the uncertainty measure.

Theorem 3.7.1. (Fano's Inequality). Given an arbitrary code (s, n) consisting of words $\mathbf{x}^{(1)}, \ldots, \mathbf{x}^{(s)}$, let $\mathbf{X} = (X_1, \ldots, X_n)$ be a random vector that equals $\mathbf{x}^{(i)}$ with probability $p(\mathbf{x}^{(i)})$, $i = 1, 2, \ldots, s$, where $\sum_{i=1}^{s} p(\mathbf{x}^{(i)}) = 1$. [In other words, we are choosing a code word at random according to the distribution $p(\mathbf{x}^{(i)})$.] Let $\mathbf{Y} = (Y_1, \ldots, Y_n)$ be the corresponding output sequence. If $p(e)$ is the probability of error of the code, computed for the given input distribution, then

$$H(\mathbf{X} \mid \mathbf{Y}) \leq H(p(e), 1 - p(e)) + p(e) \log (s - 1).$$

Proof. Consider any output sequence $\mathbf{y}$. Let $g(\mathbf{y})$ be the input selected by the decoder; thus if $\mathbf{y}$ is received, an error will occur if and only if the transmitted sequence is not equal to $g(\mathbf{y})$.

Now divide the set of values of $\mathbf{X}$ into two groups, one group consisting of $g(\mathbf{y})$ alone, and the other group consisting of all other code words. By the grouping axiom (Section 1.2, Axiom 3) we obtain

$$H(\mathbf{X} \mid \mathbf{Y} = \mathbf{y}) = H(q, 1 - q) + qH(1) + (1 - q)H(q_1, \ldots, q_{s-1})$$

where $q = P\{\mathbf{X} = g(\mathbf{y}) \mid \mathbf{Y} = \mathbf{y}\}$, and $q_1, \ldots, q_{s-1}$ are of the form

$$\frac{p(\mathbf{x} \mid \mathbf{y})}{\sum_{\mathbf{x} \neq g(\mathbf{y})} p(\mathbf{x} \mid \mathbf{y})}$$

with $\mathbf{x}$ ranging over all the code words except $g(\mathbf{y})$.

Since $H(q_1, \ldots, q_{s-1}) \le \log (s-1)$ by Theorem 1.4.2, we obtain

$$H(\mathbf{X} \mid \mathbf{Y} = \mathbf{y}) \le H(p(e \mid \mathbf{y}), 1 - p(e \mid \mathbf{y})) + p(e \mid \mathbf{y}) \log (s-1). \quad (3.7.1)$$

[The result (3.7.1) has the following intuitive interpretation. Having received the sequence $\mathbf{y}$, if we reveal whether or not a decoding error has occurred, we remove an uncertainty $H(p(e \mid \mathbf{y}), 1 - p(e \mid \mathbf{y}))$. If no error has occurred, the input sequence must have been $g(\mathbf{y})$ and the remaining uncertainty about $\mathbf{X}$ is 0. If there is an error, the input sequence could not have been $g(\mathbf{y})$; this leaves $s - 1$ possibilities, so that the remaining uncertainty about $\mathbf{X}$ cannot exceed $\log (s - 1)$. The reasoning we used when discussing the grouping axiom in Section 1.2 then suggests the relation (3.7.1).]

Now by the convexity of H (Theorem 3.3.1),

$$\begin{aligned} H(p(e), 1 - p(e)) &= H\left(\sum_{\mathbf{y}} p(\mathbf{y})p(e \mid \mathbf{y}), 1 - \sum_{\mathbf{y}} p(\mathbf{y})p(e \mid \mathbf{y})\right) \\ &= H\left(\sum_{\mathbf{y}} p(\mathbf{y})p(e \mid \mathbf{y}), \sum_{\mathbf{y}} p(\mathbf{y})[1 - p(e \mid \mathbf{y})]\right) \\ &\ge \sum_{\mathbf{y}} p(\mathbf{y})H(p(e \mid \mathbf{y}), 1 - p(e \mid \mathbf{y})). \quad (3.7.2) \end{aligned}$$

Therefore if we multiply (3.7.1) by $p(\mathbf{y})$ and sum over all $\mathbf{y}$, we find using (3.7.2) that

$$H(\mathbf{X} \mid \mathbf{Y}) \le H(p(e), 1 - p(e)) + p(e) \log (s-1),$$

as required.

We will need the following lemma.

Lemma 3.7.2. Let $X_1, \ldots, X_n$ be a sequence of inputs to a discrete memoryless channel, and $Y_1, \ldots, Y_n$ the corresponding outputs. Then $I(X_1, \ldots, X_n \mid Y_1, \ldots, Y_n) \le \sum_{i=1}^{n} I(X_i \mid Y_i)$ with equality if and only if $Y_1, \ldots, Y_n$ are independent.

Proof. $H(Y_1, \ldots, Y_n \mid X_1, \ldots, X_n) = -\sum_{(\mathbf{x},\mathbf{y})} p(\mathbf{x}, \mathbf{y}) \log p(\mathbf{y} \mid \mathbf{x})$. If $\mathbf{x} = (\alpha_1, \ldots, \alpha_n)$ and $\mathbf{y} = (\beta_1, \ldots, \beta_n)$ then $p(\mathbf{y} \mid \mathbf{x}) = p(\beta_1 \mid \alpha_1) \cdots p(\beta_n \mid \alpha_n)$; hence

$$\begin{aligned} H(Y_1, \ldots, Y_n \mid X_1, \ldots, X_n) &= -\sum_{(\mathbf{x},\mathbf{y})} p(\mathbf{x}, \mathbf{y})\left[\sum_{i=1}^{n} \log p(\beta_i \mid \alpha_i)\right] \\ &= \sum_{i=1}^{n} H(Y_i \mid X_i). \end{aligned}$$

Also (Corollary 1.4.3.1) $H(Y_1, \ldots, Y_n) \le \sum_{i=1}^n H(Y_i)$ with equality if and only if $Y_1, \ldots, Y_n$ are independent. The lemma follows.

We are now ready for the main result.

Theorem 3.7.3. (Weak Converse). The average probability of error $\overline{p(e)}$ of any code (s, n) must satisfy

$$\log s \le \frac{nC + \log 2}{1 - \overline{p(e)}}$$

where C is the channel capacity. Consequently if $s \ge 2^{n(C+\delta)}$ where $\delta > 0$, then

$$n(C + \delta) \le \frac{nC + 1}{1 - \overline{p(e)}} \quad \text{or} \quad \overline{p(e)} \ge 1 - \frac{C + 1/n}{C + \delta} \to 1 - \frac{C}{C + \delta} > 0.$$

Thus if $R > C$, no sequence of codes $([2^{nR}], n)$ can have an average probability of error which $\to 0$ as $n \to \infty$, hence no sequence of codes $([2^{nR}], n, \lambda_n)$ can exist with $\lim_{n\to\infty} \lambda_n = 0$.

Proof. Choose a code word at random with all words equally likely, that is, let $\mathbf{X}$ and $\mathbf{Y}$ be as in Theorem 3.7.1, with $p(\mathbf{x}^{(i)}) = 1/s$, $i = 1, 2, \ldots, s$. Then $H(\mathbf{X}) = \log s$ so that

$$I(\mathbf{X} \mid \mathbf{Y}) = \log s - H(\mathbf{X} \mid \mathbf{Y}) \tag{3.7.3}$$

By Lemma 3.7.2,

$$I(\mathbf{X} \mid \mathbf{Y}) \le \sum_{i=1}^{n} I(X_i \mid Y_i). \tag{3.7.4}$$

Since $I(X_i \mid Y_i) \le C$ (by definition of capacity), (3.7.3) and (3.7.4) yield

$$\log s - H(\mathbf{X} \mid \mathbf{Y}) \le nC. \tag{3.7.5}$$

By Theorem 3.7.1,

$$H(\mathbf{X} \mid \mathbf{Y}) \le H(\overline{p(e)}, 1 - \overline{p(e)}) + \overline{p(e)} \log (s - 1). \tag{3.7.6}$$

Hence

$$H(\mathbf{X} \mid \mathbf{Y}) \le \log 2 + \overline{p(e)} \log s. \tag{3.7.7}$$

The result now follows from (3.7.5) and (3.7.7).

Given a sequence of codes with at least $2^{n(C+\delta)}$ words of length n and average probability of error $\overline{p_n(e)}$, Theorem 3.7.3 implies that $\overline{p_n(e)}$ cannot be reduced to zero by allowing n to become arbitrarily large. However, $\overline{p_n(e)}$ need not be a monotone function of n, and it is at least conceivable that there might be an integer n_0 such that $\overline{p_{n_0}(e)} = 0$. In this case we could, by using the codes of word length n_0, achieve perfect reliability while maintaining a transmission rate above capacity. But this possibility is excluded, for an examination of (3.7.5) and (3.7.6) shows that the number of words in a code with zero average probability of error is at most 2^{nC}. Thus if the rate of transmission exceeds capacity, the average (hence the maximum) probability of error is bounded away from zero.

It is possible to prove a stronger result than Theorem 3.7.3, namely the *strong converse* to the fundamental theorem. The strong converse states that for any sequence of codes $([2^{nR}], n, \lambda_n)$ with average probability of error $\overline{p_n(e)}$, not only is it true that $\overline{p_n(e)}$ cannot approach zero if $R > C$, but in fact $\lim_{n\to\infty} \overline{p_n(e)} = 1$ (hence $\lim_{n\to\infty} \lambda_n = 1$).

Special cases of the strong converse will be considered in Chapters 4 and 8. For a general discussion of this subject, see Wolfowitz (1961).

3.8. *Notes and remarks*

The material in Sections 3.1 and 3.2 is standard and is discussed by Feinstein (1958) and Wolfowitz (1961). The method of calculation of channel capacity described in Section 3.3 is due to Muroga (1953); see also Fano (1961). Programming techniques and algorithms for the general problem of computing channel capacity are discussed by Eisenberg (1963).

The fundamental lemma 3.5.2 and the corresponding proof of Theorem 3.5.1 are due originally to Feinstein (1954, 1958). The present form of the fundamental lemma is due to Blackwell, Breiman, and Thomasian (1958, 1959) and Thomasian (1961). The idea of "random coding" was suggested (1948) and later made precise (1957) by Shannon. Lemma 3.5.3 was pointed out by Wolfowitz (1961).

The exponential bounding technique of Section 3.6 is due to Chernoff (1952); the technique was first applied to the calculation of error bounds by Shannon (1957). The essential content of Theorem 3.6.4 is that if $X_1, \ldots, X_n$ are independent, identically distributed random variables, each having only finitely many possible values, and if μ is the common expected value, then for every $\varepsilon > 0$, $P\{|(X_1 + \cdots + X_n)/n - \mu| \geq \varepsilon\} \to 0$ exponentially as $n \to \infty$; in other words, we have exponential convergence in the weak law of large numbers. For more general results of this type, see Katz and Thomasian (1961).

Theorem 3.7.1 was proved by Fano (1952, unpublished). The result was applied to prove the weak converse by Feinstein (1958).

Another proof of the fundamental theorem has been given by Wolfowitz (1961). Wolfowitz's approach is combinatorial and is based on a precise and general formulation of the notion of a "typical" sequence (Section 1.3). The approach of Wolfowitz seems to be better suited than the one given here for proving the strong converse, and also for dealing with questions involving the computation of the capacity of channels with memory (Chapter 7). However, the approach of this chapter seems to generalize to a broader class of channel models. [See, for example, Dobrushin (1959).]

A possible source of difficulty in understanding the physical interpretation of the fundamental theorem is the fact that if an error is made in decoding, an entire block of digits produced by the source may be reported incorrectly. However, if we have a code $(2^{nR}, n, \lambda_n)$, where each code word is assigned to a sequence of nR source digits, the average number of errors per block is $nR\lambda_n$ (making the conservative assumption that a decoding error results in a complete scrambling of the associated block of source digits.) Since the number of source digits per block is nR, the average number of errors per digit produced by the source is λ_n. Thus by reducing λ_n toward zero we are in fact achieving reliable transmission.

One final remark on terminology. The information processed by a channel is the information conveyed about a single input letter by a single output letter. Hence the units of $I(X \mid Y)$ (and the units of C) are sometimes called "bits/symbol." We have seen that if the channel can transmit 1 symbol per second, then C is the least upper bound on the number of bits (that is, binary digits) per second which can be reliably handled by the channel. Similarly if the channel can transmit α symbols per second, then the corresponding upper bound on the source rate is αC bits per second.

PROBLEMS

3.1 A discrete memoryless channel is characterized by the matrix

$$\begin{array}{c} \\ x_1 \\ x_2 \\ x_3 \end{array} \begin{array}{c} \begin{array}{ccc} y_1 & y_2 & y_3 \end{array} \\ \begin{bmatrix} \frac{1}{2} & \frac{1}{3} & \frac{1}{6} \\ \frac{1}{6} & \frac{1}{2} & \frac{1}{3} \\ \frac{1}{3} & \frac{1}{6} & \frac{1}{2} \end{bmatrix} \end{array}$$

If $p(x_1) = 1/2$, $p(x_2) = p(x_3) = 1/4$, find the ideal-observer decision scheme and calculate the associated probability of error.

3.2 Given a discrete memoryless channel with input alphabet $x_1, \ldots, x_M$, output alphabet $y_1, \ldots, y_L$, and channel matrix $[p(y_j \mid x_i)]$, a *randomized decision scheme* may be constructed by assuming that if the channel output is y_j $(j = 1, \ldots, L)$, the decoder will select x_i with probability q_{ji} $(i = 1, \ldots, M)$. For a given input distribution show that no randomized decision scheme has a lower probability of error than the ideal observer.

3.3 (Feinstein 1958). The output of a discrete memoryless channel K_1 is connected to the input of another discrete memoryless channel K_2. Show that the capacity of the cascade combination can never exceed the capacity of K_i, $i = 1, 2$. ("Information cannot be increased by data processing.")

3.4 Let X and Y be discrete random variables, with X taking on the values $x_1, \ldots, x_M$ and Y the values $y_1, \ldots, y_L$. Let g be an arbitrary function with domain $\{x_1, \ldots, x_M\}$. Define $Z = g(X)$. Show that $H(Y \mid Z) \geq H(Y \mid X)$.

3.5 A number of identical binary symmetric channels are connected in cascade, as shown below. Let X_k be the output of the kth channel (= the input of the $(k + 1)$th channel). Assume $0 < \beta < 1$, where β is the probability that a single transmitted digit is in error.

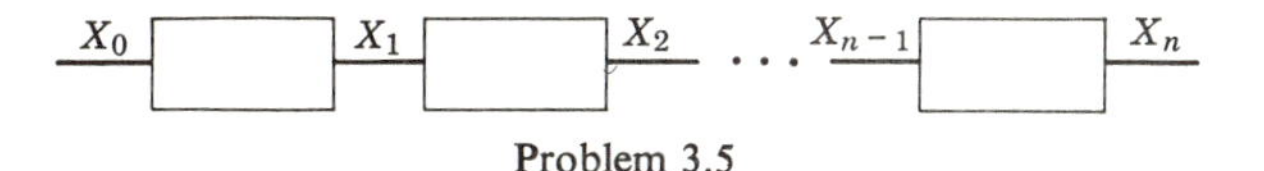

Problem 3.5

Let $p_n = P\{X_n = 0\}$, $n = 0, 1, \ldots$. Assuming p_0 known, find an explicit expression for p_n. Show that $p_n \to \frac{1}{2}$ as $n \to \infty$ regardless of p_0, so that the capacity of the cascade combination approaches zero as $n \to \infty$.

3.6 Find the channel matrix of a cascade combination in terms of the individual channel matrices

3.7 Find the capacity of the general binary channel whose channel matrix is

$$\begin{bmatrix} \alpha & 1 - \alpha \\ \beta & 1 - \beta \end{bmatrix} \quad \text{where} \quad \alpha \neq \beta.$$

3.8 (Shannon 1957). If Π_1 and Π_2 are channel matrices of discrete memoryless channels K_1 and K_2, respectively, the *sum* of K_1 and K_2 is defined as the channel whose matrix is

$$\left[\begin{array}{c|c} \Pi_1 & 0 \\ \hline 0 & \Pi_2 \end{array}\right].$$

Thus the sum channel may be operated by choosing one of the individual channels and transmitting a digit through it. The input or output symbol always identifies the particular channel used. If C_i is the capacity of K_i, $i = 1, 2$, and C is the capacity of the sum, show that

$$2^C = 2^{C_1} + 2^{C_2}.$$

3.9 (Shannon 1957). The *product* of two discrete memoryless channels K_1 and K_2 is a channel whose inputs are ordered pairs (x_i, x_j') and whose outputs are ordered pairs (y_k, y_l') where the first coordinates belong to the alphabet of K_1 and the second coordinates to the alphabet of K_2.

The channel probabilities are given by

$$p(y_k, y_l' \mid x_i, x_j') = p(y_k \mid x_i)p(y_l' \mid x_j')$$

Thus the product channel is operated by sending a digit through K_1 and then independently a digit through K_2. Show that the capacity C of the product channel is the sum $C_1 + C_2$ of the individual capacities.

3.10 (Fano 1961). Evaluate (in an effortless manner) the capacities of the channels whose matrices are given below.

a.
$$\begin{bmatrix} 1-\beta & \beta & 0 \\ \beta & 1-\beta & 0 \\ 0 & 0 & 1 \end{bmatrix}$$

b.
$$\begin{bmatrix} \frac{1-p}{2} & \frac{1-p}{2} & \frac{p}{2} & \frac{p}{2} \\ \frac{p}{2} & \frac{p}{2} & \frac{1-p}{2} & \frac{1-p}{2} \end{bmatrix}$$

c.
$$\begin{array}{c} \\ 0 \\ 1 \end{array} \begin{array}{c} \begin{matrix} 0 & \text{erase} & 1 \end{matrix} \\ \begin{bmatrix} 1-p & p & 0 \\ 0 & p & 1-p \end{bmatrix} \end{array}$$

[Channel (c) is the "binary erasure" channel.]

3.11 It is desired to construct codes for a discrete memoryless channel with the property that all code words belong to a specified set. Prove the following.

a. (Thomasian 1961). Let $p(\mathbf{x})$ be an arbitrary probability function defined on the set of all input n-sequences. For any real number a, let $A = \{(\mathbf{x}, \mathbf{y}): \log [p(\mathbf{y} \mid \mathbf{x})/p(\mathbf{y})] > a\}$. Let F be any set of input n-sequences. Then for each positive integer s there is a code (s, n, λ), *with all code words belonging to F*, such that

$$\lambda \leq s \cdot 2^{-a} + P\{(\mathbf{X}, \mathbf{Y}) \notin A\} + P\{\mathbf{X} \notin F\}.$$

b. Consider the binary symmetric channel. For each n, let F_n be the set of all binary n-sequences whose weight (that is, number of ones) is $\leq rn$, where r is a fixed number $\in [0, \frac{1}{2})$. We wish to prove a theorem of the following form.

Given $R < C_0$, there exist codes $([2^{nR}], n, \lambda_n)$, with all words belonging to F_n, such that $\lambda_n \to 0$ as $n \to \infty$.

How large can C_0 be?

3.12 Given a discrete memoryless channel with capacity C, show that there is an input distribution $p(x)$ such that the corresponding information processed is equal to C. Hence C is a true maximum rather than simply a least upper bound.

3.13 Let $X_1, \ldots, X_n$ be a sequence of inputs to a discrete memoryless channel, and let $Y_1, \ldots, Y_n$ be the corresponding output sequence. Show that it is possible to find a channel and an input distribution such that $Y_1, \ldots, Y_n$ are independent without $X_1, \ldots, X_n$ being independent.

CHAPTER FOUR

Error Correcting Codes

4.1. Introduction; minimum-distance principle

In this chapter we attack the problem of finding "good" encoding schemes to combat errors in transmission. As the fundamental theorem shows, there are codes that allow transmission at any rate below channel capacity with an arbitrarily small probability of error. However, the proofs we have given provide no clue as to how to construct the codes; in fact, there is at present no systematic procedure for constructing codes that meet the specifications of the fundamental theorem. On the other hand we do have considerable information about the structure of codes, especially for the binary symmetric channel. We will be able to give precise bounds on the error correcting ability of codes, and in addition we will exhibit encoding and decoding schemes which have desirable properties with respect to implementation.

Until the final section of the chapter, our discussion is restricted to the binary symmetric channel. The inputs to the channel will be chosen from a set of binary code words of length n, that is, a set of n-digit sequences of zeros and ones. We assume in all cases that the code words occur with equal probability, and that the ideal-observer decision scheme is used. Since an error may be made in any digit, the set of possible output sequences is the set of all 2^n binary sequences of length n. Our first problem is to find the form of the ideal-observer decision scheme for such a code. If the input code words are $\mathbf{w}_1, \mathbf{w}_2, \ldots, \mathbf{w}_s$ and the output sequences are $\mathbf{v}_1, \mathbf{v}_2, \ldots, \mathbf{v}_{2^n}$, then we wish to associate a code word with each received sequence in such a way that the probability of error is minimized. We have found that the ideal observer associates with each received sequence $\mathbf{v}$ the code word $\mathbf{w}$ for which $p(\mathbf{w} \mid \mathbf{v})$ is a maximum. Furthermore, since all code words are equally likely, the code word that maximizes $p(\mathbf{w} \mid \mathbf{v})$ is the same as the code word that maximizes $p(\mathbf{v} \mid \mathbf{w})$. We now define the *distance*† $d(\mathbf{v}_1, \mathbf{v}_2)$ between two n-digit binary sequences $\mathbf{v}_1$ and $\mathbf{v}_2$ as the number of digits in which $\mathbf{v}_1$ and $\mathbf{v}_2$ disagree. For example, if

$$\mathbf{v}_1 = 011011, \qquad \mathbf{v}_2 = 110001,$$

† Also called the "Hamming distance" after its inventor.

then

$$d(\mathbf{v}_1, \mathbf{v}_2) = 3.$$

If $d(\mathbf{w}, \mathbf{v})$ is the distance from a code word $\mathbf{w}$ to a received sequence $\mathbf{v}$ then in order that $\mathbf{w}$ be received as $\mathbf{v}$ there must be exactly $d(\mathbf{w}, \mathbf{v})$ errors in transmission, and these errors must occur in those digits where $\mathbf{w}$ and $\mathbf{v}$ disagree. Hence

$$p(\mathbf{v} \mid \mathbf{w}) = \beta^{d(\mathbf{w},\mathbf{v})}(1 - \beta)^{n-d(\mathbf{w},\mathbf{v})},$$

where β is the probability that a given digit is transmitted incorrectly. Let us now compare $p(\mathbf{v} \mid \mathbf{w}_1)$ and $p(\mathbf{v} \mid \mathbf{w}_2)$. If we write $d_i = d(\mathbf{w}_i, \mathbf{v})$ $(i = 1, 2)$ we have

$$\frac{p(\mathbf{v} \mid \mathbf{w}_1)}{p(\mathbf{v} \mid \mathbf{w}_2)} = \frac{\beta^{d_1}(1 - \beta)^{n-d_1}}{\beta^{d_2}(1 - \beta)^{n-d_2}} = \left(\frac{1 - \beta}{\beta}\right)^{d_2-d_1}.$$

We shall always assume $0 < \beta < \frac{1}{2}$, so that $(1 - \beta)/\beta > 1$. Thus

$$p(\mathbf{v} \mid \mathbf{w}_1) > p(\mathbf{v} \mid \mathbf{w}_2) \quad \text{if and only if} \quad d_1 < d_2.$$

Therefore the code word $\mathbf{w}$ that gives the largest conditional probability $p(\mathbf{v} \mid \mathbf{w})$ is that word whose distance from $\mathbf{v}$ is a minimum. We have thus established the following result.

Theorem 4.1.1. Given a code for the binary symmetric channel consisting of s equally likely code words of length n. The "ideal observer" decision scheme is a "minimum-distance" decoder, that is, the decoder selects for each received sequence $\mathbf{v}$ the code word $\mathbf{w}$ for which the distance $d(\mathbf{w}, \mathbf{v})$ is a minimum. (If there is more than one minimum, any one of the minimizing words may be chosen without affecting the probability of error.)

As an example, suppose the code words are

$$\mathbf{w}_1 = 00000$$
$$\mathbf{w}_2 = 10011$$
$$\mathbf{w}_3 = 11100$$
$$\mathbf{w}_4 = 01111.$$

If the sequence $\mathbf{v} = 01011$ is received, it is decoded as $\mathbf{w}_4$ since $d(\mathbf{w}_4, \mathbf{v}) = 1$ and $d(\mathbf{w}_i, \mathbf{v}) > 1$ for $i \neq 4$. Similarly, the sequence $\mathbf{v}' = 00110$ may be decoded as either $\mathbf{w}_1$ or $\mathbf{w}_4$ since $d(\mathbf{w}_1, \mathbf{v}') = d(\mathbf{w}_4, \mathbf{v}') = 2$ and $d(\mathbf{w}_2, \mathbf{v}') = d(\mathbf{w}_3, \mathbf{v}') = 3$.

It is easily verified that the Hamming distance has the characteristic properties of a distance function, that is,

a. $d(\mathbf{v}_1, \mathbf{v}_2) \geq 0, \qquad d(\mathbf{v}_1, \mathbf{v}_2) = 0$ if and only if $\mathbf{v}_1 = \mathbf{v}_2$.

b. $d(\mathbf{v}_1, \mathbf{v}_2) = d(\mathbf{v}_2, \mathbf{v}_1)$.

c. $d(\mathbf{v}_1, \mathbf{v}_3) \leq d(\mathbf{v}_1, \mathbf{v}_2) + d(\mathbf{v}_2, \mathbf{v}_3)$ (the triangle inequality).

4.2. Relation between distance and error correcting properties of codes; the Hamming bound

Since each received sequence $\mathbf{v}$ is decoded as the code word "closest" to $\mathbf{v}$ in the sense of Hamming distance, it appears that a "good" code is one whose code words are "far apart." If the distance between code words is large, many transmission errors must be made before a code word $\mathbf{w}$ is transformed into a received sequence $\mathbf{v}$ which is closer to some other word $\mathbf{w}'$ than to $\mathbf{w}$. As a start toward making this idea precise, we prove the following lemma.

Lemma 4.2.1. Let $\mathbf{w}_1, \mathbf{w}_2, \ldots, \mathbf{w}_s$ be binary code words of length n having the property that for a given positive integer e,

$$d(\mathbf{w}_i, \mathbf{w}_j) \geq 2e + 1 \quad \text{for} \quad i \neq j.$$

Then all single, double, . . . , e-tuple errors in transmission can be corrected. If $d(\mathbf{w}_i, \mathbf{w}_j) \geq 2e$ for $i \neq j$, then all single, double, . . . , $(e - 1)$-tuple errors can be corrected and e-tuple errors can be detected but not, in general, corrected. Conversely, any code for which all errors of magnitude $\leq e$ are corrected must satisfy $d(\mathbf{w}_i, \mathbf{w}_j) \geq 2e + 1$ for $i \neq j$, and any code for which all errors of magnitude $\leq e - 1$ are corrected and all errors of magnitude e are detected must satisfy $d(\mathbf{w}_i, \mathbf{w}_j) \geq 2e$ for $i \neq j$.

Proof. First suppose that the minimum distance between code words is at least $2e + 1$. Then in order that a code word $\mathbf{w}$ be decoded as another code word $\mathbf{w}'$ the received sequence must be at least as close to $\mathbf{w}'$ as to $\mathbf{w}$. In order for this to happen, at least $e + 1$ digits of $\mathbf{w}$ must be in error. Hence all errors involving e or fewer digits are correctible. If the distance between two code words $\mathbf{w}$ and $\mathbf{w}'$ is $2e$, then an e-tuple error in $\mathbf{w}$ will result in a received sequence $\mathbf{v}$ whose distance from $\mathbf{w}$ is e and whose distance from $\mathbf{w}'$ is at least e. [The last assertion follows from $d(\mathbf{w}, \mathbf{w}') \leq d(\mathbf{w}, \mathbf{v}) + d(\mathbf{v}, \mathbf{w}')$.] If the distance between $\mathbf{v}$ and $\mathbf{w}'$ is exactly e, then an e-tuple error in $\mathbf{w}$ and an e-tuple error in $\mathbf{w}'$ may both result in $\mathbf{v}$. Thus in general it is possible to detect the presence of an e-tuple error but not to decode with perfect accuracy. The converse follows by similar reasoning.

A geometric verification of Lemma 4.2.1 is shown in Fig. 4.2.1. If $d(\mathbf{w}_i, \mathbf{w}_j) = 9$, a change of four or fewer digits in $\mathbf{w}_i$ will leave the received sequence closer to $\mathbf{w}_i$ than to $\mathbf{w}_j$. If $d(\mathbf{w}_i, \mathbf{w}_j) = 8$, errors in three or fewer digits will be corrected, but a change of four digits in $\mathbf{w}_i$ may leave the received sequence equally far from $\mathbf{w}_i$ and $\mathbf{w}_j$. Another geometric approach to binary coding is to visualize binary sequences of length n as vertices of an n-dimensional cube. (See Fig. 4.2.2.) A change in one digit of a binary sequence corresponds to a movement from one vertex to an adjacent vertex. If dist $(\mathbf{w}_i, \mathbf{w}_j) = d$, then every path along the edges of the cube

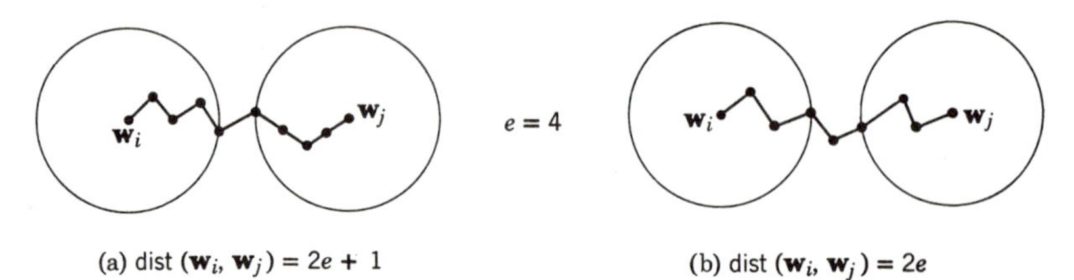

Fig. 4.2.1. Relationship between distance and error correction.

from $\mathbf{w}_i$ to $\mathbf{w}_j$ passes through at least d vertices. To form a code with minimum distance d, we select points on the cube which cannot be connected by paths of length less than d. In Fig. 4.2.2 the points 001, 010, and 111 form a code of minimum distance 2.

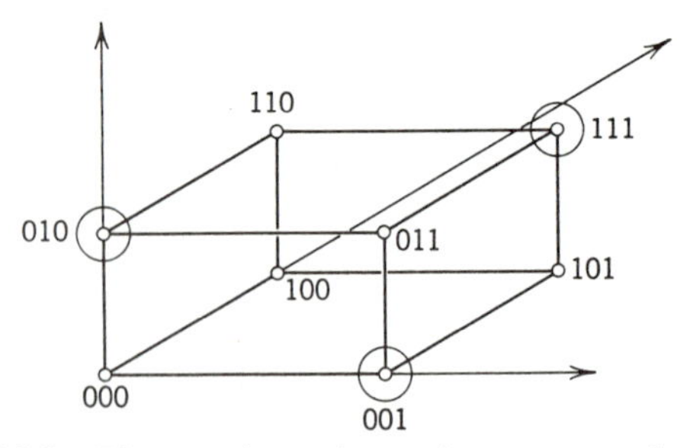

Fig. 4.2.2. Binary code words as points on an n-cube.

We saw in Chapter 3 that if a channel can transmit one symbol per second, a set of 2^{nR} code words is adequate to transmit the information produced by a source which emits R binary symbols per second. Thus a high transmission rate corresponds to a large number of code words. It is of interest to determine how many code words there can be in a code that corrects e-tuple and all smaller errors. An upper bound is provided by the following theorem.

Theorem 4.2.2. (Hamming Upper Bound on the Number of Code Words). If a code consisting of s binary sequences of length n corrects all single, double, . . . , e-tuple errors, then

$$s \leq \frac{2^n}{\sum_{i=0}^{e} \binom{n}{i}}. \tag{4.2.1}$$

Proof. Let the code words be $\mathbf{w}_1, \mathbf{w}_2, \ldots, \mathbf{w}_s$. For each code word $\mathbf{w}_i$, we can draw a "sphere" with "center" at $\mathbf{w}_i$ and "radius" e; the "sphere" is the set of all binary sequences $\mathbf{v}$ such that $d(\mathbf{w}_i, \mathbf{v}) \le e$. Since the code is e-tuple-error correcting, the spheres are disjoint. Since the sphere associated with $\mathbf{w}_i$ contains all sequences that differ from $\mathbf{w}_i$ in 0, 1, 2, ..., or e places, the number of sequences in each sphere is

$$1 + n + \binom{n}{2} + \cdots + \binom{n}{e} = \sum_{i=0}^{e} \binom{n}{i}.$$

Thus the total number of sequences included in the spheres is

$$s \sum_{i=0}^{e} \binom{n}{i}.$$

However, the number of sequences belonging to the spheres cannot exceed the total number of n-digit binary sequences, namely 2^n. Thus

$$s \sum_{i=0}^{e} \binom{n}{i} \le 2^n.$$

Note that if, for a fixed e and n, s is the largest integer satisfying (4.2.1), there may not exist an e-tuple-error correcting code containing s code words of length n. For example, if $e = 1, n = 4$, we have $2^n/(1 + n) = \frac{16}{5}$, and the largest integer that satisfies the Hamming bound is $s = 3$. However, there is no single-error correcting code with more than two words. (Try it and see.) Thus the Hamming bound is a *necessary* but *not sufficient* condition for a code to correct e-tuple and all smaller errors.

We remark again for emphasis that we associate each received sequence $\mathbf{v}$ with the closest code word $\mathbf{w}_i$. If $\mathbf{v}$ can arise from an e-tuple error in $\mathbf{w}_1$ and an e'-tuple error in $\mathbf{w}_2$, $e' > e$, then we associate $\mathbf{v}$ with $\mathbf{w}_1$ rather than $\mathbf{w}_2$. In other words, we always choose to correct the error of smaller magnitude.

4.3. *Parity check coding*

Let us summarize the results we have obtained thus far. We are considering a binary symmetric channel whose inputs are binary code words $\mathbf{w}_1, \mathbf{w}_2, \ldots, \mathbf{w}_s$. To obtain a high transmission rate, we would like to have the number of code words as large as possible; at the same time we would like to correct as many transmission errors as possible. These objectives will in general conflict. For example, if we wish to correct single errors and the code-word length is 5, we can find a code with four words, 00000, 10011, 11100, 01111. If we wish to correct double errors, we cannot use more than two words, for example, 00000, 11111. The

Hamming bound suggests that as the error correction requirements increase, the maximum possible number of code words decreases.

A consideration neglected so far is ease of decoding. The ideal-observer decision scheme is a "minimum-distance" decoder; that is, the received sequence $\mathbf{v}_j$ is decoded as the code word $\mathbf{w}_i$ that minimizes $d(\mathbf{w}_i, \mathbf{v}_j)$. Thus the decoder must contain a "table" or "code book" whose entries are all binary sequences of the specified code-word length. Each possible received sequence is assigned to the closest code word. The storage requirements for such a decoder are quite severe. In this section we investigate a coding scheme that improves the efficiency of the decoding process, and at the same time gives considerable insight into the structure of error correcting codes.

The earliest "parity check" codes were devised as a very simple and easily mechanized error detection scheme. Given a code consisting of all binary sequences of length n, an extra digit (the "parity digit") is added to each code word, the digit being chosen so as to make the total number of ones even (or odd). For example, if the original code is $\{00, 01, 10, 11\}$, and an "even parity" is used, that is, the parity digit is chosen so that the number of ones is even, the resulting code is $\{000, 011, 101, 110\}$. It is not difficult to see that if a single error (or in fact any e-tuple error where e is odd) is made in transmission, the received sequence will have an odd number of ones. Hence all errors involving an odd number of digits can be detected. The only operation the decoder must perform is to add the number of ones in the received sequence. This idea was generalized by Hamming and later by Slepian, who formulated the basic theory we present here.

If $r_1, r_2, \ldots, r_n$ are the digits of a code word, the fact that the number of ones is even may be expressed by the condition $r_1 + r_2 + \cdots + r_n = 0$ modulo 2.† Thus it appears that the binary sequences of length n which have an even number of ones are precisely the sequences which satisfy a certain modulo 2 linear equation. To generalize this idea we consider codes whose words satisfy a set of simultaneous linear equations.

Definition. Given a set of simultaneous linear equations of the form

$$\begin{aligned} a_{11}r_1 + a_{12}r_2 + \cdots + a_{1n}r_n &= 0 \\ &\vdots \\ a_{m1}r_1 + a_{m2}r_2 + \cdots + a_{mn}r_n &= 0 \end{aligned} \qquad \text{(mod 2)}. \qquad (4.3.1)$$

† Modulo 2 arithmetic is the arithmetic of the field consisting of two elements 0 and 1, with the rules $0 + 0 = 1 + 1 = 0$, $0 + 1 = 1 + 0 = 1$, $1 \cdot 1 = 1$, $0 \cdot 1 = 1 \cdot 0 = 0 \cdot 0 = 0$, $0/1 = 0$, $1/1 = 1$, $a - b = a + b$ for $a, b = 0$ or 1.

The set of solutions to (4.3.1) is called a *parity check code* (The terminology *group code* is also used, for reasons which will become clear in Section 4.4.)

The m by n matrix $A = [a_{ij}]$ is called the *parity check matrix*. If the rank of A is t, and if columns $j_1, \ldots, j_t$ of A are linearly independent, then the $n - t = k$ digits r_j, $j \neq j_1, \ldots, j \neq j_t$, may be specified arbitrarily in (4.3.1), and are called a set of *information digits* of the code. The digits $r_{j_1}, \ldots, r_{j_t}$ are then determined from (4.3.1), and are called a set of *check digits* for the code. A parity check code must contain 2^k words, where $k = n - t = n - \text{rank } A$; each assignment of values to a set of information digits determines a unique code word.

If a parity check code is to be used to process the information produced by a source which emits R binary symbols per second, then in accordance with the discussion in Chapter 3 we would set $2^{nR} = 2^k$, or $R = k/n$. Thus a high percentage of information digits corresponds to a high transmission rate; on the other hand, in general the greater the percentage of check digits the greater the error correcting ability of the code. An example of a set of parity check equations is shown below for the case $n = 6$, $t = m = 3$, $k = 3$.

$$\begin{aligned} r_1 \qquad\qquad + r_4 + r_5 \qquad\quad &= 0 \\ r_2 + r_3 + r_4 \qquad\quad + r_6 &= 0 \\ r_1 \qquad + r_3 + r_4 \qquad\quad + r_6 &= 0 \end{aligned} \quad \text{(mod 2).} \tag{4.3.2}$$

We may choose r_1, r_2, and r_3 as check digits and r_4, r_5, and r_6 as information digits. To form a code word, we first choose values for r_4, r_5, and r_6; for example, $r_4 = 0, r_5 = 1, r_6 = 1$. We then solve the parity check equation to obtain $r_1 = r_4 + r_5 = 1, r_3 = r_1 + r_4 + r_6 = 0, r_2 = r_3 + r_4 + r_6 = 1$. Thus the code word is 110011. Since there are three information digits, there are $2^3 = 8$ code words in the code; all eight code words are listed in Table 4.3.1.

The fact that the code words of a parity check code are determined by a set of linear equations allows us to use algebraic techniques to study the

Table 4.3.1. Example of a parity check code

	r_1	r_2	r_3	r_4	r_5	r_6
w_1	0	0	0	0	0	0
w_2	0	0	1	0	0	1
w_3	1	1	1	0	1	0
w_4	1	1	0	0	1	1
w_5	1	1	0	1	0	0
w_6	1	1	1	1	0	1
w_7	0	0	1	1	1	0
w_8	0	0	0	1	1	1

properties of codes. However, there is a basic question concerning the adequacy of parity check codes. It would be desirable to know whether or not such codes meet the specifications of the fundamental theorem. In other words, given a binary symmetric channel with capacity $C > 0$ and a positive number $R < C$, we would hope to find a sequence of parity check codes which maintain the transmission rate R with a probability of error which approaches zero with increasing code-word length. In fact we can find such a sequence, as we shall prove in Section 4.6. Thus from the point of view of the fundamental theorem, there is no loss of generality in considering parity check codes.

We now turn to the problem of relating the parity check equations to the error correcting ability of the code. If the digits $r_1, r_2, \ldots, r_n$ are transmitted over a binary symmetric channel, the received digits will be $r_1', r_2', \ldots, r_n'$ where $r_i' = r_i$ if no error has been made in r_i, and $r_i' = r_i + 1 \pmod 2$ if an error in r_i has occurred. Suppose that at the receiver we try to determine whether or not $r_1', r_2', \ldots, r_n'$ satisfy the parity check equations. In other words, we calculate

$$\begin{aligned} c_1 &= a_{11}r_1' + a_{12}r_2' + \cdots + a_{1n}r_n' \\ c_2 &= a_{21}r_1' + a_{22}r_2' + \cdots + a_{2n}r_n' \\ &\;\;\vdots \\ c_m &= a_{m1}r_1' + a_{m2}r_2' + \cdots + a_{mn}r_n' \end{aligned} \qquad (\text{mod } 2).\dagger$$

The binary sequence

$$\mathbf{c} = \begin{bmatrix} c_1 \\ \cdot \\ \cdot \\ \cdot \\ c_m \end{bmatrix}$$

is called the *corrector* or *syndrome* associated with the sequence $\mathbf{v} = (r_1', \ldots, r_n')$. In matrix form,

$$\mathbf{c} = A \begin{bmatrix} r_1' \\ \cdot \\ \cdot \\ \cdot \\ r_n' \end{bmatrix} = A\mathbf{v}^T.$$

(In general, code words and received sequences will be written as row vectors; $\mathbf{v}^T$ is the transpose of $\mathbf{v}$.) If no error has occurred, then $r_i' = r_i$

† For the remainder of this chapter all arithmetic will be modulo 2 unless otherwise specified.

for all i and hence $\mathbf{c}$ is the zero vector $\mathbf{0}$. We will see in the next section that the construction of the minimum-distance decoder involves the consideration of the correctors associated with various error combinations.

Before looking at the construction of the minimum-distance decoder, we develop a few properties of correctors.

Suppose that the sequence $\mathbf{w} = (r_1, \ldots, r_n)$ is transmitted and the sequence $\mathbf{v} = (r_1', \ldots, r_n')$ is received. The sequence $\mathbf{z} = \mathbf{v} - \mathbf{w} = (r_1' - r_1, \ldots, r_n' - r_n)$ is called the *error pattern vector* corresponding to $\mathbf{w}$ and $\mathbf{v}$; the error pattern vector is thus a binary sequence with ones in the positions that are in error, and zeros elsewhere. The corrector associated with $\mathbf{v}$ is $\mathbf{c} = A\mathbf{v}^T = A(\mathbf{z}^T + \mathbf{w}^T) = A\mathbf{z}^T + A\mathbf{w}^T = A\mathbf{z}^T$ ($\mathbf{w}$ is a code word so that $A\mathbf{w}^T = \mathbf{0}$). Consequently *the corrector is determined by the error pattern vector.* Now *if $\mathbf{z}$ has ones in positions $j_1, j_2, \ldots, j_e$ and zeros elsewhere, that is, digits $j_1, \ldots, j_e$ are in error, then the vector $A\mathbf{z}^T$ is simply the (mod 2) sum of columns, $j_1, j_2, \ldots, j_e$ of A.* For example, in the code described by the parity check equations (4.3.2), errors in positions 2 and 4 will yield the corrector

$$\begin{bmatrix} 0 \\ 1 \\ 0 \end{bmatrix} + \begin{bmatrix} 1 \\ 1 \\ 1 \end{bmatrix} = \begin{bmatrix} 1 \\ 0 \\ 1 \end{bmatrix}.$$

Observe that a single error in position 1, or a triple error in positions 2, 3, and 5, will yield the same corrector; in general many different error patterns will correspond to a given corrector. (In the next section we find out how many.)

4.4. *The application of group theory to parity check coding*

The object of this section is to prove the following theorem.

Theorem 4.4.1. The minimum-distance decoding scheme for a parity check code may be described as follows.

1. Given a received sequence $\mathbf{v}$, calculate the corrector $\mathbf{c}$ associated with $\mathbf{v}$.

2. Examine all error patterns whose corrector is $\mathbf{c}$, and correct the error of smallest magnitude. In other words, if among all sequences $\mathbf{z}$ such that $A\mathbf{z}^T = \mathbf{c}$, $\mathbf{z}_0$ is a sequence of minimum weight, that is, minimum number of ones, $\mathbf{v}$ is decoded as $\mathbf{v} - \mathbf{z}_0$. Thus if $\mathbf{v}$ is received, we declare that the code word $\mathbf{v} - \mathbf{z}_0$ was transmitted (note that $\mathbf{v} - \mathbf{z}_0$ is a code word since $A(\mathbf{v} - \mathbf{z}_0)^T = A\mathbf{v}^T - A\mathbf{z}_0^T = \mathbf{c} - \mathbf{c} = \mathbf{0}$) and that error pattern $\mathbf{z}_0$ occurred.

Theorem 4.4.1 implies that the minimum-distance decoder may be regarded as a function which assigns an error pattern to each corrector. Thus instead of storing a table of all 2^n binary sequences of length n and their associated code words, we need only store a table of all possible correctors and their associated error patterns. Since the correctors are linear combinations of the columns of A and the rank of A is $t = n - k$, there are $2^t = 2^{n-k}$ distinct correctors, in general a much smaller number than 2^n.

To prove Theorem 4.4.1, as well as to gain further insight into the structure of parity check codes, it is convenient to introduce some concepts from group theory.

Definition. A *group* is a set G together with an operation,† which we refer to as "addition," satisfying

1. $a \in G, b \in G$ implies $a + b \in G$ (*closure*).
2. $a + (b + c) = (a + b) + c$ for all $a, b, c, \in G$ (*associativity*).
3. There is an element $0 \in G$ such that $a + 0 = 0 + a = a$ for all $a \in G$ (*identity element*).
4. For each $a \in G$ there is an element $(-a) \in G$ such that $a + (-a) = (-a) + a = 0$ (*inverse*).

A group is *abelian* or *commutative* if $a + b = b + a$ for all $a, b \in G$.

The integers under ordinary addition and the set of binary sequences of a fixed length n under modulo 2 addition are examples of abelian groups. As far as the application to parity check coding is concerned, the basic result needed is the following:

Theorem 4.4.2. Let S be the set of code words in a parity check code. Then S is a group under modulo 2 addition.

Proof. Let A be the parity check matrix of the code. Then a binary sequence $\mathbf{w}$ is a code word if and only if $\mathbf{w}$ satisfies the parity check equations, that is, if and only if $A\mathbf{w}^T = \mathbf{0}$. To prove the closure property we note that if $\mathbf{w}_1 \in S$, $\mathbf{w}_2 \in S$, then $A\mathbf{w}_1{}^T = \mathbf{0}$, $A\mathbf{w}_2{}^T = \mathbf{0}$. Hence $A(\mathbf{w}_1 + \mathbf{w}_2)^T = A(\mathbf{w}_1{}^T + \mathbf{w}_2{}^T) = A\mathbf{w}_1{}^T + A\mathbf{w}_2{}^T = \mathbf{0}$; hence $\mathbf{w}_1 + \mathbf{w}_2 \in S$. The associative property follows from the definition of modulo 2 addition. We may take the zero vector as the identity element, since the zero vector always satisfies the parity check equations and is therefore a code word. We may take each element of S as its own inverse, again by definition of modulo 2 addition. This completes the proof.

† By an operation we mean a function that assigns to every pair a, b of elements in G, another element $a + b$.

The set S, of course, is an abelian group, since modulo 2 addition is commutative. If the code words of S are of length n, S is said to be a *subgroup* of the group $\mathcal{B}_n$ of all binary sequences of length n, since S is contained in the larger group $\mathcal{B}_n$.

We now prove the converse of Theorem 4.4.2.

Theorem 4.4.3. Let S be a set of binary sequences of length n that forms a group under modulo 2 addition. Then S is a parity check code, that is, there is a parity check matrix A such that the code determined by A is S.

Proof. Let us arrange the code words in an s by n matrix M, where s is the total number of code words. An example is given in Table 4.4.1. It may be verified that the code does in fact form a group; for example, $\mathbf{w}_1 + \mathbf{w}_2 = \mathbf{w}_4$, $\mathbf{w}_5 + \mathbf{w}_7 = \mathbf{w}_2$.

Table 4.4.1. Matrix of code words

$$\begin{array}{c|cccccc} \mathbf{w}_0 & 0 & 0 & 0 & 0 & 0 & 0 \\ \mathbf{w}_1 & 1 & 0 & 1 & 0 & 0 & 1 \\ \mathbf{w}_2 & 1 & 1 & 0 & 0 & 1 & 0 \\ \mathbf{w}_3 & 0 & 1 & 0 & 1 & 0 & 1 \\ \mathbf{w}_4 & 0 & 1 & 1 & 0 & 1 & 1 \\ \mathbf{w}_5 & 1 & 1 & 1 & 1 & 0 & 0 \\ \mathbf{w}_6 & 1 & 0 & 0 & 1 & 1 & 1 \\ \mathbf{w}_7 & 0 & 0 & 1 & 1 & 1 & 0 \end{array} = M$$

Let k be the rank of M, and $m = n - k$. Then k is the maximum number of linearly independent rows of M, and also the maximum number of linearly independent columns of M. Assume for convenience that $\mathbf{w}_1, \mathbf{w}_2, \ldots, \mathbf{w}_k$ are linearly independent. (If not, we may rearrange the rows.) Then every code word in S is a modulo 2 linear combination of $\mathbf{w}_1, \mathbf{w}_2, \ldots, \mathbf{w}_k$. But every linear combination of $\mathbf{w}_1, \mathbf{w}_2, \ldots, \mathbf{w}_k$ is a member of S since S is a group. Thus S coincides with the set of all binary sequences $\mathbf{w}$ which can be expressed as

$$\mathbf{w} = \lambda_1\mathbf{w}_1 + \lambda_2\mathbf{w}_2 + \cdots + \lambda_k\mathbf{w}_k, \qquad \lambda_i = 0 \text{ or } 1. \tag{4.4.1}$$

By the linear independence of $\mathbf{w}_1, \ldots, \mathbf{w}_k$, each distinct choice of $(\lambda_1, \ldots, \lambda_k)$ produces a different $\mathbf{w}$. Thus the number of code words in S is 2^k. Now assume for convenience that the last k columns of M (columns $m + 1, m + 2, \ldots, m + k$ where $m = n - k$) are linearly independent. Then the remaining columns $1, 2, \ldots, m$ may be expressed as linear combinations of columns $m + 1, m + 2, \ldots, m + k$. In other words, each digit $r_1, r_2, \ldots, r_m$ of the code words of S is expressible as a linear combination of digits $r_{m+1}, r_{m+2}, \ldots, r_{m+k}$. Hence we may write a set of linear equations relating $r_1, r_2, \ldots, r_m$ to $r_{m+1}, r_{m+2}, \ldots, r_{m+k}$, that is,

$$\begin{aligned} r_1 &= b_{11}r_{m+1} + b_{12}r_{m+2} + \cdots + b_{1k}r_{m+k} \\ r_2 &= b_{21}r_{m+1} + b_{22}r_{m+2} + \cdots + b_{2k}r_{m+k} \\ &\;\;\vdots \\ r_m &= b_{m1}r_{m+1} + b_{m2}r_{m+2} + \cdots + b_{mk}r_{m+k}. \end{aligned} \tag{4.4.2}$$

The equations (4.4.2) are of the form

$$A\begin{bmatrix} r_1 \\ \cdot \\ \cdot \\ \cdot \\ r_n \end{bmatrix} = 0,$$

where

$$A = \left[\begin{array}{c|cccc} & -b_{11} & -b_{12} & \cdots & -b_{1k} \\ & & \cdot & & \\ I_m & & \cdot & & \\ & & \cdot & & \\ & -b_{m1} & -b_{m2} & \cdots & -b_{mk} \end{array}\right] \tag{4.4.3†}$$

and I_m is an identity matrix of order m.

Thus we have shown that every code word $\mathbf{w} \in S$ satisfies $A\mathbf{w}^T = \mathbf{0}$, where A is the matrix (4.4.3). To show that every solution of $A\mathbf{w}^T = \mathbf{0}$ is in S we note that in solving the equations (4.4.2) the variables r_{m+1}, $r_{m+2}, \ldots, r_{m+k}$ may be specified arbitrarily. Thus $A\mathbf{w}^T = \mathbf{0}$ has 2^k solutions. But we have shown that S has 2^k elements, each of which satisfies $A\mathbf{w}^T = \mathbf{0}$. Hence the set of row vectors $\mathbf{w}$ satisfying $A\mathbf{w}^T = \mathbf{0}$ is identical to the set S of code words, that is, the code determined by A is S. The proof is complete.

To illustrate the process of constructing the parity check equations (4.4.2), consider the code of Table 4.4.1. Since $\mathbf{w}_1$, $\mathbf{w}_2$, and $\mathbf{w}_3$ form a maximal set of linearly independent code words, any parity check equations satisfied by $\mathbf{w}_1$, $\mathbf{w}_2$, and $\mathbf{w}_3$ will necessarily be satisfied by all code words. [To see this, we observe that $A\mathbf{w}_i^T = \mathbf{0}$ for $i = 1, 2, \ldots, k$ implies $A(\lambda_1\mathbf{w}_1^T + \cdots + \lambda_k\mathbf{w}_k^T) = \mathbf{0}$ for all possible choices of the λ_i.] Thus

† Addition and subtraction are the same in modulo 2 arithmetic, but the use of the minus sign in certain arguments facilitates generalizations to nonbinary coding (see Section 4.9).

we restrict our attention to the submatrix

$$Q = \begin{bmatrix} 1 & 0 & 1 & 0 & 0 & 1 \\ 1 & 1 & 0 & 0 & 1 & 0 \\ 0 & 1 & 0 & 1 & 0 & 1 \end{bmatrix}$$

formed by $\mathbf{w}_1$, $\mathbf{w}_2$, and $\mathbf{w}_3$. Since the last three columns of Q are linearly independent, we should be able to express the remaining three columns in terms of these. For example, to express column 1 in terms of columns 4, 5, and 6, we have to find numbers a_1, a_2, and a_3 such that

$$\begin{bmatrix} 1 \\ 1 \\ 0 \end{bmatrix} = a_1 \begin{bmatrix} 0 \\ 0 \\ 1 \end{bmatrix} + a_2 \begin{bmatrix} 0 \\ 1 \\ 0 \end{bmatrix} + a_3 \begin{bmatrix} 1 \\ 0 \\ 1 \end{bmatrix}, \qquad a_i = 0 \text{ or } 1. \tag{4.4.4}$$

Equation (4.4.4) yields $a_3 = 1$, $a_2 = 1$, $a_1 + a_3 = 0$ or $a_1 = 1$. Similarly, to express columns 2 and 3 in terms of columns 4, 5, and 6 we have to solve

$$\begin{bmatrix} 0 \\ 1 \\ 1 \end{bmatrix} = b_1 \begin{bmatrix} 0 \\ 0 \\ 1 \end{bmatrix} + b_2 \begin{bmatrix} 0 \\ 1 \\ 0 \end{bmatrix} + b_3 \begin{bmatrix} 1 \\ 0 \\ 1 \end{bmatrix} \tag{4.4.5}$$

$$\begin{bmatrix} 1 \\ 0 \\ 0 \end{bmatrix} = d_1 \begin{bmatrix} 0 \\ 0 \\ 1 \end{bmatrix} + d_2 \begin{bmatrix} 0 \\ 1 \\ 0 \end{bmatrix} + d_3 \begin{bmatrix} 1 \\ 0 \\ 1 \end{bmatrix}. \tag{4.4.6}$$

Solving (4.4.5) and (4.4.6) we obtain $b_1 = 1$, $b_2 = 1$, $b_3 = 0$, $d_1 = 1$, $d_2 = 0$, $d_3 = 1$. From (4.4.4), (4.4.5), and (4.4.6) we may write

$$\begin{aligned} r_1 &= r_4 + r_5 + r_6 \\ r_2 &= r_4 + r_5 \\ r_3 &= r_4 \phantom{{}+r_5} + r_6. \end{aligned}$$

Hence the parity check matrix is

$$A = \begin{bmatrix} 1 & 0 & 0 & 1 & 1 & 1 \\ 0 & 1 & 0 & 1 & 1 & 0 \\ 0 & 0 & 1 & 1 & 0 & 1 \end{bmatrix}.$$

Because of Theorems 4.4.2 and 4.4.3 a parity check code is often called a "group code." The term "(n, k) code" is also used to refer to a parity check code with code words of length n and k information digits.

We now turn to the problem of constructing the minimum-distance decoder for a parity check code. As an example, consider the code defined by the parity check equations

$$\begin{bmatrix} 1 & 0 & 1 & 0 \\ 0 & 1 & 1 & 1 \end{bmatrix} \begin{bmatrix} r_1 \\ r_2 \\ r_3 \\ r_4 \end{bmatrix} = \begin{bmatrix} 0 \\ 0 \end{bmatrix}. \tag{4.4.7}$$

The four code words corresponding to (4.4.7) are $\mathbf{w}_0 = 0000$, $\mathbf{w}_1 = 0101$, $\mathbf{w}_2 = 1110$, and $\mathbf{w}_3 = 1011$.

Table 4.4.2. Error pattern vectors and received sequences for a group code

$\mathbf{w}_0$	$\mathbf{w}_1$	$\mathbf{w}_2$	$\mathbf{w}_3$
0000	0101	1110	1011
0110	0011	1000	1101
1111	1010	0001	0100
0010	0111	1100	1001

Various error pattern vectors and their associated received sequences are shown in Table 4.4.2. The leftmost element in any row is an error pattern vector. The element $\mathbf{v}_i$ ($i = 0, 1, 2, 3$) in a given row represents the received sequence when code word $\mathbf{w}_i$ is transmitted and the error pattern vector is $\mathbf{v}_0$. For example, if $\mathbf{w}_2$ is transmitted and the error pattern vector is 0110, that is, errors are made in positions 2 and 3, the received sequence is 1000. The rows of Table 4.4.2 are examples of what are called *cosets* in group theory.

Definition. Given a subgroup S of a group G, the *coset* associated with an element $\mathbf{z} \in G$ is the set of all elements $\mathbf{z} + \mathbf{w}$, where $\mathbf{w}$ ranges over S. The coset associated with $\mathbf{z}$ is written $\mathbf{z} + S$.

In the example we are considering, G is the group of all sixteen binary sequences of length 4 under modulo 2 addition, S is the set of code words $\{\mathbf{w}_0, \mathbf{w}_1, \mathbf{w}_2, \mathbf{w}_3\}$. Various cosets are

$$\begin{aligned} 0110 + S &= \{0110, 0011, 1000, 1101\} \\ 1000 + S &= \ 0110 + S \\ 1111 + S &= \{1111, 1010, 0001, 0100\} \\ 0000 + S &= \ S. \end{aligned}$$

The key property of cosets is the following.

Lemma 4.4.4. The cosets $\mathbf{z}_1 + S$ and $\mathbf{z}_2 + S$ are either disjoint or identical. (Thus in Table 4.4.2 no sequence appears in more than one row.)

Proof. Suppose $\mathbf{z}_1 + S$ and $\mathbf{z}_2 + S$ are not disjoint. Then there is an element $\mathbf{v}'$ such that $\mathbf{v}' \in \mathbf{z}_1 + S$ and $\mathbf{v}' \in \mathbf{z}_2 + S$. Hence $\mathbf{v}' = \mathbf{z}_1 + \mathbf{w}_1$ for some $\mathbf{w}_1 \in S$ and $\mathbf{v}' = \mathbf{z}_2 + \mathbf{w}_2$ for some $\mathbf{w}_2 \in S$. To show that $\mathbf{z}_1 + S$ and $\mathbf{z}_2 + S$ are identical, suppose $\mathbf{v} \in \mathbf{z}_1 + S$. Then $\mathbf{v} = \mathbf{z}_1 + \mathbf{w}_3$ for some $\mathbf{w}_3 \in S$. Therefore $\mathbf{v} = \mathbf{v}' - \mathbf{w}_1 + \mathbf{w}_3 = \mathbf{z}_2 + (\mathbf{w}_2 - \mathbf{w}_1 + \mathbf{w}_3)$. Hence $\mathbf{v}$ is of the form $\mathbf{z}_2 + \mathbf{w}$ where $\mathbf{w} = \mathbf{w}_2 - \mathbf{w}_1 + \mathbf{w}_3 \in S$. Thus $\mathbf{v} \in \mathbf{z}_2 + S$. A symmetrical argument shows that $\mathbf{v} \in \mathbf{z}_2 + S$ implies $\mathbf{v} \in \mathbf{z}_1 + S$, completing the proof.

Each coset $\mathbf{z} + S$ has exactly as many elements as S since all the elements $\mathbf{z} + \mathbf{w}\ (\mathbf{w} \in S)$ are distinct. (If $\mathbf{z} + \mathbf{w}_1 = \mathbf{z} + \mathbf{w}_2$, then $\mathbf{w}_1 = \mathbf{w}_2$.) Hence the number of cosets is the number of elements of G divided by the number of elements of S. If G is the group of 2^n binary sequences of length n and S is a group code with 2^k code words of length n, then the number of cosets is 2^{n-k}. Thus the number of cosets and the number of possible corrector values coincide. This is not a coincidence, as we shall see shortly.

We now establish a symmetry property of the minimum-distance decoder for a group code.

Lemma 4.4.5. Let S be a group code. Suppose that for a particular code word $\mathbf{w}_i$ and a particular error pattern vector $\mathbf{z}$,

$$d(\mathbf{w}_i + \mathbf{z}, \mathbf{w}_i) \leq d(\mathbf{w}_i + \mathbf{z}, \mathbf{w}_j) \tag{4.4.8}$$

for all code words $\mathbf{w}_j$. Then

$$d(\mathbf{w} + \mathbf{z}, \mathbf{w}) \leq d(\mathbf{w} + \mathbf{z}, \mathbf{w}') \tag{4.4.9}$$

for all code words $\mathbf{w}, \mathbf{w}'$.

The interpretation of Lemma 4.4.5 is as follows. Suppose that the error pattern $\mathbf{z}$ is corrected by the minimum-distance decoder when the code word $\mathbf{w}_i$ is transmitted, that is, the minimum-distance decoder assigns to the received sequence $\mathbf{w}_i + \mathbf{z}$ the code word $\mathbf{w}_i$, so that condition (4.4.8) is satisfied. Then by (4.4.9) we may without loss of generality assume that the error pattern $\mathbf{z}$ is corrected no matter which code word is sent. In other words, without changing the probability of error we may adjust the minimum-distance decoder so that *a given error pattern is either always corrected or never corrected.* From now on, we assume that the minimum-distance decoder always has this property; thus the statement "the error pattern $\mathbf{z}$ is corrected" is unambiguous.

Proof. First observe that if $\mathbf{v}_1$, $\mathbf{v}_2$, and $\mathbf{v}_3$ are binary sequences of the same length, then $d(\mathbf{v}_1 + \mathbf{v}_3, \mathbf{v}_2 + \mathbf{v}_3) = d(\mathbf{v}_1, \mathbf{v}_2)$. For example,

$$d(0111, 1010) = 3 \quad \text{and} \quad d(0111 + 1100, 1010 + 1100) = d(1011, 0110) = 3.$$

We therefore have

$$\begin{aligned} d(\mathbf{z} + \mathbf{w}, \mathbf{w}') &= d(\mathbf{z} + \mathbf{w} + \mathbf{w}_i - \mathbf{w}, \mathbf{w}' + \mathbf{w}_i - \mathbf{w}) \\ &= d(\mathbf{z} + \mathbf{w}_i, \mathbf{w}' + \mathbf{w}_i - \mathbf{w}) \geq d(\mathbf{z} + \mathbf{w}_i, \mathbf{w}_i) \qquad \text{(by 4.4.8)} \\ &= d(\mathbf{z} + \mathbf{w}_i + \mathbf{w} - \mathbf{w}_i, \mathbf{w}_i + \mathbf{w} - \mathbf{w}_i) = d(\mathbf{z} + \mathbf{w}, \mathbf{w}). \end{aligned}$$

Another consequence of Lemma 4.4.5 is that if we wish to correct an error pattern $\mathbf{z}_1$, we cannot correct any other error pattern $\mathbf{z}_2$ in the coset $\mathbf{z}_1 + S$. For if $\mathbf{z}_2 = \mathbf{z}_1 + \mathbf{w}$ where $\mathbf{w} \in S$, $\mathbf{w} \neq \mathbf{0}$, then the transmission of $\mathbf{w}$ together with the error pattern $\mathbf{z}_1$ will yield the same received sequence as the transmission of the zero vector together with error pattern $\mathbf{z}_2$. Thus it is impossible to correct both $\mathbf{z}_1$ and $\mathbf{z}_2$ for all possible transmitted sequences. Hence *exactly one sequence in each coset can serve as a correctible error pattern.* Since the probability of an error pattern increases as the number of digits involved in the error decreases, it appears that among the sequences in a given coset, we should choose as the correctible error pattern that sequence with the fewest number of ones. This result is established by the following theorem.

Theorem 4.4.6. To construct the minimum-distance decoding scheme for a group code S, it is sufficient to choose as the correctible error pattern for each coset a sequence with minimum weight, that is, minimum number of ones. If $\mathbf{z}$ is such a sequence, then the received sequence $\mathbf{z} + \mathbf{w}$ ($\mathbf{w} \in S$) is decoded as $\mathbf{w}$.

Proof. If $\mathbf{z}$ is a sequence of minimum weight in the coset $\mathbf{z} + S$, then $d(\mathbf{z}, \mathbf{w}) = d(\mathbf{z} - \mathbf{w}, \mathbf{w} - \mathbf{w}) = d(\mathbf{z} - \mathbf{w}, \mathbf{0})$. Since $\mathbf{z}$ has minimum weight among all sequences $\mathbf{z} + \mathbf{w}$, $\mathbf{w} \in S$, we have $d(\mathbf{z} - \mathbf{w}, \mathbf{0}) \geq d(\mathbf{z}, \mathbf{0})$. Hence the sequence $\mathbf{z}$ is at least as close to zero as to any other code word, so that in the minimum-distance decoder we may correct the error pattern $\mathbf{z}$ when $\mathbf{0}$ is transmitted, and hence by Lemma 4.4.5 we may correct $\mathbf{z}$ for all possible transmitted sequences. The theorem is proved.

It follows from Theorem 4.4.6 that the minimum-distance decoder may be constructed in the following way.

Write down a table of cosets (such as Table 4.4.2); choose a sequence $\mathbf{z}$ of minimum weight in the coset $\mathbf{z} + S$; rearrange the table so that the entries in the row corresponding to $\mathbf{z} + S$ are $\mathbf{z} + \mathbf{0}$, $\mathbf{z} + \mathbf{w}_1, \ldots,$ $\mathbf{z} + \mathbf{w}_{2^k-1}$, where the $\mathbf{w}_i$ are the code words. (Thus $\mathbf{z}$ is the leftmost element of its coset, or the so-called "*coset leader*.") If a received sequence $\mathbf{v}$ is in the column of the coset table headed by $\mathbf{w}_i$, then $\mathbf{v} = \mathbf{w}_i + \mathbf{z}$ for some correctible error pattern $\mathbf{z}$; hence $\mathbf{v}$ is decoded as $\mathbf{w}_i$. The minimum-distance decoding scheme for the code defined by (4.4.7) and Table 4.4.2 is shown in Table 4.4.3. Notice that row 3 has two sequences of weight one, namely 0001 and 0100. Either sequence can be chosen as the coset leader.

Table 4.4.3. Minimum-distance decoding scheme for a group code

$\mathbf{w}_0$	$\mathbf{w}_1$	$\mathbf{w}_2$	$\mathbf{w}_3$
0000	0101	1110	1011
1000	1101	0110	0011
0001	0100	1111	1010
0010	0111	1100	1001

We now show that there is a one-to-one correspondence between the 2^{n-k} correctible error patterns of a group code and the 2^{n-k} possible values of the corrector.

Theorem 4.4.7. All sequences in the same coset of a group code S have the same corrector. Two sequences in different cosets have different correctors.

Proof. If $\mathbf{z}_1$ and $\mathbf{z}_2$ are in the same coset, then $\mathbf{z}_2 = \mathbf{z}_1 + \mathbf{w}$ for some $\mathbf{w} \in S$. Thus $A\mathbf{z}_2^T = A\mathbf{z}_1^T$ so that $\mathbf{z}_1$ and $\mathbf{z}_2$ have the same corrector. On the other hand, if $\mathbf{z}_1$ and $\mathbf{z}_2$ have the same corrector then $A\mathbf{z}_1^T = A\mathbf{z}_2^T$ or $A(\mathbf{z}_2 - \mathbf{z}_1)^T = \mathbf{0}$. But then $\mathbf{z}_2 - \mathbf{z}_1$ satisfies the parity check equations, hence $\mathbf{z}_2 - \mathbf{z}_1$ is a code word $\mathbf{w}$. Therefore $\mathbf{z}_1$ and $\mathbf{z}_2 = \mathbf{z}_1 + \mathbf{w}$ are in the same coset.

Finally, we prove the main theorem of this section.

Proof of Theorem 4.4.1. Let $\mathbf{c}$ be the corrector associated with a given received sequence $\mathbf{v}$; let $\mathbf{v}$ belong to the coset $\mathbf{z}_0 + S$ where $\mathbf{z}_0$ is a sequence of minimum weight in $\mathbf{z}_0 + S$. By Theorem 4.4.7, the sequences of the coset $\mathbf{z}_0 + S$ are precisely the sequences whose corrector is $\mathbf{c}$. It follows that among all sequences $\mathbf{z}$ such that $A\mathbf{z}^T = \mathbf{c}$, $\mathbf{z}_0$ is a sequence of minimum weight. If $\mathbf{v} = \mathbf{z}_0 + \mathbf{w}$, then, by Theorem 4.4.6, $\mathbf{v}$ is decoded as $\mathbf{w} = \mathbf{v} - \mathbf{z}_0$. This agrees exactly with the procedure given in Theorem 4.4.1.

We conclude this section with an example. Consider the code whose parity check matrix is

$$A = \begin{bmatrix} 1 & 0 & 0 & 0 & 1 & 1 \\ 0 & 1 & 0 & 0 & 1 & 0 \\ 0 & 0 & 1 & 0 & 1 & 1 \\ 0 & 0 & 0 & 1 & 0 & 1 \end{bmatrix}.$$

The minimum-distance decoding table is shown in Table 4.4.4. Notice that all single errors are corrected, along with some (but not all) double errors and some triple errors.

Table 4.4.4. Minimum-distance decoding table for a parity check code

Code Words	000000	111010	101101	010111	(Transposed) Correctors 0000
Single errors	100000	011010	001101	110111	1000
	010000	101010	111101	000111	0100
	001000	110010	100101	011111	0010
	000100	111110	101001	010011	0001
	000010	111000	101111	010101	1110
	000001	111011	101100	010110	1011
Double errors	110000	001010	011101	100111	1100
	101000	010010	000101	111111	1010
	100100	011110	001001	110011	1001
	100010	011000	001111	110101	0110
	100001	011011	001100	110110	0011
	010100	101110	111001	000011	0101
	010001	101011	111100	000110	1111
Triple errors	110100	001110	011001	100011	1101
	110001	001011	011100	100110	0111

To calculate the probability of error for a group code, we note that correct transmission will occur if and only if the error pattern is a coset leader and hence correctible. Thus if N_i is the number of coset leaders of weight i, or equivalently the number of i-tuple errors corrected by the code, then the probability of correct transmission is

$$p(e') = \sum_{i=0}^{n} N_i \beta^i (1 - \beta)^{n-i}.$$

For the code of Table 4.4.4,

$$p(e') = (1-\beta)^6 + 6\beta(1-\beta)^5 + 7\beta^2(1-\beta)^4 + 2\beta^3(1-\beta)^3.$$

Note that the probability of correct transmission $p(e' \mid \mathbf{w})$ is the same for each transmitted code word $\mathbf{w}$; hence for a parity check code the average and maximum probability of error coincide.

4.5. Upper and lower bounds on the error correcting ability of parity check codes

In this section we investigate the relation between the number of code words in a parity check code and its error correcting ability. As we have seen previously, an increase in the number of words in a code will in general be accompanied by a reduction in the number of errors corrected. We will try to obtain quantitative results to this effect.

We first propose the following problem. We wish to construct a parity check code that has certain error correcting properties; for example, we might want a code that corrects e-tuple (and all smaller) errors. How do we choose the parity check matrix so that the resulting code has the desired properties? To begin with, let us consider the construction of a single-error correcting code. Since the corrector associated with an error in a single digit r_j is the jth column of the parity check matrix, we need only choose n distinct (nonzero) columns; the resulting code must be single-error correcting, since no two distinct single errors can result in the same corrector. (Also, since no column is zero, a single error cannot be confused with the "no error" condition.) For example, if $n = 7$, $k = 4$, the code defined by the following parity check equations will correct single errors.

$$\begin{bmatrix} 1 & 0 & 0 & 1 & 1 & 0 & 1 \\ 0 & 1 & 0 & 1 & 0 & 1 & 1 \\ 0 & 0 & 1 & 0 & 1 & 1 & 1 \end{bmatrix} \begin{bmatrix} r_1 \\ r_2 \\ r_3 \\ r_4 \\ r_5 \\ r_6 \\ r_7 \end{bmatrix} = \begin{bmatrix} 0 \\ 0 \\ 0 \end{bmatrix}$$

r_1, r_2, r_3 are check digits.
r_4, r_5, r_6, r_7 are information digits.

Note that if the parity check matrix A is m by n, we may as well assume that the rank of A is m. For if the rank is $t < m$, then $m - t$ rows are linearly dependent on the remaining t rows and may be eliminated. We may also assume for convenience that the m by m submatrix in the left-hand corner of A is nonsingular; in fact we may assume without loss of generality that the submatrix is an identity matrix. If the submatrix is not originally an identity matrix the equations may be diagonalized by elementary row transformations. For example, consider the parity check equations (4.3.2):

$$\begin{aligned} r_1 \qquad\quad + r_4 + r_5 \qquad &= 0 \\ r_2 + r_3 + r_4 \qquad + r_6 &= 0 \\ r_1 \quad + r_3 + r_4 \qquad + r_6 &= 0. \end{aligned}$$

Let us add equation 1 to equation 3, obtaining

$$\begin{aligned} r_1 \qquad\quad + r_4 + r_5 \qquad &= 0 \\ r_2 + r_3 + r_4 \qquad + r_6 &= 0 \\ r_3 \qquad + r_5 + r_6 &= 0. \end{aligned}$$

Now let us add equation 3 to equation 2, obtaining

$$\begin{aligned} r_1 \qquad + r_4 + r_5 \qquad &= 0 \\ r_2 \quad + r_4 + r_5 \qquad &= 0 \\ r_3 \qquad + r_5 + r_6 &= 0. \end{aligned}$$

The submatrix in the lower left-hand corner is now an identity matrix, and the new equations have exactly the same solution as the original equations; in other words, both sets of equations define the same code. The diagonalization process will be successful as long as the original submatrix is nonsingular.

The general relation between the parity check matrix and the error correcting ability of the code is contained in the following theorem.

Theorem 4.5.1. The parity check code defined by the matrix A will correct e-tuple (and all smaller) errors if and only if every set of $2e$ columns of A is linearly independent.

Proof. By Theorem 4.4.1, e-tuple and all smaller errors will be corrected if and only if all error patterns of weight $\leq e$ yield distinct correctors, that is, if and only if no linear combination of e or fewer columns of A equals another such linear combination. But this is precisely the condition that each set of $2e$ columns of A be linearly independent.

As an example, the following parity check matrix describes a double-error correcting code with 7 check digits and 3 information digits.

$$A = \begin{bmatrix} 1 & 0 & 0 & 0 & 0 & 0 & 0 & 1 & 0 & 1 \\ 0 & 1 & 0 & 0 & 0 & 0 & 0 & 0 & 0 & 1 \\ 0 & 0 & 1 & 0 & 0 & 0 & 0 & 1 & 0 & 1 \\ 0 & 0 & 0 & 1 & 0 & 0 & 0 & 0 & 1 & 1 \\ 0 & 0 & 0 & 0 & 1 & 0 & 0 & 1 & 1 & 0 \\ 0 & 0 & 0 & 0 & 0 & 1 & 0 & 0 & 1 & 0 \\ 0 & 0 & 0 & 0 & 0 & 0 & 1 & 1 & 1 & 0 \end{bmatrix}. \qquad (4.5.1)$$

It may be verified that every set of 4 columns of the matrix (4.5.1) is linearly independent. However, we may find triple errors which are not correctible. For example,

$$\mathbf{c}(r_1) + \mathbf{c}(r_8) + \mathbf{c}(r_9) = \begin{bmatrix} 0 \\ 0 \\ 1 \\ 1 \\ 0 \\ 1 \\ 0 \end{bmatrix} = \mathbf{c}(r_3) + \mathbf{c}(r_4) + \mathbf{c}(r_6)$$

$[\mathbf{c}(r_j) = j\text{th column of } A]$.

Thus a triple error in r_1, r_8, and r_9 will be confused with a triple error in r_3, r_4, and r_6.

Next we shall consider the following problem. Suppose we wish to construct a parity check code that corrects e-tuple and all smaller errors, with a fixed code-word length n. We ask how many check digits are necessary to construct the code. We wish to use as few check digits as possible, since a small number of check digits corresponds to a large number of code words. In general, we cannot determine the minimum number of check digits exactly, but we can establish some useful lower and upper bounds, as follows.

Theorem 4.5.2. (Hamming Lower Bound on the Number of Check Digits). The number of check digits in an e-tuple-error correcting parity check code must satisfy

$$2^m \geq \sum_{i=0}^{e} \binom{n}{i} \tag{4.5.2}$$

where n = code word length
m = number of check digits $= n - k$.

Proof. In order that a parity check code be capable of correcting e-tuple and all smaller errors, each error combination involving e or fewer digits must result in a distinct corrector. The number of possible i-tuple errors in an n-digit code word is equal to the number of ways of choosing i digits out of n, namely $\binom{n}{i}$. The number of possible corrector values is 2^m; hence in order that there be a unique corrector for each error combination, we must have

$$2^m \geq 1 + \binom{n}{1} + \binom{n}{2} + \cdots + \binom{n}{e},$$

which proves (4.5.2). (The term "1" appears on the right since if no error is made, the associated corrector must be the zero vector.)

The Hamming lower bound on the number of check digits is in fact identical to the Hamming upper bound on the number of code words, which we proved for general binary codes in Section 4.2. To see this, note that the number of code words is $2^k = 2^{n-m} = 2^n/2^m$. Hence (4.5.2) is equivalent to

$$2^k \leq \frac{2^n}{\sum_{i=0}^{e} \binom{n}{i}},$$

which is (4.2.1) for the special case of a parity check code.

The Hamming lower bound is *necessary but not sufficient* for the constructibility of an e-tuple-error correcting parity check code. In other words, if m_0 is the smallest positive integer satisfying (4.5.2) for a given n and e, there may not be a parity check code using m_0 check digits. For example, if $n = 10, e = 2$, we find from (4.5.2) that $m_0 = 6$. However, there is no double-error correcting parity check code with fewer than seven check digits, as the reader may verify by trial and error.

An upper bound on the number of check digits required for the construction of a code is provided by the following theorem.

Theorem 4.5.3. (Varsharmov-Gilbert-Sacks Condition). An e-tuple-error correcting parity check code with words of length n may be constructed if the number of check digits m satisfies

$$2^m > \sum_{i=0}^{2e-1} \binom{n-1}{i}. \tag{4.5.3}$$

The condition (4.5.3) is *sufficient but not necessary* for constructibility of a code. In other words, if for a fixed n and e, m_0 is the smallest positive integer satisfying (4.5.3), then it is guaranteed that an e-tuple-error correcting parity check code with m_0 check digits exists. However, there may be an e-tuple-error correcting code with fewer check digits. For example, if $n = 10$ and $e = 2$, then (4.5.3) yields $m_0 = 8$. However, it is possible to construct a double-error correcting code with seven check digits. [See the parity check matrix (4.5.1).]

Proof. The proof is accomplished by giving a synthesis procedure for the required code. We shall choose the columns $\mathbf{c}(r_1), \mathbf{c}(r_2), \ldots, \mathbf{c}(r_n)$ of the parity check matrix successively. By Theorem 4.5.1 we must satisfy the condition that every set of $2e$ columns be linearly independent. First we choose $\mathbf{c}(r_1)$ arbitrarily. [Of course we must have $\mathbf{c}(r_1) \neq \mathbf{0}$ since the zero vector must correspond to perfect transmission.] We now choose $\mathbf{c}(r_2)$ such that $\mathbf{c}(r_2) \neq \mathbf{0}$ and $\mathbf{c}(r_2) \neq \mathbf{c}(r_1)$. This can be done provided $2^m > 2$. We now choose $\mathbf{c}(r_3)$ such that

$$\mathbf{c}(r_3) \neq \mathbf{0}, \qquad \mathbf{c}(r_3) \neq \mathbf{c}(r_1), \qquad \mathbf{c}(r_3) \neq \mathbf{c}(r_2), \qquad \mathbf{c}(r_3) \neq \mathbf{c}(r_1) + \mathbf{c}(r_2),$$

thus guaranteeing that $\mathbf{c}(r_1)$, $\mathbf{c}(r_2)$ and $\mathbf{c}(r_3)$ will be linearly independent. A choice for $\mathbf{c}(r_3)$ can be made if $2^m > 4$. Let us digress for a moment to consider the special case $e = 2$. Then every set of four columns must be linearly independent. Thus we choose $\mathbf{c}(r_4)$ such that $\mathbf{c}(r_4)$ is not equal to any of the following:

$$\mathbf{0}, \quad \mathbf{c}(r_1), \quad \mathbf{c}(r_2), \quad \mathbf{c}(r_3), \quad \mathbf{c}(r_1) + \mathbf{c}(r_2),$$
$$\mathbf{c}(r_1) + \mathbf{c}(r_3), \quad \mathbf{c}(r_2) + \mathbf{c}(r_3), \quad \mathbf{c}(r_1) + \mathbf{c}(r_2) + \mathbf{c}(r_3).$$

Such a choice can be made if $2^m > 8$. We now choose $\mathbf{c}(r_5)$ such that

$$\mathbf{c}(r_5) \neq \mathbf{0}$$
$$\mathbf{c}(r_5) \neq \mathbf{c}(r_i), \qquad i = 1, 2, 3, 4$$
$$\mathbf{c}(r_5) \neq \mathbf{c}(r_i) + \mathbf{c}(r_j), \qquad i, j = 1, 2, 3, 4, \quad i \neq j$$
$$\mathbf{c}(r_5) \neq \mathbf{c}(r_i) + \mathbf{c}(r_j) + \mathbf{c}(r_k), \qquad i, j, k, = 1, 2, 3, 4, \quad i \neq k, j \neq k.$$

Note that $\mathbf{c}(r_5)$ *can* equal $\mathbf{c}(r_1) + \mathbf{c}(r_2) + \mathbf{c}(r_3) + \mathbf{c}(r_4)$ since we do not require that every set of five columns be linearly independent. Therefore $\mathbf{c}(r_5)$ can be chosen if

$$2^m > 1 + 4 + \binom{4}{2} + \binom{4}{3} = 15.$$

Now let us return to the general case. Suppose we have determined $\mathbf{c}(r_1), \mathbf{c}(r_2), \ldots, \mathbf{c}(r_{n-1})$. We then choose $\mathbf{c}(r_n)$ such that

$$\mathbf{c}(r_n) \neq \mathbf{0}$$
$$\mathbf{c}(r_n) \neq \mathbf{c}(r_i), \qquad i = 1, 2, \ldots, n-1$$
$$\mathbf{c}(r_n) \neq \mathbf{c}(r_i) + \mathbf{c}(r_j), \qquad i, j = 1, 2, \ldots, n-1, \quad i \neq j$$
$$\mathbf{c}(r_n) \neq \mathbf{c}(r_i) + \mathbf{c}(r_j) + \mathbf{c}(r_k), \qquad i, j = 1, 2, \ldots, n-1, \quad i, j, k \text{ distinct}$$
$$\vdots$$
$$\mathbf{c}(r_n) \neq \mathbf{c}(r_{i_1}) + \mathbf{c}(r_{i_2}) + \cdots + \mathbf{c}(r_{i_{2e-1}}),$$
$$i_1, i_2, \ldots, i_{2e-1} = 1, 2, \ldots, n-1, \quad i_1, i_2, \ldots, i_{2e-1} \text{ distinct}$$

At worst, all of the above combinations are distinct.† The number of

† It is not necessary that all the combinations be distinct; for example, in the parity check matrix (4.5.1) we have $2e = 4$ but several combinations involving three or fewer columns are equal, for example $\mathbf{c}(r_1) + \mathbf{c}(r_8) + \mathbf{c}(r_9) = \mathbf{c}(r_3) + \mathbf{c}(r_4) + \mathbf{c}(r_6)$; $\mathbf{c}(r_2) + \mathbf{c}(r_6) = \mathbf{c}(r_8) + \mathbf{c}(r_9) + \mathbf{c}(r_{10})$.

combinations is

$$1 + \binom{n-1}{1} + \binom{n-1}{2} + \cdots + \binom{n-1}{2e-1};$$

hence the theorem.

In general the smallest value of m (say m_1) satisfying the Varsharmov-Gilbert-Sacks condition exceeds the minimum value of m (say m_2) satisfying the Hamming lower bound. However, it was proved by Sacks (1958) that $m_1/m_2 \leq 2$. It may also be seen from (4.5.2) and (4.5.3) that the upper and lower bounds coincide when $e = 1$. In other words, a single-error correcting code is constructible *if and only if* $2^m \geq 1 + n$. (This fact may be seen directly without reference to Theorem 4.5.3, since for single-error correction we merely need n distinct nonzero columns.)

The bounds of this section say nothing about the probability of error of a code, which is a more basic quantity than the number of errors corrected. However, the results will be useful in Section 4.7, where we give precise bounds on the probability of error for general binary codes.

4.6. *Parity check codes are adequate*

In this section we show that parity check codes meet the specifications of the fundamental theorem. More precisely, we will prove the following result.

Theorem 4.6.1. Given a binary symmetric channel with capacity $C > 0$ and a positive number $R < C$, there is a sequence of parity check codes (s_i, n_i, λ_i), $i = 1, 2, \ldots,$ where $n_1 < n_2 < \ldots, s_i = 2^{k_i}, R \leq k_i/n_i < C$, and $\lim_{i\to\infty} \lambda_i = 0$.

Thus for each i we can construct a parity check code with at least $2^{n_i R}$ words and a (maximum) probability of error λ_i which approaches zero as $i \to \infty$. Hence, even if we are restricted to the class of parity check codes, we can still maintain the transmission rate $R < C$ with an arbitrarily small probability of error.

Proof. First observe that given a parity check code with 2^k words, we may find a *set of generators* for the code, that is, a set consisting of k code words (say) $\mathbf{w}_1, \ldots, \mathbf{w}_k$, such that each word of the code can be written as a modulo 2 linear combination of $\mathbf{w}_1, \ldots, \mathbf{w}_k$. (See the proof of Theorem 4.4.3.) The idea of the proof is to form a parity check code by selecting generators at random, and then to show that the "random-coding" proof of Section 3.5 goes through almost word for word. In particular we consider the following random experiment:

1. Fix R' such that $R < R' < C$. Let k and n be positive integers such that $R \leq k/n \leq R'$. Independently select binary sequences $\mathbf{w}_1, \mathbf{w}_2, \ldots, \mathbf{w}_k$ of length n, choosing the components of each sequence independently with 0 and 1 equally likely.

2. Form a code with $s = 2^k$ words† $\mathbf{w}_j, j = 0, 1, \ldots, 2^k - 1$, by taking all possible modulo 2 linear combinations of $\mathbf{w}_1, \ldots, \mathbf{w}_k$; in other words, set $\mathbf{w}_j = b_{1j}\mathbf{w}_1 + \cdots + b_{kj}\mathbf{w}_k$ (mod 2), where $b_{ij} = 0$ or 1, $i = 1, 2, \ldots, k$. (Let $\mathbf{w}_0$ be the zero vector, corresponding to $b_{i0} = 0$ for all i.)

Now if a_{tj} is the tth component of $\mathbf{w}_j$, then

$$a_{tj} = b_{1j}a_{t1} + b_{2j}a_{t2} + \cdots + b_{kj}a_{tk}. \tag{4.6.1}$$

Since the components of the words $\mathbf{w}_1, \ldots, \mathbf{w}_k$ are chosen independently, the a_{tj}, $t = 1, 2, \ldots, n$, are independent. Furthermore $P\{a_{tj} = 0\} = P\{a_{tj} = 1\} = 1/2$ for all t, provided $j \neq 0$. To see this, we observe that there are as many sequences $(a_{t1}, \ldots, a_{tk})$ that produce $a_{tj} = 0$ in (4.6.1) as there are sequences that produce $a_{tj} = 1$. For if $j \neq 0$, at least one coefficient b_{rj} is not zero. If $\sum_{m=1}^{k} b_{mj}a_{tm} = 0$, then define $a_{tm}{}^* = a_{tm}$, $m \neq r$; $a_{tr}{}^* = 1 + a_{tr}$. This yields $\sum_{m=1}^{k} b_{mj}a_{tm}{}^* = 1$. In this way we establish a one-to-one correspondence between sequences that produce $a_{tj} = 0$ in (4.6.1) and sequences that produce $a_{tj} = 1$.

Thus the components of the words $\mathbf{w}_j, j \neq 0$, are (statistically) independent and have the distribution that achieves channel capacity (that is, 0 and 1 are equally probable).

By a similar argument, we may show that the words $\mathbf{w}_j$ and $\mathbf{w}_{j'}$, j, $j' \neq 0$, $j \neq j'$, are independent. To accomplish this, we will prove that

$$P\{a_{tj} = \gamma, a_{tj'} = \gamma'\} = 1/4 \quad \text{for} \quad \gamma, \gamma' = 0{,}1, \quad t = 1, 2, \cdots n. \tag{4.6.2}$$

---

† Although the words $\mathbf{w}_1, \ldots, \mathbf{w}_k$ chosen in step 1 are statistically independent, a particular choice may result in $\mathbf{w}_1, \ldots, \mathbf{w}_k$ being *linearly* dependent. In this case, the words $\mathbf{w}_j, j = 0, 1, \ldots, 2^k - 1$, will not be distinct, and we obtain a code which strictly speaking is not a parity check code as defined in Section 4.3. If, however, the number of distinct words of the code is 2^r, $r < k$, and $\mathbf{w}_1', \ldots, \mathbf{w}_r'$ is a set of r linearly independent words, we may select sequences $\mathbf{w}_{r+1}', \ldots, \mathbf{w}_k'$ such that $\mathbf{w}_1', \ldots, \mathbf{w}_k'$ are linearly independent, and in this way construct a parity check code with 2^k distinct words. It can be shown (see Problem 4.11) that the average probability of error of the new code cannot exceed the average probability of error of the original code. Hence if the random experiment described above yields a code with average probability of error ε, we may always find a parity check code with the same number of words and an average probability of error $\leq \varepsilon$.

We note also that in this proof and elsewhere, "independent" will mean "statistically independent"; the term "linearly independent" will not be abbreviated.

It then follows that $P\{a_{tj} = \gamma, a_{tj'} = \gamma'\} = P\{a_{tj} = \gamma\} P\{a_{tj'} = \gamma'\}$; hence a_{tj} and $a_{tj'}$ are independent for each t, which implies the independence of $\mathbf{w}_j$ and $\mathbf{w}_{j'}$.

To prove (4.6.2), we note that since $j \neq j'$, we have $b_{rj} \neq b_{rj'}$ for some r. Say $b_{rj} = 1$ and $b_{rj'} = 0$. Then given any sequence $(a_{t1}, \ldots, a_{tk})$ such that

$$\sum_{m=1}^{k} b_{mj} a_{tm} = 0, \qquad \sum_{m=1}^{k} b_{mj'} a_{tm} = 0,$$

we define $a_{tm}^* = a_{tm}$, $m \neq r$; $a_{tr}^* = 1 + a_{tr}$. It follows that

$$\sum_{m=1}^{k} b_{mj} a_{tm}^* = 1, \qquad \sum_{m=1}^{k} b_{mj'} a_{tm}^* = 0.$$

This establishes a one-to-one correspondence between sequences that produce $a_{tj} = a_{tj'} = 0$ in (4.6.1) and sequences that produce $a_{tj} = 1$, $a_{tj'} = 0$. Thus

$$P\{a_{tj} = 0, a_{tj'} = 0\} = P\{a_{tj} = 1, a_{tj'} = 0\}. \tag{4.6.3}$$

But

$$P\{a_{tj} = 0, a_{tj'} = 0\} + P\{a_{tj} = 1, a_{tj'} = 0\} = P\{a_{tj'} = 0\} = 1/2,$$

so that both terms of (4.6.3) are 1/4.

In a similar fashion, we establish the other relations of (4.6.2). Note that although the $\mathbf{w}_j$ are *pairwise* independent, it is not true that the entire set $(\mathbf{w}_0, \mathbf{w}_1, \ldots, \mathbf{w}_{s-1})$ is independent. However, $\mathbf{w}_1, \ldots, \mathbf{w}_k$ do form an independent set.

Now let us try to reproduce the "random-coding" proof of Section 3.5. The role of the code words $\mathbf{x}^{(1)}, \ldots, \mathbf{x}^{(s)}$ in that proof is played by $\mathbf{w}_1, \ldots, \mathbf{w}_{s-1}$. The constant a in the definition of the set A is taken as $\frac{1}{2}n(R' + C)$. If $\mathbf{x}$ is any binary sequence of length n, we have shown that $P\{\mathbf{w}_j = \mathbf{x}\} = (1/2)^n$, which is the probability determined by selecting independently the components of the input sequence, each component chosen with the distribution that achieves capacity. The random-coding proof may be taken over word for word from Section 3.5, except for two difficulties. In the original proof, the code words were independent. However, as we observed at the time [see the discussion preceding (3.5.11)], only pairwise independence is required.

A slight difficulty is also caused by our leaving out the zero code word. Let S_1 be the code consisting of the words $\mathbf{w}_1, \ldots, \mathbf{w}_{s-1}$, with the decoding sets B_i, $1 \leq i \leq s - 1$, determined by the minimum-distance decoding scheme. Let $\bar{\varepsilon}_1$ be the average probability of error of S_1. Let S_2 be the code formed by adding the zero word to S_1. However, instead of using the minimum-distance decoder for S_2, take the decoding set of $\mathbf{w}_i$ to be B_i

when $i \neq 0$, and take the decoding set of $\mathbf{w}_0$ to be the empty set (so that an error is always made when $\mathbf{w}_0$ is transmitted). The average probability of error of S_2 is then

$$\bar{\varepsilon}_2 = \frac{1}{s}\sum_{i=0}^{s-1} p(e \mid \mathbf{w}_i) = \frac{1}{s} + \frac{s-1}{s}\bar{\varepsilon}_1.$$

Finally, let $S_2{}^*$ be the code S_2 with the minimum-distance decoder, and let ε^* be the average probability of error of $S_2{}^*$. Since the minimum-distance decoder corresponds to the ideal observer with all words equally likely, $\varepsilon^* \leq \bar{\varepsilon}_2$. The random-coding proof shows that $E[\bar{\varepsilon}_1] \to 0$ as $n \to \infty$.

Since $s = 2^k \geq 2^{nR} \to \infty$ as $n \to \infty$, $E[\varepsilon^*] \to 0$ as $n \to \infty$. Since $R < C$, we can find integers k_i and n_i, $i = 1, 2, \ldots$, such that $n_1 < n_2 < \cdots$ and $R \leq k_i/n_i < R' < C$. Thus the above procedure yields codes that maintain the transmission rate R and have an arbitrarily small probability of error. The theorem is proved.

4.7. *Precise error bounds for general binary codes*

In Chapter 3 we showed that we could construct a sequence of codes $([2^{nR}], n, \lambda_n)$ with $\lambda_n \to 0$ at an exponential rate. In this section we obtain more precise information about the rate of approach of λ_n to zero. We shall need a few combinatorial results.

Lemma 4.7.1. If $0 < p < 1$, $q = 1 - p$, and np is an integer, then

$$(8npq)^{-1/2} \cdot 2^{nH(p,q)} \leq \binom{n}{np} \leq (2\pi npq)^{-1/2} \cdot 2^{nH(p,q)}. \tag{4.7.1}$$

Proof. We use *Stirling's formula:*

$$n! = \sqrt{2\pi n}\, n^n e^{-n} \exp\left(\frac{1}{12n} - \frac{1}{360n^3} + \cdots\right).$$

If no terms of the exponential series are taken, $n!$ is underestimated; if one term is taken, $n!$ is overestimated, etc. To establish the left-hand inequality, we write

$$\binom{n}{np} = \frac{n!}{(np)!\,(nq)!} \geq \frac{\sqrt{2\pi n}\, n^n e^{-n} \exp(-1/12np - 1/12nq)}{\sqrt{2\pi np}\,(np)^{np} e^{-np} \sqrt{2\pi nq}\,(nq)^{nq} e^{-nq}}.$$

[We have underestimated $n!$ and overestimated $(np)!$ and $(nq)!$.] Assume that $np \geq 1$ and $nq \geq 3$. [By the symmetry in p and q of the expression (4.7.1), the only remaining cases to consider are $np = 1$, $nq = 1$; $np = 1$,

$nq = 2$; $np = 2, nq = 2$; for these cases, the inequality may be verified by direct substitution.] Then

$$\frac{1}{12np} + \frac{1}{12nq} \leq \frac{1}{12} + \frac{1}{36} = \frac{1}{9},$$

and

$$\exp\left(-\frac{1}{12np} - \frac{1}{12nq}\right) \geq e^{-1/9} = 0.895 \geq \tfrac{1}{2}\sqrt{\pi}.$$

Hence

$$\binom{n}{np} \geq (8npq)^{-1/2} p^{-np} q^{-nq} = (8npq)^{-1/2} 2^{-n(p \log p + q \log q)}$$

which proves the left-hand inequality. To establish the right-hand side, we overestimate $n!$ and underestimate $(np)!$ and $(nq)!$, as follows:

$$\binom{n}{np} \leq \frac{\sqrt{2\pi n}\, n^n e^{-n} e^{1/12n}}{\sqrt{2\pi np}\,(np)^{np} e^{-np} \sqrt{2\pi nq}(nq)^{nq} e^{-nq}} \times \exp\left[\frac{1}{12np} - \frac{1}{360(np)^3} + \frac{1}{12nq} - \frac{1}{360(nq)^3}\right]$$

Assume $p \geq q$; then $1/360(np)^3 \leq 1/360np$ since $np \geq 1$; similarly $1/360(nq)^3 \leq 1/360nq$. Thus

$$\frac{1}{12n} - \frac{1}{12np} - \frac{1}{12nq} + \frac{1}{360(np)^3} + \frac{1}{360(nq)^3} \leq -\frac{1}{12nq} + \frac{1}{360np} + \frac{1}{360nq} \leq (\text{since } p \geq q) - \frac{1}{12nq} + \frac{1}{180nq} \leq 0.$$

Thus

$$\binom{n}{np} \leq (2\pi npq)^{-1/2} p^{-np} q^{-nq} = (2\pi npq)^{-1/2} \cdot 2^{nH(p,q)}.$$

Since the upper bound is symmetrical in p and q, the restriction $p \geq q$ may be dropped; this completes the proof.

Lemma 4.7.2. Let $\Psi_p(n, \lambda)$ be the "tail" of a binomial distribution (with parameters n and p) from λn to n, that is,

$$\Psi_p(n, \lambda) = \sum_{k=\lambda n}^{n} \binom{n}{k} p^k q^{n-k}$$

where λn is an integer and $0 < p < \lambda < 1$. Then

$$[8n\lambda(1-\lambda)]^{-\frac{1}{2}} \cdot 2^{-nB(\lambda,p)} \le \Psi_p(n, \lambda) \le 2^{-nB(\lambda,p)} \tag{4.7.2}$$

where
$$B(\lambda, p) = -H(\lambda, 1-\lambda) - \lambda \log p - (1-\lambda)\log q$$
$$= \lambda \log \frac{\lambda}{p} + (1-\lambda) \log \frac{1-\lambda}{q}.$$

It follows that

$$\lim_{n\to\infty} -\frac{1}{n} \log \Psi_p(n, \lambda_n) = B(\lambda, p) \tag{4.7.3}$$

provided $n\lambda_n$ is an integer and $\lambda_n \to \lambda$. In particular if $p = q = \frac{1}{2}$ we obtain

$$\frac{2^{nH(\lambda,1-\lambda)}}{\sqrt{8n\lambda(1-\lambda)}} \le \sum_{k=\lambda n}^{n} \binom{n}{k} \le 2^{nH(\lambda,1-\lambda)}, \qquad \lambda > \tfrac{1}{2}, \tag{4.7.4}$$

so that

$$\lim_{n\to\infty} \left\{-\frac{1}{n} \log \left[\sum_{k=n\lambda_n}^{n} \binom{n}{k}\right]\right\} = H(\lambda, 1-\lambda) \quad \text{provided} \quad \lambda_n \to \lambda. \tag{4.7.5}$$

Proof. Let X be a binomially distributed random variable with parameters n and p. Since X may be regarded as the number of successes in a sequence of n Bernoulli trials, X may be represented as $X_1 + X_2 + \cdots + X_n$ where $X_1, \ldots, X_n$ are independent. X_i, interpreted as the number of successes on trial i, is 1 with probability p and 0 with probability q. Thus

$$E[e^{rX}] = E\left[\prod_{k=1}^{n} e^{rX_k}\right] = \prod_{k=1}^{n} E[e^{rX_k}] = (q + pe^r)^n.$$

Now by Lemma 3.6.2,

$$P\{X \ge n\lambda\} \le e^{-nr\lambda} E[e^{rX}] = e^{-nr\lambda} e^{n \log_e (q+pe^r)}. \tag{4.7.6}$$

In (4.7.6), r is any positive real number. We may try to find an appropriate value of r in (4.7.6) by differentiating the exponent:

$$\frac{d}{dr}[-r\lambda + \log_e(q + pe^r)] = -\lambda + \frac{pe^r}{q + pe^r}. \tag{4.7.7}$$

Expression (4.7.7) becomes zero if r is chosen so that

$$e^r = \frac{\lambda q}{(1-\lambda)p}. \tag{4.7.8}$$

If $\lambda > p$, then $\lambda q > (1 - \lambda)p$ so that a positive r satisfying (4.7.8) can be found. With this value of r, (4.7.6) becomes

$$\Psi_p(n, \lambda) = P\{X \geq n\lambda\}$$
$$\leq \exp\left[-n\lambda \log_e \frac{\lambda q}{(1-\lambda)p} + n \log_e \left(q + \frac{\lambda q}{1-\lambda}\right)\right],$$

or

$$\Psi_p(n, \lambda) \leq \exp_2 \{[-\lambda \log \lambda - \lambda \log q + \lambda \log (1 - \lambda) + \lambda \log p + \log q - \log (1 - \lambda)]n\}.$$

Thus

$$\Psi_p(n, \lambda) \leq 2^{nH(\lambda, 1-\lambda)} 2^{n\lambda \log p} 2^{n(1-\lambda) \log q}$$

which establishes the right side of (4.7.2). To prove the left side, note that

$$\Psi_p(n, \lambda) = \sum_{k=\lambda n}^{n} \binom{n}{k} p^k q^{n-k} \geq \binom{n}{\lambda n} p^{\lambda n} q^{n(1-\lambda)}.$$

The result then follows immediately from Lemma 4.7.1; thus (4.7.2) is proved. Since (4.7.3), (4.7.4), and (4.7.5) follow from (4.7.2), the proof is finished.

It may be verified from the definition of $B(\lambda, p)$ that

$$B(\lambda, p) = H(p, 1 - p) + (\lambda - p) \frac{d}{dp} H(p, 1 - p) - H(\lambda, 1 - \lambda). \quad (4.7.9)$$

Thus $B(\lambda, p_0)$ is the difference between the tangent

$$T(\lambda, p_0) = H(p_0, 1 - p_0) + (\lambda - p_0) \left[\frac{d}{dp} H(p, 1 - p)\right]_{p=p_0}$$

to $H(p,1 - p)$ at the point p_0 and the value of H at the point λ. (See Fig. 4.7.1.)

Now consider a binary symmetric channel with probability of error $\beta < 1/2$. Given $R < C = 1 - H(\beta, 1 - \beta)$, we know that there exist codes $([2^{nR}], n, \varepsilon_n)$ with $\varepsilon_n \to 0$ as $n \to \infty$. We will try to obtain precise information about the rate of approach of ε_n to zero.

Let us define a parameter λ $(\beta \leq \lambda \leq 1/2)$ by setting

$$R = 1 - H(\lambda, 1 - \lambda) = 1 + \lambda \log \lambda + (1 - \lambda) \log (1 - \lambda). \quad (4.7.10)$$

Then as R increases from 0 to C, λ decreases from 1/2 to β. Define the *critical rate* R_c as $1 - H(\lambda_c, 1 - \lambda_c)$, where

$$\frac{\lambda_c}{1 - \lambda_c} = \left(\frac{\beta}{1 - \beta}\right)^{1/2},$$

or equivalently

$$\lambda_c = \frac{\sqrt{\beta}}{\sqrt{\beta} + \sqrt{1 - \beta}}.$$

We have the following theorem.

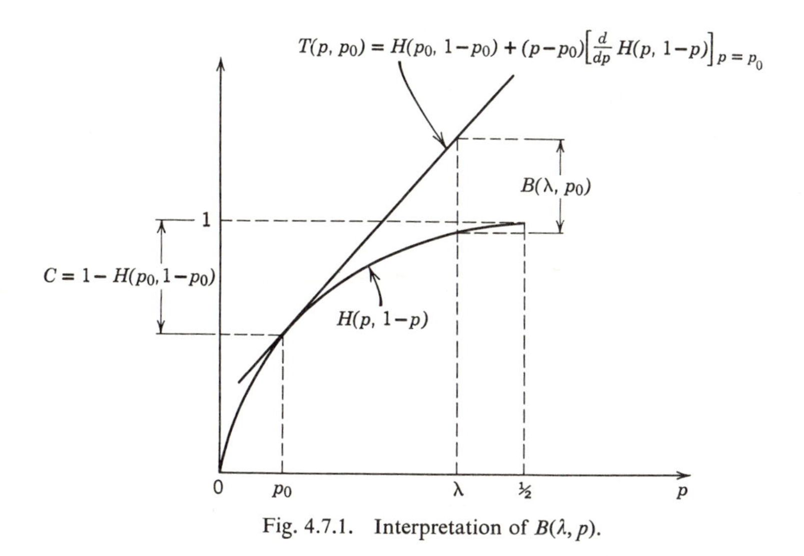

Fig. 4.7.1. Interpretation of $B(\lambda, p)$.

Theorem 4.7.3. Let $\varepsilon_n{}^*$ be the lowest average probability of error among all codes $([2^{nR}], n)$ for the binary symmetric channel (where R is a fixed positive number). If $R_c \leq R < C$, then $\alpha = \lim_{n\to\infty} [-(1/n) \log \varepsilon_n{}^*]$ exists and equals $B(\lambda, \beta)$ where λ is determined from R by (4.7.10). Thus for large n, $\varepsilon_n{}^* \sim 2^{-\alpha n}$ where $\alpha = B(\lambda, \beta)$.

Proof. We first establish an upper bound on $\varepsilon_n{}^*$. As in the "random-coding" proof of Section 3.5, we choose code words $\mathbf{w}_1, \ldots, \mathbf{w}_s$, $s = [2^{nR}]$, independently and with $P\{\mathbf{w} = \mathbf{x}\} = (1/2)^n$ for all code words $\mathbf{w}$ and all binary n-sequences $\mathbf{x}$. We then calculate the average probability of error ε for the code selected, using the ideal observer with all words equally likely, that is, minimum-distance decoding.

Suppose that the word $\mathbf{w}_i$ of the code chosen above is transmitted through the channel. Let us calculate the probability $p^*(e \mid \mathbf{w}_i, \mathbf{x}, \mathbf{y};\ t)$ of a decoding error given that

a. $\mathbf{w}_i$ is transmitted
b. $\mathbf{w}_i = \mathbf{x}$
c. $\mathbf{y}$ is received, where dist $(\mathbf{x}, \mathbf{y}) = t$.

An error in decoding implies that at least one code word $\mathbf{w}_j, j \neq i$, is at least as close to the received sequence as $\mathbf{w}_i$. Thus

$$\begin{aligned} p^*(e \mid \mathbf{w}_i, \mathbf{x}, \mathbf{y}; t) &\leq P\{\text{dist}\,(\mathbf{w}_j, \mathbf{y}) \leq t \text{ for at least one } j \neq i\} \\ &\leq \sum_{j \neq i} P\{\text{dist}\,(\mathbf{w}_j, \mathbf{y}) \leq t\}. \end{aligned}$$

Now the probability that $\mathbf{w}_j$ and $\mathbf{y}$ agree in any specific digit is 1/2. If $t = np$ where $p < 1/2$, then

$$\begin{aligned} P\{\text{dist}\,(\mathbf{w}_j, \mathbf{y}) \leq t\} &= \sum_{k=0}^{np} \binom{n}{k} \left(\frac{1}{2}\right)^n \\ &= \sum_{k=n(1-p)}^{n} \binom{n}{k} \left(\frac{1}{2}\right)^n \leq \left(\frac{1}{2}\right)^n 2^{nH(p,1-p)} = 2^{-n[1-H(p,1-p)]} \end{aligned}$$

(by Lemma 4.7.2).
Thus

$$p^*(e \mid \mathbf{w}_i, \mathbf{x}, \mathbf{y}; t) \leq 2^{nR} 2^{-n[1-H(p,1-p)]} \quad \text{if} \quad t = np, \quad p < 1/2. \tag{4.7.11}$$

Clearly $p^*(e \mid \mathbf{w}_i, \mathbf{x}, \mathbf{y};\ t) \leq 1$ under any circumstances, in particular if $t = np$ where $p \geq 1/2$. As in the random-coding proof of Section 3.5 we are looking for $E[\varepsilon_i \mid \mathbf{w}_i = \mathbf{x}]$, where ε_i is the probability of error when $\mathbf{w}_i$ is transmitted. We have [see (3.5.9)]

$$E[\varepsilon_i \mid \mathbf{w}_i = \mathbf{x}] \leq \sum_{t=0}^{n} P\{N = t\} h(t) \tag{4.7.12}$$

where N is the number of errors in transmission, and

$$\begin{aligned} h(t) &= 2^{nR} 2^{-n[1-H(p,1-p)]} && \text{if} \quad t = np, \quad p < 1/2 \\ &= 1 && \text{if} \quad t = np, \quad p \geq 1/2. \end{aligned}$$

We break (4.7.12) into three parts, first summing from 0 to $[n\beta]$, then from $[n\beta + 1]$ to $[n\lambda]$ and finally from $[n\lambda + 1]$ to n. [Recall that λ and R

are related by (4.7.10)]. First we have, using (4.7.12) and the fact that $H(p, 1-p) \leq H(\beta, 1-\beta)$ for $0 \leq p \leq \beta < 1/2$,

$$\sum_{t=0}^{[n\beta]} P\{N = t\}h(t) \leq 2^{nR}2^{-n[1-H(\beta,1-\beta)]} \sum_{t=0}^{[n\beta]} P\{N = t\}$$
$$\leq 2^{-n[H(\lambda,1-\lambda)-H(\beta,1-\beta)]} \quad \text{(by 4.7.10).} \tag{4.7.13}$$

For the second contribution to (4.7.12) we have

$$\sum_{t=[n\beta+1]}^{[n\lambda]} P\{N = t\}h(t) \leq \sum_{t=[n\beta+1]}^{[n\lambda]} P\{N \geq t\}h(t) \tag{4.7.14}$$

By Lemma 4.7.2,

$$P\{N \geq t\} = \Psi_\beta(n, p) \leq 2^{-nB(p,\beta)} \quad \text{if} \quad t = np, \quad p > \beta. \tag{4.7.15}$$

Also, by (4.7.12),

$$h(t) \leq 2^{nR}2^{-n[1-H(p,1-p)]} \quad \text{if} \quad t = np, \quad p \leq \lambda < \tfrac{1}{2}. \tag{4.7.16}$$

Using (4.7.10), (4.7.15), and (4.7.16), we find that (4.7.14) becomes

$$\sum_{t=[n\beta+1]}^{[n\lambda]} P\{N = t\}h(t) \leq \sum_{t=np=[n\beta+1]}^{[n\lambda]} 2^{-n[H(\lambda,1-\lambda)+B(p,\beta)-H(p,1-p)]}. \tag{4.7.17}$$

Finally we estimate the third contribution to (4.7.12):

$$\begin{aligned} \sum_{t=[n\lambda+1]}^{n} P\{N = t\}h(t) &\leq \sum_{t=[n\lambda+1]}^{n} P\{N = t\} \\ &= P\{N \geq [n\lambda + 1]\} \leq 2^{-nB(\lambda,\beta)} \quad \text{(by 4.7.15)} \end{aligned} \tag{4.7.18}$$

From (4.7.12), (4.7.13), (4.7.17), and (4.7.18) we obtain

$$E[\varepsilon_i \mid \mathbf{w}_i = \mathbf{x}] \leq 2^{-n[H(\lambda,1-\lambda)-H(\beta,1-\beta)]} + \sum_{t=np=[n\beta+1]}^{[n\lambda]} 2^{-n[H(\lambda,1-\lambda)+B(p,\beta)-H(p,1-p)]} + 2^{-nB(\lambda,\beta)}. \tag{4.7.19}$$

Consider the term

$$\begin{aligned} g(p) &= H(\lambda, 1-\lambda) + B(p, \beta) - H(p, 1-p) \\ &= H(\lambda, 1-\lambda) + H(\beta, 1-\beta) \\ &\quad + (p-\beta)\left[\frac{d}{dp} H(p, 1-p)\right]_{p=\beta} - 2H(p, 1-p). \end{aligned}$$

Differentiating with respect to p, we obtain

$$g'(p) = \log \frac{1-\beta}{\beta} - 2 \log \frac{1-p}{p}$$

$$g'(p) = 0 \quad \text{when} \quad \frac{p}{1-p} = \left(\frac{\beta}{1-\beta}\right)^{1/2} \quad \text{or } p = \lambda_c$$

$$g'(p) < 0 \quad \text{for} \quad p < \lambda_c$$

$$g'(p) > 0 \quad \text{for} \quad p > \lambda_c$$

We have the following two possibilities.

CASE 1. $R \geq R_c$, hence $\lambda \leq \lambda_c$. Then $g(p)$ is a decreasing function of p for $0 \leq p \leq \lambda$. Consequently the largest term in (4.7.19) is the last.

CASE 2. $R < R_c$, hence $\lambda > \lambda_c$. Since $g(p)$ is a minimum when $p = \lambda_c$, we may overbound (4.7.19) by replacing p by λ_c in all terms. Since the number of terms is $\leq [n\lambda] + 1 \leq n/2$ for large enough n, we have:

If $R \geq R_c$, then $E[\varepsilon_i \mid \mathbf{w}_i = \mathbf{x}] \leq \frac{n}{2} 2^{-nB(\lambda,\beta)}$.
If $R < R_c$, then

$$E[\varepsilon_i \mid \mathbf{w}_i = \mathbf{x}] \leq \frac{n}{2} 2^{-n[H(\lambda,1-\lambda)+B(\lambda_c,\beta)-H(\lambda_c,1-\lambda_c)]} \tag{4.7.20}$$

We need only the first case in the present proof. Since the bound of (4.7.20) does not depend on i or $\mathbf{x}$,

$$\begin{aligned} \varepsilon_n^* &\leq \frac{n}{2} 2^{-nB(\lambda,\beta)} \quad \text{if} \quad R \geq R_c \\ &\leq \frac{n}{2} 2^{-n[H(\lambda,1-\lambda)+B(\lambda_c,\beta)-H(\lambda_c,1-\lambda_c)]} \quad \text{if} \quad R < R_c. \end{aligned} \tag{4.7.21}$$

We now establish a lower bound on ε_n^*. The average probability of correct transmission of a code S with words $\mathbf{w}_1, \ldots, \mathbf{w}_s$ is

$$1 - \bar{\varepsilon} = s^{-1} \sum_{i=1}^{s} [1 - p(e \mid \mathbf{w}_i)] = s^{-1} \sum_{i=1}^{s} \sum_{j=1}^{k_i} P\{\mathbf{z} = \mathbf{z}_{ij}\}, \tag{4.7.22}$$

where $\mathbf{z}$ is the error pattern vector and the $\mathbf{z}_{ij}$, $j = 1, 2, \ldots, k_i$, are the distinct error patterns that are corrected when $\mathbf{w}_i$ is transmitted.

The number k_i of error patterns corrected when $\mathbf{w}_i$ is transmitted is the number of elements in the decoding set associated with $\mathbf{w}_i$. Since the decoding sets are disjoint and together exhaust the entire space of 2^n possible received sequences, it follows that $\sum_{i=1}^{s} k_i = 2^n$. (Note that although $\mathbf{z}_{i1}, \ldots, \mathbf{z}_{ik_i}$ are distinct, $\mathbf{z}_{im_1}$ may equal $\mathbf{z}_{jm_2}$ for $i \neq j$.) Since the probability associated with a given error pattern increases as the weight of the pattern decreases, it follows that $\bar{\varepsilon}$ cannot be smaller than the number obtained when the $\mathbf{z}_{ij}$ have minimum weight in (4.7.22). To minimize the weights of the $\mathbf{z}_{ij}$ subject to the requirements that $\sum_{i=1}^{s} k_i = 2^n$ and that $\mathbf{z}_{i1}, \ldots, \mathbf{z}_{ik_i}$ be distinct, we can do no better than to correct, for each word $\mathbf{w}_i$, all error patterns of weight $0, 1, \ldots, t-2$, where $t-2$ is the largest integer satisfying the Hamming bound $\sum_{i=0}^{t-2} \binom{n}{i} \leq \frac{2^n}{s}$ (Theorem 4.2.2), and to complete the decoding sets by correcting $(t-1)$-tuple errors for some (but not all) of the code words. The average probability of error of this (possibly hypothetical) code S' cannot exceed the original figure $\bar{\varepsilon}$; this provides a lower bound on $\varepsilon_n{}^*$.

If N is the number of errors made in transmission, then $N = t$ implies a decoding error in S'; hence

$$\varepsilon_n{}^* \geq P\{N = t\} = \binom{n}{t} \beta^t (1-\beta)^{n-t}. \tag{4.7.23}$$

Now by Lemma 4.7.1, with $np_1 = t - 2$,

$$[8np_1(1-p_1)]^{-1/2} 2^{nH(p_1, 1-p_1)} \leq \binom{n}{np_1} \leq \sum_{i=0}^{t-2} \binom{n}{i} \leq \frac{2^n}{s}$$

$$= \frac{2^n}{[2^{nR}]} \leq (2) 2^{n(1-R)} = (2) 2^{nH(\lambda, 1-\lambda)}.$$

Thus $H(p_1, 1-p_1) \leq H(\lambda, 1-\lambda) + (2n)^{-1} \log 2n + n^{-1}$. It follows that for large enough n and any fixed $\delta > 0$, we have $p_1 < \lambda + \delta$, hence

$$p = \frac{t}{n} \leq \lambda + \delta \quad \text{for} \quad n \text{ sufficiently large.} \tag{4.7.24}$$

By (4.7.23) and Lemma 4.7.1,

$$\begin{aligned} \varepsilon_n{}^* &\geq [8np(1-p)]^{-1/2} 2^{nH(p, 1-p)} \beta^{np} (1-\beta)^{n(1-p)} \\ &= [8np(1-p)]^{-1/2} 2^{-nB(p,\beta)} \geq (2n)^{-1/2} 2^{-nB(p,\beta)}. \end{aligned} \tag{4.7.25}$$

Thus by (4.7.24) and (4.7.25),

$$\varepsilon_n{}^* \geq (2n)^{-1/2} 2^{-nB(p,\beta)} \geq (2n)^{-1/2} 2^{-nB(\lambda+\delta, \beta)} \tag{4.7.26}$$

where $\delta > 0$ is arbitrary and n is sufficiently large. The theorem now follows from (4.7.21) and (4.7.26).

Theorem 4.7.3 gives very precise information about the asymptotic behavior of binary codes with $[2^{nR}]$ words of length n as long as R is not below the critical rate R_c. For values of R below R_c, the asymptotic behavior of $-n^{-1} \log \varepsilon_n{}^*$ is not known exactly, although it may be estimated from above and below. We define

$$\alpha^* = \limsup_{n\to\infty} (-n^{-1} \log \varepsilon_n{}^*)$$
$$\alpha_* = \liminf_{n\to\infty} (-n^{-1} \log \varepsilon_n{}^*).$$

From (4.7.26) we have

$$\alpha^* \leq B(\lambda, \beta) \tag{4.7.27}$$

and from (4.7.21) we have, for $R < R_c$,

$$\alpha_* \geq H(\lambda, 1 - \lambda) + B(\lambda_c, \beta) - H(\lambda_c, 1 - \lambda_c). \tag{4.7.28}$$

Thus for large n, $\varepsilon_n{}^* \sim 2^{-\alpha n}$ where α lies between the limits (4.7.27) and (4.7.28).

Both the upper and lower bounds above can be improved. We will prove one result concerning the lower bound.

Theorem 4.7.4. $\alpha_* \geq \lambda B(\frac{1}{2}, \beta) - R$, where λ and R are related, as in (4.7.10.), by

$$R = 1 - H(\lambda, 1 - \lambda).$$

Proof. We make use of the Varsharmov-Gilbert-Sacks condition (Theorem 4.5.3). We know that there exists a code with 2^k words of length n correcting e-tuple and all smaller errors, provided

$$2^{n-k} > \sum_{i=0}^{2e-1} \binom{n-1}{i}.$$

Now by Lemma 4.7.2,

$$\sum_{i=0}^{2e-1} \binom{n-1}{i} \leq 2^{(n-1)H((2e-1)/(n-1),1-(2e-1)/(n-1))} \quad \text{if} \quad \frac{2e-1}{n-1} < \frac{1}{2}$$
$$< 2^{nH((2e+1)/n,1-(2e+1)/n)} \quad \text{if} \quad \frac{2e+1}{n} \leq \frac{1}{2}.$$

[Since $(2e - 1)/(n - 1) < (2e + 1)/n$, the second requirement implies the first.] Hence, setting $k = nR$ in the Varsharmov-Gilbert-Sacks condition, we find that there is an e-tuple-error correcting code with 2^{nR} words of length n, if

$$1 - R \geq H\left(\frac{2e+1}{n}, 1 - \frac{2e+1}{n}\right). \tag{4.7.29}$$

In particular, since $1 - R = H(\lambda, 1 - \lambda)$ we may take $2e + 1 = \lambda n$ (thus $(2e + 1)/n \leq \frac{1}{2}$ and the application of Lemma 4.7.2 above is legitimate).

We may bound the average probability of error $\overline{p(e)}$ of such a code as follows. If $\mathbf{w}_i$ is transmitted then the occurrence of an error implies that the received sequence is at least as close to one of the words $\mathbf{w}_j, j \neq i$, as to $\mathbf{w}_i$. Hence

$$p(e \mid \mathbf{w}_i) \leq \sum_{j \neq i} P\left\{\begin{matrix}\text{received sequence is at least} \\ \text{as close to } \mathbf{w}_j \text{ as to } \mathbf{w}_i\end{matrix} \;\middle|\; \mathbf{w}_i \text{ transmitted}\right\}.$$

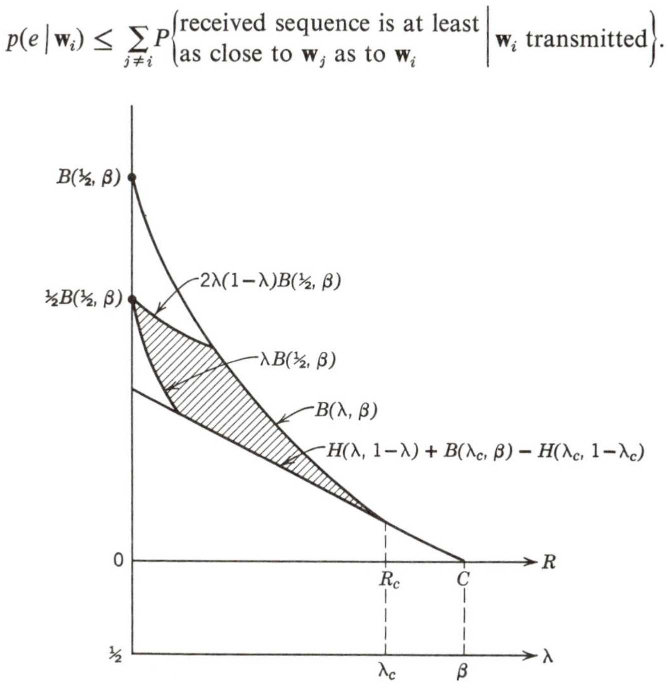

Fig. 4.7.2. Error bounds for the binary symmetric channel; $\varepsilon_n^* \sim 2^{-\alpha n}$ where α lies in the shaded area.

Suppose that $\mathbf{w}_i$ and $\mathbf{w}_j$ differ in u positions. By Lemma 4.2.1, for a code to correct e-tuple and all smaller errors, the minimum distance between code words must be at least $2e + 1$; therefore $u \geq 2e + 1 = \lambda n$. Now $P\{$received sequence is at least as close to $\mathbf{w}_j$ as to $\mathbf{w}_i \mid \mathbf{w}_i$ transmitted$\}$ = $P\{$at least half of the u digits in which $\mathbf{w}_i$ and $\mathbf{w}_j$ differ are in error$\}$ =

$$\sum_{u/2}^{u} \binom{u}{k} \beta^k (1 - \beta)^{u-k} = \Psi_\beta(u, \tfrac{1}{2}).$$

By Lemma 4.7.2, $\Psi_\beta(u, \frac{1}{2}) \leq 2^{-uB(\frac{1}{2},\beta)} \leq 2^{-\lambda n B(\frac{1}{2},\beta)}$. Thus

$$p(e \mid \mathbf{w}_i) \leq \sum_{j \neq i} 2^{-\lambda n B(\frac{1}{2},\beta)} \leq 2^{nR} 2^{-\lambda n B(\frac{1}{2},\beta)}.$$

Consequently $\overline{p(e)} \leq 2^{-n[\lambda B(\frac{1}{2},\beta) - R]}$ and the theorem is proved.

Certain quantities (such as $\frac{1}{2}u$) in the above argument have been treated as integers in order to avoid laborious details; a completely precise analysis would require the use of the greatest integer function.

The bound of Theorem 4.7.4 may be improved to $\alpha_* \geq \lambda B(\frac{1}{2}, \beta)$ as was shown by Gallager (1963). The proof in the text is due to Wyner (1964), who also proved that $\alpha^* \leq 2\lambda(1 - \lambda)B(\frac{1}{2}, \beta)$.

The bounds described in this section are shown in Fig. 4.7.2.

4.8. *The strong converse for the binary symmetric channel*

In this section we prove, for the special case of the binary symmetric channel, that if the transmission rate is maintained above channel capacity, not only is the probability of error bounded away from zero, as we have established in Section 3.7, but in fact the probability of error must approach one with increasing code-word length. Specifically, we will prove the following result.

Theorem 4.8.1. Consider a binary symmetric channel with capacity C. If ε is any fixed positive number, and λ is any fixed nonnegative number less than one, then for sufficiently large n, any code (s, n, λ) must satisfy

$$s < 2^{n(C+\varepsilon)}$$

In particular, given any sequence of codes $([2^{nR}], n, \lambda_n)$ with $R > C$, for sufficiently large n we must have $\lambda_n > \lambda$; in other words,

$$\lambda_n \to 1 \text{ as } n \to \infty.$$

Proof. The idea of the proof is to estimate the number of sequences in the decoding sets of the code. There must be enough sequences in each decoding set to make the probability of correct transmission at least $1 - \lambda$ for each possible transmitted code word; on the other hand, the number of sequences is limited by the fact that there are only 2^n binary sequences of length n.

Given a code (s, n, λ) with $0 \leq \lambda < 1$, let r be the smallest integer such that

$$\sum_{j=0}^{r} \binom{n}{j} \beta^j(1 - \beta)^{n-j} \geq 1 - \lambda. \tag{4.8.1}$$

It follows that each decoding set has more than $\sum_{j=0}^{r-1} \binom{n}{j}$ sequences. For suppose that the decoding set B associated with the code word $\mathbf{w}$ has fewer sequences. Since the probability of an error pattern increases as its weight decreases, the largest possible probability of correct transmission corresponds to the correction of errors of the smallest possible weight. Now if B in fact had $\sum_{j=0}^{r-1}\binom{n}{j}$ sequences, $p(e \mid \mathbf{w})$ would be minimized if the correctible error patterns corresponding to the points of B were the zero pattern together with all single, double, ..., $(r-1)$-tuple errors. In this case the probability of correct transmission when $\mathbf{w}$ is transmitted would be

$$p(e' \mid \mathbf{w}) = \sum_{j=0}^{r-1}\binom{n}{j}\beta^j(1-\beta)^{n-j} < 1 - \lambda.$$

Since the number of sequences of B is at most

$$\sum_{j=0}^{r-1}\binom{n}{j},$$

$p(e' \mid \mathbf{w}) < 1 - \lambda$ in any event, contradicting the fact that we have assumed a code with maximum probability of error λ.

Thus we have a lower bound on the size of each decoding set. However, since there are s decoding sets and 2^n sequences altogether, we have

$$s\sum_{j=0}^{r-1}\binom{n}{j} < 2^n. \tag{4.8.2}$$

Let N be the number of errors in transmission. Then

$$P\{N \le r\} = \sum_{j=0}^{r}\binom{n}{j}\beta^j(1-\beta)^{n-j} \ge 1 - \lambda$$

by definition of r. Now N/n converges in probability to β as $n \to \infty$ by the weak law of large numbers, so for any fixed $\delta > 0$, $P\{N \le n(\beta - \delta)\} \to 0$ as $n \to \infty$. In particular, $P\{N \le n(\beta - \delta)\} < 1 - \lambda$ for sufficiently large n (remember that $\lambda < 1$). Thus for n sufficiently large we have

$$r \ge n(\beta - \delta). \tag{4.8.3}$$

From (4.8.2) and (4.8.3) we have

$$s\sum_{j=0}^{[n(\beta-\delta)-1]}\binom{n}{j} < 2^n.$$

We may write $[n(\beta - \delta) - 1] = n(\beta - \delta')$ where $\delta' \to \delta$ as $n \to \infty$. Lemma 4.7.2 then yields

$$s\frac{2^{nH(\beta-\delta',\,1-(\beta-\delta'))}}{[8n(\beta-\delta')(1-(\beta-\delta'))]^{1/2}} < 2^n,$$

or

$$s < 2^{n\{1-H(\beta-\delta',1-(\beta-\delta'))+(2n)^{-1}\log[8n(\beta-\delta')(1-(\beta-\delta'))]\}}.$$

Since $C = 1 - H(\beta, 1 - \beta)$ and δ' may be taken arbitrarily small, the strong converse is proved.

We have shown that if the transmission rate is maintained above capacity, the *maximum* probability of error approaches 1. In fact, the *average* probability of error also approaches 1. (See Problem 4.10.)

4.9. Nonbinary coding

Most of the results in this chapter can be extended to more general situations. In this section we present some of the extensions without proof. In many cases, the proofs are very close to the arguments given in the binary case; where this is not so, we give a reference.

For a given positive integer q, the *q-ary symmetric channel* is a discrete memoryless channel with input and output alphabet $\{0, 1, \ldots, q - 1\}$ and channel probabilities $p(i \mid i) = 1 - \beta, p(j \mid i) = \beta/(q - 1)$ for $i \neq j$. Thus the probability of receiving digit j given that digit i is transmitted is the same for all $j \neq i$. As in the binary case, define the distance between two n-sequences as the number of digits in which the two sequences disagree. Then the ideal observer with equally likely inputs is a minimum-distance decoder, as in Theorem 4.1.1. Lemma 4.2.1 goes through word for word, but Theorem 4.2.2 becomes

$$s \leq \frac{q^n}{\sum_{i=0}^{e} \binom{n}{i}(q - 1)^i}.$$

If q is a prime or a power of a prime, then the elements $0, 1, \ldots, q - 1$ form a finite field under appropriate definitions of addition and multiplication. Finite fields will be discussed in Chapter 5 (see page 142); for a more complete treatment see Peterson (1961) or Albert (1956). A parity check code for the q-ary symmetric channel may be defined as a set of solutions to a system of homogeneous linear equations over the field with q elements. The discussion of Section 4.3 applies to the more general situation. Theorem 4.4.1 carries over word for word. The discussion of Section 4.4 must be modified. A parity check code forms a *vector space*† over the field with q elements; conversely, any set of

† A *vector space* V over a field F is an abelian group together with an operation of "scalar multiplication," that is, a function which assigns to each pair $(\lambda, \mathbf{v})$, $\lambda \in F$, $\mathbf{v} \in V$, an element $\lambda\mathbf{v} \in V$, such that for all elements $\lambda, \mu \in F$ and all elements $\mathbf{v}, \mathbf{w} \in V$ we have $\lambda(\mathbf{v} + \mathbf{w}) = \lambda\mathbf{v} + \lambda\mathbf{w}$, $(\lambda + \mu)\mathbf{v} = \lambda\mathbf{v} + \mu\mathbf{v}$, $(\lambda\mu)\mathbf{v} = \lambda(\mu\mathbf{v})$, and $1 \cdot \mathbf{v} = \mathbf{v}$, where 1 is the multiplicative identity of F. Now if F is a finite field with q elements, where q is a prime, then the elements of F are the integers modulo q. If V is any abelian group consisting of sequences of length n whose components belong to F, where the group operation is componentwise modulo q addition, we may define $\lambda\mathbf{v} = \mathbf{v} + \mathbf{v} + \cdots + \mathbf{v}$ (λ times). With this definition, V becomes a vector space over F.

q-ary sequences that forms a vector space is a parity check code. If q is a prime, any abelian group of q-ary sequences is also a vector space over the field modulo q; thus only the group structure is needed when $q = 2$. With this modification, Section 4.4 survives essentially intact. Theorem 4.5.1 is unchanged, but Theorems 4.5.2 and 4.5.3 become, respectively:

$$q^m \geq \sum_{i=0}^{e} \binom{n}{i}(q-1)^i \quad \text{and} \quad q^m > \sum_{i=0}^{2e-1} \binom{n-1}{i}(q-1)^i.$$

The proof of Theorem 4.6.1 parallels that of the binary case.

Error bounds for nonbinary channels are significantly harder to find than those for binary channels. Blackwell, Breiman, and Thomasian (1959) proved that given any discrete memoryless channel with capacity $C > 0$ and given any $R < C$ such that $C - R \leq \frac{1}{2} \log_2 e$, there is a code $([2^{nR}], n, \varepsilon_n)$ with

$$\varepsilon_n < 2e^{-[(C-R)^2/16ab(\log_2 e)^2]n}$$

where a is the number of input symbols and b the number of output symbols. Sharper bounds may be found in Fano (1961).

4.10. *Notes and remarks*

A general reference for error correcting codes is Peterson (1961). Single-error correcting parity check codes were introduced by Hamming (1950), who also pointed out the geometric interpretation of the binary coding problem. The general notion of a parity check code and the results of Section 4.4 are due to Slepian (1956). Theorem 4.5.3 is due to Varsharmov (1957) and Gilbert (1952); the argument in the text follows Sacks (1958). Theorem 4.6.1 was stated by Elias (1955). Lemma 4.7.2 is an application of the exponential bounding technique of Chernoff (1952). Theorem 4.7.3 and the bounds (4.7.21) are due to Elias (1955, 1956); see also Wozencraft and Reiffen (1961). For a further discussion of error bounds on codes with low transmission rates, see Gallager (1963) and Wyner (1965). In particular, Wyner proves the following results, which provide a strengthening of the Hamming bound.

Lemma 4.10.1. Let $\mathbf{w}_i = (x_{i1}, x_{i2}, \ldots, x_{in})$, $i = 1, 2, \ldots, s$, be code words of an arbitrary binary code. Let d be the minimum distance of the code, that is, $d = \min_{i \neq j} \text{dist}(\mathbf{w}_i, \mathbf{w}_j)$. Let $\mathbf{y}$ be any binary n-sequence and d_i the distance between $\mathbf{w}_i$ and $\mathbf{y}$ $(i = 1, 2, \ldots, s)$. Then

$$\left(\frac{1}{n}\sum_{i=1}^{s} d_i\right)^2 - \frac{s}{n}\sum_{i=1}^{s} d_i + s(s-1)\frac{d}{2n} \leq 0. \tag{4.10.1}$$

Proof. Without loss of generality we may assume $\mathbf{y} = \mathbf{0}$. [If $\mathbf{y} \neq \mathbf{0}$, consider a new code with words $\mathbf{w}_i + \mathbf{y}$ (mod 2).] Since the minimum distance is d,

$$\binom{s}{2} d \leq \sum_{1 \leq i < j \leq s} d(\mathbf{w}_i, \mathbf{w}_j).$$

Now

$$d(\mathbf{w}_i, \mathbf{w}_j) = \sum_{k=1}^{n} (x_{ik} - x_{jk})^2$$

with *ordinary* (*not modulo 2*) arithmetic. Thus

$$\binom{s}{2} d \leq \sum_{k=1}^{n} \sum_{1 \leq i < j \leq s} (x_{ik} - x_{jk})^2.$$

In the expansion of the inner summation, we have terms of the form $x_{rk}^2 + x_{jk}^2, j > r$, and terms of the form $x_{ik}^2 + x_{rk}^2, i < r$; thus each term x_{rk}^2 appears $s - 1$ times. Hence

$$\binom{s}{2} d \leq \sum_{k=1}^{n} \left[(s-1) \sum_{i=1}^{s} x_{ik}^2 - 2 \sum_{1 \leq i < j \leq s} x_{ik} x_{jk} \right]$$
$$= \sum_{k=1}^{n} \left[s \sum_{i=1}^{s} x_{ik}^2 - \left(\sum_{i=1}^{s} x_{ik} \right)^2 \right].$$

Since $\mathbf{y} = \mathbf{0}$, we have $d_i = \sum_{k=1}^{n} x_{ik}^2$; thus

$$\binom{s}{2} d \leq s \sum_{i=1}^{s} d_i - \sum_{k=1}^{n} g_k^2 \quad \text{where} \quad g_k = \sum_{i=1}^{s} x_{ik}.$$

Now by the Schwarz inequality for sums,

$$\left(\sum_{k=1}^{n} 1 \cdot g_k \right)^2 \leq \sum_{k=1}^{n} 1^2 \sum_{k=1}^{n} g_k^2 = n \sum_{k=1}^{n} g_k^2.$$

Thus

$$\binom{s}{2} d \leq s \sum_{i=1}^{s} d_i - \frac{1}{n} \left(\sum_{k=1}^{n} g_k \right)^2$$

or

$$\binom{s}{2} d \leq s \sum_{i=1}^{s} d_i - \frac{1}{n} \left(\sum_{i=1}^{s} \sum_{k=1}^{n} x_{ik} \right)^2.$$

Since $x_{ik}^2 = x_{ik}$, we have $d_i = \sum_{k=1}^{n} x_{ik}$, so that

$$\binom{s}{2} d \leq s \sum_{i=1}^{s} d_i - n \left(\frac{\sum_{i=1}^{s} d_i}{n} \right)^2$$

On dividing through by n, the lemma follows. Note that $\mathbf{w}_1, \ldots, \mathbf{w}_s$ need not be the entire code.

Now let $M(n, d)$ be the maximum possible number of code words in a binary code with words of length n and minimum distance d. We have the following theorem.

Theorem 4.10.2. If $d/n < \frac{1}{2}$, then

$$M(n, d) \leq \frac{n \cdot 2^n K(p)}{\sum_{r=0}^{[tpn]} \binom{n}{r}\left(\frac{td}{2} - r\right)} \tag{4.10.2}$$

where $p = d/2n$, $t = (1/2p)(1 - \sqrt{1 - 4p})$, $K(p) = p/\sqrt{1 - 4p}$.

Proof. Consider a code consisting of $M(n, d)$ words of length n, with minimum distance d.

In the space of all 2^n n-sequences, consider, for each code word $\mathbf{w}_i$, the sphere S_i consisting of all sequences $\mathbf{y}$ such that $d(\mathbf{w}_i, \mathbf{y}) \leq [td/2]$. Now within the sphere, assign to each sequence $\mathbf{y}$ whose distance from $\mathbf{w}_i$ is r the mass $\mu_i(\mathbf{y}) = td/2 - r$. The total mass of each sphere is therefore

$$\mu = \sum_{r=0}^{[td/2]} \binom{n}{r}\left(\frac{td}{2} - r\right) = \sum_{r=0}^{[tpn]} \binom{n}{r}(tpn - r). \tag{4.10.3}$$

If a sequence $\mathbf{y}$ belongs to each of the spheres S_i, $i = 1, 2, \ldots, s$, then we assign a mass $\mu(\mathbf{y})$ to $\mathbf{y}$ given by the sum of the masses assigned to $\mathbf{y}$ in each sphere; that is,

$$\mu(\mathbf{y}) = \sum_{i=1}^{s} \mu_i(\mathbf{y}) = \frac{std}{2} - \sum_{i=1}^{s} d(\mathbf{w}_i, \mathbf{y}) \tag{4.10.4}$$

Consequently the total mass assigned to all sequences is

$$\sum_{\mathbf{y}} \mu(\mathbf{y}) = M(n, d)\mu. \tag{4.10.5}$$

We shall bound $M(n, d)$ by finding a bound on $\sum_{\mathbf{y}} \mu(\mathbf{y})$. Let $b = b(\mathbf{y}) = \mu(\mathbf{y})/n$; then (4.10.4) becomes

$$\frac{1}{n}\sum_{i=1}^{s} d_i = \frac{std}{2n} - b = stp - b \quad \text{where} \quad d_i = d(\mathbf{w}_i, \mathbf{y}). \tag{4.10.6}$$

Substituting (4.10.6) into (4.10.1) we obtain

$$s^2t^2p^2 - 2stpb + b^2 - s^2tp + sb + s^2p - sp \leq 0$$

or

$$0 \leq b^2 \leq s[p - sp(t^2p - t + 1) - b(1 - 2tp)]. \tag{4.10.7}$$

Since $t^2p - t + 1 = 0$ and $1 - 2tp > 0$ for $p < 1/4$, (4.10.7) can be satisfied only when

$$b \leq \frac{p}{1 - 2tp} = \frac{p}{\sqrt{1 - 4p}} = K(p).$$

Thus

$$\sum_{\mathbf{y}} \mu(\mathbf{y}) = n \sum_{\mathbf{y}} b(\mathbf{y}) \leq n \cdot 2^n K(p).$$

By (4.10.3) and (4.10.5) we have

$$M(n, d) \leq \frac{n \cdot 2^n K(p)}{\sum_{r=0}^{[tpn]} \binom{n}{r} (tpn - r)},$$

as asserted.

It is unfortunate that the synthesis procedure given in the proof of the Varsharmov-Gilbert-Sacks condition is not practical, for codes that satisfy the condition have desirable asymptotic properties. In particular, we may construct codes that correct e-tuple and all smaller errors, provided

$$1 - R \geq H\left(\frac{2e + 1}{n}, 1 - \frac{2e + 1}{n}\right)$$

where $R = k/n$ is the transmission rate, and n is the code-word length (see (4.7.29)). By the weak law of large numbers, the number of errors in transmission will be of the order of $n\beta$ for large n, where β is the probability of a single digit being received incorrectly. Thus, roughly speaking, by setting $e = n\beta$ we can achieve arbitrarily high reliability provided $1 - R \geq H(2\beta, 1 - 2\beta)$, or $R \leq 1 - H(2\beta, 1 - 2\beta)$. [Note that β must be less than $\frac{1}{4}$ for this to be valid; see the discussion preceding (4.7.29).] Hence if we could synthesize codes meeting the Varsharmov-Gilbert-Sacks condition, such codes could be used to maintain any transmission rate up to $1 - H(2\beta, 1 - 2\beta)$ (which is close to capacity for small β) with an arbitrarily small probability of error.

Theorem 4.8.1 is a weakened version of a result proved by Weiss (1960); see also Wolfowitz (1961).

PROBLEMS

4.1 A *Hamming code* may be defined as a group code with the property that the jth column of the parity check matrix is the binary number j. For example, if A is 3 by 6 then

$$A = \begin{bmatrix} 1 & 0 & 1 & 0 & 1 & 0 \\ 0 & 1 & 1 & 0 & 0 & 1 \\ 0 & 0 & 0 & 1 & 1 & 1 \end{bmatrix}$$

where, for example, the column $\begin{bmatrix} 1 \\ 1 \\ 0 \end{bmatrix}$

represents the binary number $011 = 3$. Digits 1, 2, 4, 8, . . . may be chosen as check digits (to avoid degeneracy, assume that there are no zero rows). If a Hamming code has m check digits, what is the maximum code-word length n? Give the parity check matrix for $m = 4$ and n as large as possible.

4.2 a. Show that a parity check code will correct e-tuple (and all smaller) errors and detect (but not necessarily correct) $(e + 1)$-tuple errors if and only if every set of $2e + 1$ columns of the parity check matrix is linearly independent.

b. Given an e-tuple-error correcting parity check code with parity check matrix $A = [I_m \mid A_1]$, where I_m is an identity matrix of order m. Show that the code defined by the matrix

$$A_0 = \left[\begin{array}{c|c|c} I_m & \begin{matrix} 0 \\ 0 \\ \cdot \\ \cdot \\ \cdot \\ 0 \end{matrix} & A_1 \\ \hline 1 \quad 1 \quad \cdots \quad 1 & 1 & 1 \quad 1 \quad \cdots \quad 1 \end{array}\right]$$

will correct e-tuple errors and detect $(e + 1)$-tuple errors. (This corresponds to adding one check digit which performs a parity check on *all* the digits of a code word.)

c. Find a parity check matrix of a single-error correcting, double-error detecting code with 5 check digits and 8 information digits.

d. What is the analog of the Varsharmov-Gilbert-Sacks condition for codes that correct e-tuple and detect $(e + 1)$-tuple errors?

4.3 A group code has the following parity check matrix.

$$A = \begin{bmatrix} 1 & 0 & 0 & 0 & 1 & a_{16} \\ 1 & 1 & 0 & 0 & 0 & a_{26} \\ 1 & 0 & 1 & 0 & 0 & a_{36} \\ 0 & 1 & 1 & 1 & 0 & a_{46} \end{bmatrix}.$$

a. If $a_{16} = a_{26} = a_{36} = a_{46} = 1$, list the code words in the resulting code.

b. Show that $a_{16}, a_{26}, a_{36}, a_{46}$ can be chosen so that the code will correct single errors and detect double errors. Show also that not all double errors can be corrected.

4.4 Show that the minimum distance between code words of a parity check code is d if and only if every nonzero code word has at least d ones and at least one code word has exactly d ones. (In other words, the minimum distance equals the minimum weight over all nonzero code words.)

4.5 It is desired to construct a parity check code to *detect* e-tuple and all smaller errors. If an error of magnitude $\leq e$ is made, the decoder must decide that an error has occurred; it is not required that the decoder provide any information about the nature of the error.

a. Find the necessary and sufficient conditions on the parity check matrix for this type of error detection.

b. Find a convenient condition on the code words which is equivalent to the condition of part (a).

c. Given a parity check code with 2^k words of length n, find the number of error patterns that can be detected by the code. [Note that this number is the same for all (n, k) codes.]

4.6 A *generator matrix* for an (n, k) binary code S is a matrix G whose rows consist of k linearly independent code words of S. Every code word of S may be obtained as a linear combination of the rows of G. Assuming that the last k columns of G are linearly independent (if not we can interchange columns), we may reduce G by elementary row transformations to a new generator matrix

$$G^* = [B \mid I_k]$$

where B is k by m (m = rank of the parity check matrix $= n - k$) and I_k is an identity matrix of order k.

a. Show that the matrix $A = [I_m \mid B^T]$ is a parity check matrix for S.

b. Find parity check matrices for the codes whose generator matrices are

i.
$$G = \begin{bmatrix} 1 & 1 & 0 & 0 & 1 & 0 \\ 0 & 0 & 1 & 1 & 0 & 1 \\ 0 & 1 & 0 & 1 & 1 & 1 \\ 0 & 1 & 0 & 0 & 0 & 1 \end{bmatrix}$$

ii.
$$G = [1 \quad 1 \quad 1 \quad 1 \quad 1].$$

4.7 Given a single-error correcting, double-error detecting parity check code characterized by the parity check matrix A. Show that it is possible to delete a row and a column of A such that the resulting matrix is the parity check matrix of a single-error correcting code.

4.8 A binary code is said to be *close-packed* if for some e, all single, double, . . . , e-tuple errors, possibly some $(e + 1)$-tuple errors, but no errors of greater weight are corrected by the minimum-distance decoder, no matter which code word is transmitted. The code is said to be *lossless* if, for each transmitted word, the code corrects all single, double, . . . , e-tuple errors, but nothing else. If a code is lossless we may surround each word $\mathbf{w}_i$ by a sphere of radius

e consisting of all sequences whose distance from $\mathbf{w}_i$ does not exceed e, such that the spheres are disjoint and cover the entire space of 2^n binary sequences. Equivalently, a lossless code meets the Hamming bound. Note that by the argument after (4.7.22), a close-packed or lossless code has the lowest average probability of error among all codes having the same number of code words and the same code-word length.

a. Find a single-error correcting lossless parity check code with sixteen code words.

b. Find a double-error correcting lossless parity check code with four check digits.

c. Show that the Hamming code (Problem 4.1) is close-packed.

4.9 Show that the lower bound $H(\lambda, 1 - \lambda) + B(\lambda_c, \beta) - H(\lambda_c, 1 - \lambda_c)$ of (4.7.28), plotted as a function of R [where $R = 1 - H(\lambda, 1 - \lambda)$; see Fig. 4.7.2] is tangent to $B(\lambda, \beta)$ at $\lambda = \lambda_c$.

4.10 Consider a binary symmetric channel with capacity C. If $R > C$ and $0 \leq \lambda < 1$, show that for sufficiently large n, no code $([2^{nR}], n)$ can have an average probability of error less than λ. Hence given a sequence of codes $([2^{nR}], n)$, $R > C$, with average probability of error $\overline{p_n(e)}$, we have $\overline{p_n(e)} \to 1$ as $n \to \infty$.

4.11 Consider an arbitrary code (s, n) with not all words distinct. Assume that $\mathbf{w}_1 = \mathbf{w}_2$, and form a new code by replacing $\mathbf{w}_1$ by any sequence $\mathbf{w}_1' \neq \mathbf{w}_1$. Assume that in each case the decision scheme is determined by the ideal observer with all words equally likely. Show that the average probability of error of the new code cannot exceed the average probability of error of the original code.

CHAPTER FIVE

Further Theory of Error Correcting Codes

5.1. *Feedback shift registers and cyclic codes*

In Chapter 4 we talked about general properties of codes but we never came to grips with the synthesis problem. If we wish to construct an e-tuple-error correcting code for a specified word length n, the only

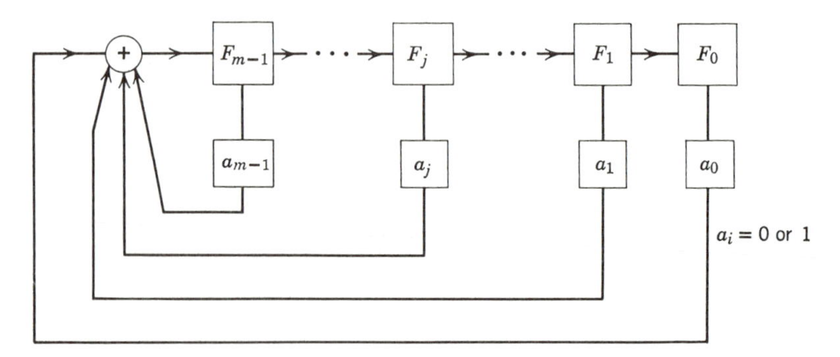

Fig. 5.1.1. Feedback shift register.

general method we have available at this point is the very laborious procedure described in the proof of the Varsharmov-Gilbert-Sacks bound. (In the special case $e = 1$ there is no problem, since all we need to do is to take a parity check matrix with distinct nonzero columns.) It would be desirable to be able to write down a parity check matrix which we knew to be the matrix of an e-tuple-error correcting code. In this chapter we attack the synthesis problem and in addition study some of the algebraic properties of codes.

The class of codes (the "cyclic codes") that we shall examine may be thought of intuitively as being generated by a physical device called a *feedback shift register.*

A feedback shift register (see Fig. 5.1.1) is a system of binary storage devices $F_0, F_1, \ldots, F_{m-1}$; switches; and a modulo 2 adder. The boxes labeled $a_0, a_1, \ldots, a_{m-1}$ represent switches. If $a_i = 1$, the corresponding switch is closed and the contents of F_i are fed back to the input of F_{m-1}.

If $a_i = 0$, the switch is open. The system is controlled by a clock pulse; at each pulse the contents of F_i are shifted to F_{i-1} ($i = 1, 2, \ldots, m - 1$), and the new number stored in F_{m-1} is $a_0x_0 + a_1x_1 + \cdots + a_{m-1}x_{m-1}$ (mod 2), where x_i is the number stored in F_i before the clock pulse.

The operation of a feedback shift register may be described by a matrix equation. If x_i is the number stored in F_i after the tth clock pulse and $x_i{'}$ is the number stored in F_i after the $(t + 1)$th clock pulse,

$$\begin{aligned} x_0{'} &= x_1 \\ x_1{'} &= x_2 \\ &\vdots \\ x'_{m-2} &= x_{m-1} \\ x'_{m-1} &= a_0x_0 + a_1x_1 + \cdots + a_{m-1}x_{m-1} \qquad \text{(mod 2)} \end{aligned}$$

or in matrix form,

$$\mathbf{x}' = T\mathbf{x} \qquad \text{(mod 2),†}$$

where

$$\mathbf{x} = \begin{bmatrix} x_0 \\ x_1 \\ \cdot \\ \cdot \\ \cdot \\ x_{m-1} \end{bmatrix}, \qquad \mathbf{x}' = \begin{bmatrix} x_0{'} \\ x_1{'} \\ \cdot \\ \cdot \\ \cdot \\ x'_{m-1} \end{bmatrix}$$

$$T = \begin{bmatrix} 0 & 1 & 0 & 0 & \cdots & 0 & 0 \\ 0 & 0 & 1 & 0 & \cdots & 0 & 0 \\ 0 & 0 & 0 & 1 & \cdots & 0 & 0 \\ & & \cdot & & & & \\ & & \cdot & & & & \\ & & \cdot & & & & \\ 0 & 0 & 0 & 0 & \cdots & 0 & 1 \\ a_0 & a_1 & a_2 & a_3 & \cdots & a_{m-2} & a_{m-1} \end{bmatrix} \tag{5.1.1}$$

The (column) vector $\mathbf{x}$ is called the *state* or *content* of the shift register, and T is called the *characteristic matrix* of the shift register. It is easily

† All arithmetic in this chapter is modulo 2; all codes to be studied will be binary. If the results of the chapter are to be generalized to codes whose words have components in an arbitrary finite field, the coefficients a_i in the last row of the matrix T should be replaced by $-a_i$ in order that the characteristic (and minimal) polynomial of T still be given by $a_0 + a_1\lambda + \cdots + a_{m-1}\lambda^{m-1} + \lambda^m$ (see Section 5.2).

seen that the determinant of T is a_0; we assume that $a_0 = 1$ so that T is nonsingular.

If $\mathbf{x}_t$ is the state of the shift register after the tth clock pulse ($t = 0, 1, 2, \ldots$), then as the clock pulses appear the register content traces out a sequence $\mathbf{x}_0, \mathbf{x}_1, \mathbf{x}_2, \ldots$, where $\mathbf{x}_1 = T\mathbf{x}_0$, $\mathbf{x}_2 = T^2\mathbf{x}_0$, etc. We claim that

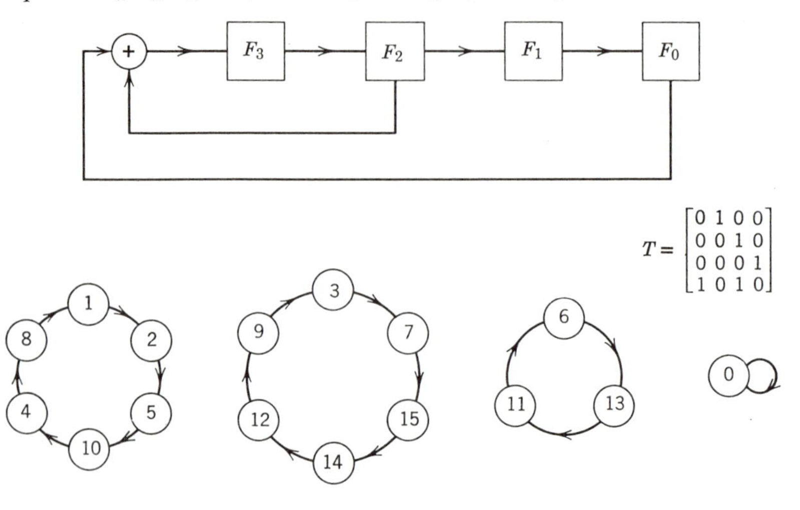

Fig. 5.1.2. Feedback shift register and associated cycle set.

if $\mathbf{x}_0 \neq \mathbf{0}$, then $\mathbf{x}_n$ must equal $\mathbf{x}_0$ for some $n > 0$; that is, the state must eventually return to $\mathbf{x}_0$. To prove this, note that the shift register has only a finite number 2^m of possible states. If the state never returns to $\mathbf{x}_0$, then eventually one of the other states must occur twice in the sequence, that is, there must be two distinct states $\mathbf{x}_i$ and $\mathbf{x}_j$ such that $T\mathbf{x}_i = \mathbf{x}_{i+1}$ and $T\mathbf{x}_j = \mathbf{x}_{i+1}$. Thus $T(\mathbf{x}_i - \mathbf{x}_j) = \mathbf{0}$, and hence by the nonsingularity of T we have $\mathbf{x}_i = \mathbf{x}_j$, a contradiction. Thus the shift register sequences are periodic. If the initial state of the register is $\mathbf{x}_0$ and a sequence of clock pulses is applied, the resulting sequence of states forms a cycle $\mathbf{x}_0, \mathbf{x}_1, \ldots, \mathbf{x}_k, \mathbf{x}_0, \ldots$, with the states $\mathbf{x}_0, \mathbf{x}_1, \ldots, \mathbf{x}_k$ distinct. The set of all possible cycles is called the *cycle set* of the feedback shift register, or of the associated matrix T.

A feedback shift register and its associated cycle set is shown in Fig. 5.1.2. For simplicity, we have identified state vectors with binary numbers, for example,

$$\begin{bmatrix} 0 \\ 0 \\ 1 \\ 1 \end{bmatrix} \leftrightarrow 3, \qquad \begin{bmatrix} 1 \\ 1 \\ 0 \\ 1 \end{bmatrix} \leftrightarrow 13.$$

There are two cycles of length 6, one cycle of length 3, and one cycle of length 1 consisting of the zero vector alone. The cycle consisting of zero alone will always be present in any cycle set, since $T \cdot \mathbf{0} = \mathbf{0}$.

We now establish a connection between feedback shift registers and a certain type of coding system.

Definition. A parity check code is said to be *cyclic* if it has the property that whenever $\mathbf{w} = (r_0, r_1, \ldots, r_{n-2}, r_{n-1})$ is a code word, the sequence $\mathbf{w}' = (r_{n-1}, r_0, r_1, \ldots, r_{n-2})$ obtained from $\mathbf{w}$ by a "cyclic shift" is also a code word.

A feedback shift register "generates" a cyclic code, in the following sense.

Theorem 5.1.1. Let a parity check code be formed by taking, as the columns of the parity check matrix, one or more repetitions of a cycle generated by a feedback shift register. In other words, consider a parity check code with matrix $A = [\mathbf{x} \mid T\mathbf{x} \mid T^2\mathbf{x} \mid \cdots \mid T^{n-1}\mathbf{x}]$ where $\mathbf{x}$ is a binary m-vector, T is a nonsingular matrix of the form (5.1.1), and $T^n\mathbf{x} = \mathbf{x}$. The resulting code is cyclic.

We may use the feedback shift register of Fig. 5.1.2 to generate various cyclic codes; for example, we may construct codes with parity check matrices

$$A_1 = \begin{bmatrix} 0 & 0 & 0 & 1 & 0 & 1 \\ 0 & 0 & 1 & 0 & 1 & 0 \\ 0 & 1 & 0 & 1 & 0 & 0 \\ 1 & 0 & 1 & 0 & 0 & 0 \end{bmatrix}, \qquad A_2 = \begin{bmatrix} 0 & 0 & 1 & 1 & 1 & 1 \\ 0 & 1 & 1 & 1 & 1 & 0 \\ 1 & 1 & 1 & 1 & 0 & 0 \\ 1 & 1 & 1 & 0 & 0 & 1 \end{bmatrix},$$

$$A_3 = \begin{bmatrix} 0 & 1 & 1 \\ 1 & 1 & 0 \\ 1 & 0 & 1 \\ 0 & 1 & 1 \end{bmatrix}.$$

It may be verified that A_1 and A_2 define the same code, namely the code whose words are 000000, 101010, 010101, and 111111. (This is no accident; see Lemma 5.3.3b). The matrix A_3 is degenerate in the sense that the rank of A_3 is two while the number of rows is 4.

Since the last two rows of A_3 may be expressed as linear combinations of the first two rows, the parity check equations associated with A_3 become:

$$r_2 + r_3 = 0$$
$$r_1 + r_2 = 0$$

Thus we may take one digit, say r_3, as an information digit; the remaining digits r_1 and r_2 will be check digits. The code defined by A_3 is therefore {000, 111}.

We may also form a cyclic code by taking two repetitions of the cycle of length 3. The corresponding parity check matrix is

$$A_4 = \begin{bmatrix} 0 & 1 & 1 & 0 & 1 & 1 \\ 1 & 1 & 0 & 1 & 1 & 0 \\ 1 & 0 & 1 & 1 & 0 & 1 \\ 0 & 1 & 1 & 0 & 1 & 1 \end{bmatrix}.$$

Since A_4 is of rank 2, the code defined by A_4 has 16 words, and thus differs from the code defined by A_1 and A_2.

Proof of Theorem 5.1.1. If $\mathbf{w} = (r_0, r_1, \ldots, r_{n-1})$ is a code word, then $A\mathbf{w}^T = \mathbf{0}$, hence $r_0\mathbf{x} + r_1T\mathbf{x} + \cdots + r_{n-1}T^{n-1}\mathbf{x} = \mathbf{0}$. Multiplying by T, we obtain (since $T^n\mathbf{x} = \mathbf{x}$) $r_{n-1}\mathbf{x} + r_0T\mathbf{x} + \cdots + r_{n-2}T^{n-1}\mathbf{x} = \mathbf{0}$; hence $(r_{n-1}, r_0, r_1, \ldots, r_{n-2})$ is a code word.

The converse of Theorem 5.1.1 is also true; namely, for every cyclic code there is a corresponding feedback shift register. We will not be able to prove this until we have developed some general properties of binary matrices.

5.2. *General properties of binary matrices and their cycle sets*

Before proceeding to a study of cyclic codes, we must apply some results from linear algebra to binary matrices and their cycle sets.

Let T be a non-zero m by m matrix of zeros and ones [not necessarily of the form (5.1.1)]. The *characteristic polynomial* of the matrix T is defined by

$$\varphi(\lambda) = |T - \lambda I|,$$

where I is an identity matrix of the same order as T. For the matrix T of Fig. 5.1.2, we have

$$\varphi(\lambda) = \begin{vmatrix} \lambda & 1 & 0 & 0 \\ 0 & \lambda & 1 & 0 \\ 0 & 0 & \lambda & 1 \\ 1 & 0 & 1 & \lambda \end{vmatrix} = \lambda^4 + \lambda^2 + 1 \qquad (\text{mod } 2).$$

In general, if the matrix T is given by (5.1.1), it may be verified by expanding $|T - \lambda I|$ by minors of the last row that

$$\varphi(\lambda) = a_0 + a_1\lambda + a_2\lambda^2 + \cdots + a_{m-1}\lambda^{m-1} + \lambda^m.$$

Note that if T is the characteristic matrix of a feedback shift register, then $\varphi(\lambda)$ determines T uniquely.

Now we define the *minimal polynomial* of T as the monic† polynomial $f(\lambda)$ of lowest degree such that $f(T) = 0$.

The reader may recall that $\varphi(T) = 0$ (this is the Cayley-Hamilton theorem, which we shall use without proof) and that $f(\lambda)$ is a divisor of $\varphi(\lambda)$ (see Lemma 5.2.2). In general, $f(\lambda) \neq \varphi(\lambda)$. For example, for any square matrix T with minimal polynomial $f(\lambda)$ and characteristic polynomial $\varphi(\lambda)$, let

$$T_0 = \left[\begin{array}{c|c} T & 0 \\ \hline 0 & T \end{array}\right].$$

Then since

$$p(T_0) = \left[\begin{array}{c|c} p(T) & 0 \\ \hline 0 & p(T) \end{array}\right]$$

for any polynomial p, the minimal polynomial of T_0 is $f(\lambda)$, but the characteristic polynomial of T_0 is $[\varphi(\lambda)]^2$, which is always of higher degree than $f(\lambda)$. However, we do have the following result.

Lemma 5.2.1. If a matrix T is of the form (5.1.1), then its minimal and characteristic polynomial are the same.

Proof. Let $r_1(T^i)$ denote the first row of the matrix T^i. By a direct computation it follows that

$$r_1(T^0) = r_1(I) = [1 \quad 0 \quad 0 \quad 0 \quad \cdots \quad 0], \qquad r_1(T) = [0 \quad 1 \quad 0 \quad \cdots \quad 0],$$
$$r_1(T^2) = [0 \quad 0 \quad 1 \quad 0 \quad \cdots \quad 0], \ldots,$$
$$r_1(T^{m-1}) = [0 \quad 0 \quad 0 \quad \cdots \quad 0 \quad 1].$$

---

† A polynomial $p(x) = b_0 + b_1x + b_2x^2 + \cdots + b_nx^n$, $b_n \neq 0$, is said to be monic if $b_n = 1$. In modulo 2 arithmetic, all nonzero polynomials are monic.

If $f(\lambda) \neq \varphi(\lambda)$ then since $\varphi(T) = 0$, the degree of $f(\lambda)$ is at most $m - 1$. But if $f(T) = \sum_{k=0}^{m-1} b_k T^k = 0$, the linear independence of the $r_1(T^i)$, $i = 0, 1, \ldots, m - 1$, implies that all $b_k = 0$, a contradiction. Thus $f(\lambda) = \varphi(\lambda)$.

In view of Lemma 5.2.1, our main interest will be in matrices whose minimal and characteristic polynomials are the same. However, the results to follow will be stated in general terms.

Let T be a binary matrix whose minimal polynomial has degree q (note that if T is m by m and $\varphi(\lambda) = f(\lambda)$ then necessarily $q = m$). Let Z be the set of *matric polynomials* in T of degree $\leq q - 1$, that is, the set of all matrices

$$b_0 I + b_1 T + \cdots + b_{q-1} T^{q-1} \qquad (\text{mod } 2;\ b_i = 0 \text{ or } 1). \qquad (5.2.1)$$

We observe that all matrices of the form (5.2.1) are distinct, that is, $I, T, \ldots, T^{q-1}$ are linearly independent. For if

$$\sum_{i=0}^{q-1} b_i T^i = \sum_{i=0}^{q-1} b_i' T^i$$

then

$$p(T) = \sum_{i=0}^{q-1} (b_i - b_i') T^i = 0.$$

Thus we have a polynomial $p(\lambda)$ of degree $<q$ such that $p(T) = 0$. Since the minimal polynomial has degree q, the only possible conclusion is that $p(\lambda) \equiv 0$, that is, $b_i = b_i'$ for all i.

Thus Z has 2^q members. Before deriving further properties of the set Z, we must recall some facts about the minimal polynomial.

Lemma 5.2.2. Let $f(\lambda)$ be the minimal polynomial of a matrix T. If $p(\lambda)$ is any polynomial such that $p(T) = 0$, then $f(\lambda)$ divides $p(\lambda)$; that is, $p(\lambda) = g(\lambda) f(\lambda)$ for some polynomial $g(\lambda)$. In particular, by the Cayley-Hamilton theorem $f(\lambda)$ divides the characteristic polynomial $\varphi(\lambda)$.

Proof. We may divide $p(\lambda)$ by $f(\lambda)$ to obtain a quotient and a remainder, that is, $p(\lambda) = g(\lambda) f(\lambda) + h(\lambda)$ where the degree of $h(\lambda)$ is less than the degree of $f(\lambda)$. Replacing λ by T we have $p(T) = g(T) f(T) + h(T)$, whence $h(T) = 0$. Since $f(\lambda)$ is the nonzero polynomial of least degree which is annulled by T, it follows that $h(\lambda) = 0$ and $f(\lambda)$ divides $p(\lambda)$.

Lemma 5.2.3. Let $f(\lambda)$ be the minimal polynomial of a matrix T. Suppose that $f(\lambda)$ is *irreducible*, that is, the only polynomials which divide $f(\lambda)$ are $f(\lambda)$ itself and the constant polynomial 1. Let $p(T)$ be any matric polynomial in T such that $p(T)$ is not the zero matrix. Then $p(T)$ is nonsingular, and the inverse of $p(T)$ is also a matric polynomial in T. (In particular, T itself is nonsingular.)

Proof. Let S be the set of all polynomials of the form $a(\lambda)f(\lambda) + b(\lambda)p(\lambda)$, where $a(\lambda)$ and $b(\lambda)$ range over all polynomials. Let $d(\lambda)$ be the monic polynomial of lowest degree in S. Then:

a. $d(\lambda)$ divides both $f(\lambda)$ and $p(\lambda)$.

b. Any polynomial that divides $f(\lambda)$ and $p(\lambda)$ is a divisor of $d(\lambda)$, thus $d(\lambda)$ is the polynomial of highest degree which divides both $f(\lambda)$ and $p(\lambda)$; $d(\lambda)$ is called the *greatest common divisor* of $f(\lambda)$ and $p(\lambda)$.

Since $d(\lambda) \in S$, $d(\lambda) = a(\lambda)f(\lambda) + b(\lambda)p(\lambda)$ for some polynomials $a(\lambda)$ and $b(\lambda)$; (b) follows immediately.

To prove (a), write $f(\lambda) = g(\lambda)\, d(\lambda) + h(\lambda)$ where the degree of $h(\lambda)$ is less than the degree of $d(\lambda)$. Observe that the set S is closed under the operations of addition, subtraction, and multiplication by a polynomial; hence $h(\lambda) = f(\lambda) - g(\lambda)\, d(\lambda) \in S$. But by the minimality of $d(\lambda)$, $h(\lambda) = 0$ so that $d(\lambda)$ divides $f(\lambda)$. Similarly $d(\lambda)$ divides $p(\lambda)$.

Now since $f(\lambda)$ is irreducible, the greatest common divisor of $f(\lambda)$ and $p(\lambda)$ is either 1 or $f(\lambda)$. If $d(\lambda) = f(\lambda)$, then $f(\lambda)$ divides $p(\lambda)$. But $p(\lambda) = g(\lambda)f(\lambda)$ implies $p(T) = g(T)f(T) = 0$, contradicting the hypothesis. Thus $d(\lambda) = 1$.

Since $d(\lambda) = 1 \in S$, it follows that there are polynomials $a(\lambda)$ and $b(\lambda)$ such that $a(\lambda)f(\lambda) + b(\lambda)p(\lambda) = 1$. Replacing λ by T we have $a(T)f(T) + b(T)p(T) = I$; but $f(T) = 0$, so that $b(T)p(T) = I$ and $p(T)$ is nonsingular. The inverse of $p(T)$ is $b(T)$, a matric polynomial.

We now return to the set Z.

Theorem 5.2.4. Let Z be the set of all 2^q matric polynomials in T of degree $\leq q - 1$ where q is the degree of the minimal polynomial $f(\lambda)$ of T. Then

a. Z is an abelian group under addition.

b. $p_1(p_2 + p_3) = p_1p_2 + p_1p_3$, and $(p_1 + p_2)p_3 = p_1p_3 + p_2p_3$ for all $p_1, p_2, p_3 \in Z$.

In addition, if $f(\lambda)$ is irreducible, then

c. The set Z_0 of all nonzero elements of Z is an abelian group under multiplication.

Proof. The modulo 2 sum of a matric polynomial of degree r and a matric polynomial of degree $s \geq r$ is a polynomial of degree at most s. Hence Z is closed under addition. Addition of matric polynomials is associative since matrix addition is associative. The zero polynomial serves as the additive identity and each element of Z is its own additive inverse. Commutativity of addition is clear, and hence (a) is proved. Part (b) is a consequence of the distributivity of matrix multiplication over addition.

To prove (c), note that the product of two matric polynomials p_1 and p_2 of degree $\leq q - 1$ may be expressed as another matric polynomial p_3 of degree $\leq q - 1$ using the minimal polynomial. For example, let

$$T = \begin{bmatrix} 0 & 1 & 0 & 0 \\ 0 & 0 & 1 & 0 \\ 0 & 0 & 0 & 1 \\ 1 & 0 & 0 & 1 \end{bmatrix}; \qquad \varphi(\lambda) = f(\lambda) = 1 + \lambda^3 + \lambda^4.$$

Let $p_1(T) = T^3 + T + I$, $p_2(T) = T^2 + I$; then

$$p_1(T)p_2(T) = T^5 + 2T^3 + T^2 + T + I = T^5 + T^2 + T + I.$$

Now $f(T) = T^4 + T^3 + I = 0$, hence

$$T^4 = T^3 + I, \qquad T^5 = TT^4 = T^4 + T = T^3 + T + I;$$

thus $p_1(T)p_2(T) = T^3 + T^2 = p_3(T)$. We must show that $p_1, p_2 \in Z_0$ implies $p_1p_2 \in Z_0$, that is, Z_0 is closed under multiplication. But if $p_1(T)$ and $p_2(T)$ are nonzero matric polynomials, then $p_1(T)$ and $p_2(T)$ are nonsingular by Lemma 5.2.3. Hence $p_1(T)p_2(T)$ is nonsingular, and hence nonzero. [Alternately, if $p_1(T)p_2(T) = 0$ then $f(\lambda)$ divides $p_1(\lambda)p_2(\lambda)$ by Lemma 5.2.2; since $f(\lambda)$ is irreducible, the unique factorization theorem for polynomials implies that either $f(\lambda)$ divides $p_1(\lambda)$ or $f(\lambda)$ divides $p_2(\lambda)$; this contradicts the assumption that the degrees of $p_1(\lambda)$ and $p_2(\lambda)$ are $\leq q - 1$.]

Thus Z_0 is closed under multiplication. Multiplication of matric polynomials, being ordinary matrix multiplication, is associative. The identity matrix I serves as the multiplicative identity. By Lemma 5.2.3, any polynomial $p(T) \in Z_0$ has an inverse $b(T)$ whch is also a matric polynomial. As above, $b(T)$ may be expressed as a polynomial of degree $\leq q - 1$, hence $b(T) \in Z_0$. Since $T^iT^j = T^jT^i$ for all i, j, multiplication of matric polynomials is commutative. The proof is complete.

A set containing at least two members that is closed under two operations (called "addition" and "multiplication") and that satisfies conditions (a), (b), and (c) of Theorem 5.2.4 is called a *field*. Roughly speaking, a field is a set in which we can do addition, subtraction, multiplication, and division without leaving the set. The field Z of matric polynomials $p(T)$ of degree $\leq q - 1$ has 2^q elements;

Z is called the *Galois field* of order 2^q, written GF(2^q)

(Any finite field is referred to as a Galois field; it can be shown that a finite field must contain p^q elements where p is a prime and q is a positive integer; furthermore, any two finite fields of the same size are isomorphic.)

Now consider any nonsingular m by m binary matrix T and its powers T^i, $i = 0, 1, 2, \ldots$. We claim that for some r, $T^r = I$. For since T has only finitely many components, there must eventually be repetition in the

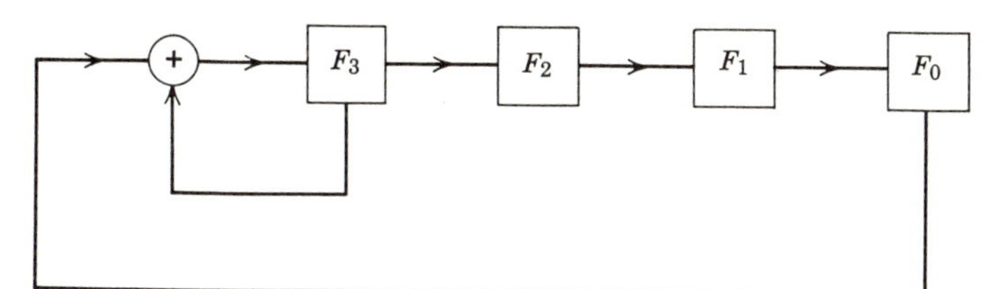

$$T = \begin{bmatrix} 0 & 1 & 0 & 0 \\ 0 & 0 & 1 & 0 \\ 0 & 0 & 0 & 1 \\ 1 & 0 & 0 & 1 \end{bmatrix} \qquad \varphi(\lambda) = 1 + \lambda^3 + \lambda^4$$

Parity check matrix of associated cyclic code:

$$A = \begin{bmatrix} 0 & 0 & 0 & 1 & 1 & 1 & 1 & 0 & 1 & 0 & 1 & 1 & 0 & 0 & 1 \\ 0 & 0 & 1 & 1 & 1 & 1 & 0 & 1 & 0 & 1 & 1 & 0 & 0 & 1 & 0 \\ 0 & 1 & 1 & 1 & 1 & 0 & 1 & 0 & 1 & 1 & 0 & 0 & 1 & 0 & 0 \\ 1 & 1 & 1 & 1 & 0 & 1 & 0 & 1 & 1 & 0 & 0 & 1 & 0 & 0 & 0 \end{bmatrix}$$

Fig. 5.2.1. A matrix with maximal period and its associated feedback shift register and cyclic code.

sequence $I, T, T^2, \ldots$, that is, for some i and j we must have $T^i = T^j$. If say $i > j$, then the nonsingularity of T implies $T^{i-j} = I$. We define the *period* of T as the smallest integer r such that $T^r = I$. (The period cannot exceed $2^q - 1$, where q is the degree of the minimal polynomial of T. See Problem 5.10.)

Now if $\mathbf{x}_0$ is any nonzero binary m-vector, then just as in Section 5.1 the vectors $\mathbf{x}_0, \mathbf{x}_1 = T\mathbf{x}_0, \mathbf{x}_2 = T\mathbf{x}_1 = T^2\mathbf{x}_0, \ldots$ form a cycle, that is, $\mathbf{x}_n = \mathbf{x}_0$ for some $n > 0$. We have the following theorem.

Theorem 5.2.5. Let T be a nonsingular m by m binary matrix with period r. If the minimal polynomial $f(\lambda)$ is irreducible, then every nonzero vector $\mathbf{x}$ lies on a cycle of length r. Since the number of nonzero m-vectors is $2^m - 1$, it follows that $r \leq 2^m - 1$, and in fact r is a divisor of $2^m - 1$.

Proof. Suppose $\mathbf{x}$ lies on a cycle of length $j < r$. (Observe that $j > r$ is impossible since $T^r = I$, hence $T^r\mathbf{x} = \mathbf{x}$ for all $\mathbf{x}$.) Then $T^j\mathbf{x} = \mathbf{x}$, or

$(T^j - I)\mathbf{x} = \mathbf{0}$. Since $j < r$, $T^j - I \neq 0$, so that $T^j - I$ is nonsingular by Lemma 5.2.3. But then $\mathbf{x} = \mathbf{0}$, a contradiction.

If the period of T is $2^m - 1$, T is said to have *maximal period.* Equivalently, T has maximal period if and only if its cycle set consists of one cycle of length $2^m - 1$ together with the trivial cycle of length 1. The maximal-period case is of interest in coding theory because if a matrix T of the form (5.1.1) generates a cyclic code, then if T has maximal period the resulting code has the largest possible word length. For an example of a matrix with maximal period, see Fig. 5.2.1.

Necessary and sufficient conditions for T to have maximal period are given in the following theorem.

Theorem 5.2.6. Let T be a nonsingular m by m binary matrix with minimal polynomial $f(\lambda)$. Then T has maximal period if and only if $f(\lambda)$ is irreducible and $f(\lambda)$ is not a divisor of $\lambda^k - 1$ for any $k < 2^m - 1$.

Proof. First suppose $f(\lambda)$ is irreducible and does not divide $\lambda^k - 1$ for any $k < 2^m - 1$. Let r be the period of T. By Theorem 5.2.5, $r \leq 2^m - 1$. Dividing $\lambda^r - 1$ by $f(\lambda)$ we obtain

$$\lambda^r - 1 = g(\lambda)f(\lambda) + h(\lambda)$$

where degree $h(\lambda) <$ degree $f(\lambda)$. Replacing λ by T we have $0 = T^r - I = g(T)f(T) + h(T) = h(T)$. By the minimality of $f(\lambda)$, $h(\lambda) = 0$; hence $f(\lambda)$ divides $\lambda^r - 1$. But by hypothesis, r cannot be less than $2^m - 1$; hence $r = 2^m - 1$ and T has maximal period.

Conversely, suppose T has maximal period. If $\lambda^k - 1 = g(\lambda)f(\lambda)$ for some $k < 2^m - 1$, then $T^k - I = g(T)f(T) = 0$, contradicting the fact that the period of T is $2^m - 1$. It remains to show that $f(\lambda)$ is irreducible. If not, there are two possibilities:

CASE 1. $f(\lambda) = g(\lambda)h(\lambda)$ where $g(\lambda)$ and $h(\lambda)$ are relatively prime [that is, the greatest common divisor of $g(\lambda)$ and $h(\lambda)$ is unity], and the degrees of $g(\lambda)$ and $h(\lambda)$ are each greater than 1.

CASE 2. $f(\lambda)$ is $[g(\lambda)]^k$ where $g(\lambda)$ is irreducible and $k \geq 2$.

In Case 1, there are polynomials $a(\lambda)$ and $b(\lambda)$ such that $a(\lambda)g(\lambda) + b(\lambda)h(\lambda) = 1$ (see the argument of Lemma 5.2.3).

Now let S_1 be the set of all m-vectors $\mathbf{x}$ such that $g(T)\mathbf{x} = \mathbf{0}$, and S_2 the set of all $\mathbf{x}$ such that $h(T)\mathbf{x} = \mathbf{0}$. We will show the following:

$$S_1 \cap S_2 = \{\mathbf{0}\}. \tag{5.2.2}$$

Any m-vector $\mathbf{x}$ may be written as $\mathbf{y}_1 + \mathbf{y}_2$

$$\text{where} \quad \mathbf{y}_1 \in S_1, \ \mathbf{y}_2 \in S_2. \tag{5.2.3}$$

(Note that $\mathbf{y}_1$ and $\mathbf{y}_2$ are unique for a given $\mathbf{x}$, for if $\mathbf{y}_1 + \mathbf{y}_2 = \mathbf{y}_1' + \mathbf{y}_2'$ where $\mathbf{y}_1, \mathbf{y}_1' \in S_1$ and $\mathbf{y}_2, \mathbf{y}_2' \in S_2$, then $\mathbf{y}_1 - \mathbf{y}_1' = \mathbf{y}_2' - \mathbf{y}_2 \in S_1 \cap S_2$, so by (5.2.2), $\mathbf{y}_1 = \mathbf{y}_1'$, $\mathbf{y}_2 = \mathbf{y}_2'$.)

$$\begin{aligned} \mathbf{x} \in S_1 &\Rightarrow T\mathbf{x} \in S_1 \\ \mathbf{x} \in S_2 &\Rightarrow T\mathbf{x} \in S_2. \end{aligned} \tag{5.2.4}$$

To establish (5.2.2), (5.2.3), and (5.2.4), we first note that $a(\lambda)g(\lambda) + b(\lambda)h(\lambda) = 1$ implies

$$a(T)g(T) + b(T)h(T) = I. \tag{5.2.5}$$

If $\mathbf{x}$ is any binary m-vector, then if $\mathbf{x} \in S_1 \cap S_2$ we have $g(T)\mathbf{x} = h(T)\mathbf{x} = \mathbf{0}$; hence, by (5.2.5), $\mathbf{x} = \mathbf{0}$. This proves (5.2.2). Again by (5.2.5),

$$\mathbf{x} = I\mathbf{x} = a(T)g(T)\mathbf{x} + b(T)h(T)\mathbf{x} = \mathbf{y}_2 + \mathbf{y}_1.$$

Since

$$\begin{aligned} h(T)\mathbf{y}_2 &= h(T)a(T)g(T)\mathbf{x} = a(T)h(T)g(T)\mathbf{x} \\ &= a(T)f(T)\mathbf{x} = \mathbf{0}, \end{aligned}$$

and similarly $g(T)\mathbf{y}_1 = \mathbf{0}$, we have $\mathbf{y}_1 \in S_1$ and $\mathbf{y}_2 \in S_2$, proving (5.2.3).

To prove (5.2.4), observe that if $\mathbf{x} \in S_1$, then $g(T)\mathbf{x} = \mathbf{0}$ so that $g(T)T\mathbf{x} = Tg(T)\mathbf{x} = T\mathbf{0} = \mathbf{0}$, whence $T\mathbf{x} \in S_1$; similarly $\mathbf{x} \in S_2$ implies $T\mathbf{x} \in S_2$.

Finally, we note that S_1 and S_2 each contain at least one nonzero vector. For suppose $S_2 = \{\mathbf{0}\}$. Then by (5.2.3), S_1 is the entire space V of all binary m-vectors, that is, $g(T)\mathbf{x} = \mathbf{0}$ for all $\mathbf{x}$. It follows that $g(T) = 0$, contradicting the minimality of $f(\lambda)$.

Now let $\mathbf{x}_1 \in S_1$, $\mathbf{x}_2 \in S_2$, with $\mathbf{x}_1 \neq \mathbf{0}$, $\mathbf{x}_2 \neq \mathbf{0}$. By (5.2.2) and (5.2.4), $\mathbf{x}_1$ and $\mathbf{x}_2$ must lie on different cycles, so T cannot possibly have maximal period.

In Case 2 we must appeal to a theorem in Birkhoff and MacLane (1953, p. 329) which states that if the minimal polynomial of T is $[g(\lambda)]^k$ where $g(\lambda)$ is irreducible, then the entire space V may be written as a direct sum of nontrivial subspaces $S_1, \ldots, S_r$, $r \geq 2$ (that is, each $\mathbf{x} \in V$ has a unique expression $\mathbf{x} = \mathbf{y}_1 + \cdots + \mathbf{y}_r$, $\mathbf{y}_i \in S_i$) where $\mathbf{x} \in S_i \Rightarrow T\mathbf{x} \in S_i$, $i = 1, 2, \ldots, r$. Now just as above, if we pick $\mathbf{x}_1 \in S_1$, $\mathbf{x}_2 \in S_2$ with $\mathbf{x}_1 \neq \mathbf{0}$, $\mathbf{x}_2 \neq \mathbf{0}$, then $\mathbf{x}_1$ and $\mathbf{x}_2$ must lie on different cycles. The proof is complete.

If T is nonsingular and has maximal period, the set of nonzero matric polynomials of degree $\leq m - 1$ coincides with the set of powers T^i, $0 \leq i \leq 2^m - 2$. For the matrices T^i, $0 \leq i \leq 2^m - 2$ are distinct. (If $T^i = T^j$ where $0 \leq j < i \leq 2^m - 2$, then $T^{i-j} = I$ where $0 < i - j \leq 2^m - 2$, contradicting the fact that the period of T is $2^m - 1$.) Since

Table 5.2.1. Binary irreducible polynomials and their periods (Elspas, 1959)

Polynomial	Period	Coefficients a_0	a_1	a_2	a_3	a_4
$\lambda + 1$	1	1				
$\lambda^2 + \lambda + 1$	3	1	1			
$\lambda^3 + \lambda + 1$	7	1	1	0		
$\lambda^3 + \lambda^2 + 1$	7	1	0	1		
$\lambda^4 + \lambda^3 + \lambda^2 + \lambda + 1$	5	1	1	1	1	
$\lambda^4 + \lambda + 1$	15	1	1	0	0	
$\lambda^4 + \lambda^3 + 1$	15	1	0	0	1	
$\lambda^5 + \lambda^2 + 1$	31	1	0	1	0	0
$\lambda^5 + \lambda^3 + 1$	31	1	0	0	1	0
$\lambda^5 + \lambda^4 + \lambda^3 + \lambda + 1$	31	1	1	0	1	1
$\lambda^5 + \lambda^4 + \lambda^2 + \lambda + 1$	31	1	1	1	0	1
$\lambda^5 + \lambda^4 + \lambda^3 + \lambda^2 + 1$	31	1	0	1	1	1
$\lambda^5 + \lambda^3 + \lambda^2 + \lambda + 1$	31	1	1	1	1	0

$f(T) = 0$, each T^i can be expressed as a (nonzero) matric polynomial of degree $\leq m - 1$. (See the proof of Theorem 5.2.4.) Since there are exactly $2^m - 1$ nonzero matric polynomials of degree $\leq m - 1$, these polynomials must be in one-to-one correspondence with the powers of T.

For example, consider the matrix

$$T = \begin{bmatrix} 0 & 1 & 0 & 0 \\ 0 & 0 & 1 & 0 \\ 0 & 0 & 0 & 1 \\ 1 & 0 & 0 & 1 \end{bmatrix} \quad \text{with} \quad f(\lambda) = \varphi(\lambda) = 1 + \lambda^3 + \lambda^4.$$

(See Fig. 5.2.1.) Using the fact that $T^4 + T^3 + I = 0$, we may express the powers of T in terms of matric polynomials of degree ≤ 3 as follows:

$$\begin{aligned}
I &= I & T^8 &= T^3 + T^2 + T \\
T &= T & T^9 &= T^2 + I \\
T^2 &= T^2 & T^{10} &= T^3 + T \\
T^3 &= T^3 & T^{11} &= T^3 + T^2 + I \\
T^4 &= T^3 + I & T^{12} &= T + I \\
T^5 &= TT^4 = T^4 + T = T^3 + T + I & T^{13} &= T^2 + T \\
T^6 &= T^4 + T^2 + T & T^{14} &= T^3 + T^2 \\
&= T^3 + T^2 + T + I & T^{15} &= I, \\
T^7 &= T^2 + T + I & &
\end{aligned}$$

and so on.

Polynomials $f(\lambda)$ satisfying the conditions of Theorem 5.2.6 exist for all possible degrees. (See, for example, Peterson, 1961.) A brief table of irreducible polynomials and the periods of the corresponding matrices is shown in Table 5.2.1. Note that $\lambda^4 + \lambda^3 + \lambda^2 + \lambda + 1$ is irreducible but is a divisor of $\lambda^5 + 1$; the associated matrix does not have maximal period.

We remark that if T is m by m and has maximal period, then the minimal polynomial $f(\lambda)$ is of degree m, and furthermore $f(\lambda)$ divides $\lambda^{2^m-1} - 1$. For if $f(\lambda)$ is of degree q, then the period of T is at most $2^q - 1$ (Problem 5.10), which proves the first assertion. In general, if T is a nonsingular binary matrix with minimal polynomial $f(\lambda)$ and period r, then we have seen (at the beginning of the proof of Theorem 5.2.6) that $f(\lambda)$ divides $\lambda^r - 1$; this proves the second assertion.

5.3. *Properties of cyclic codes*

In this section we apply some of the results of Section 5.2 to the study of cyclic codes. First we develop a convenient algebraic characterization of such codes. If $b(x) = b_0 + b_1x + \cdots + b_{n-1}x^{n-1}$ ($b_i = 0$ or 1) is a polynomial of degree $\leq n - 1$, then we say that $b(x)$ *represents* the sequence $(b_0, b_1, \ldots, b_{n-1})$. Now let K be the set of all binary polynomials of degree $\leq n - 1$. It is clear that K is an abelian group under (componentwise) modulo 2 addition (in fact K is a vector space over the modulo 2 field). We define a multiplication operation "$\cdot$" in K as follows. If $b(x)$ and $c(x)$ are in K, let $b(x) \cdot c(x)$ be the ordinary product (with arithmetic modulo 2), subject to the relation $x^n = 1$. More precisely, $b(x) \cdot c(x)$ is the remainder after the ordinary product of $b(x)$ and $c(x)$ is divided by $x^n - 1$. For example, if $n = 4$ then

$$(x^3 + x + 1) \cdot (x^2 + x) = x^5 + x^4 + x^3 + x^2 + x^2 + x = x^3 + 1$$

since $x^4 = 1$, $x^5 = x$, $x^2 + x^2 = 0$. (For those familiar with algebra, what we are actually doing here is considering the set K^* of residue classes of binary polynomials modulo $x^n - 1$; if $[b(x)]$ is the residue class of the polynomial $b(x)$, then addition and multiplication in K^* are defined by

$$[b(x)] + [c(x)] = [b(x) + c(x)], \qquad [b(x)] \cdot [c(x)] = [b(x)c(x)].$$

With this definition of addition and multiplication, K^* (or K) is a vector space that is also a ring; that is, K^* is an *algebra* over the modulo 2 field.)

For the remainder of this section, the notation $b(x) \cdot c(x)$ will always signify multiplication in K; the notation $b(x)c(x)$ will always indicate ordinary multiplication.

Now in K,

$$x \cdot (b_0 + b_1x + \cdots + b_{n-1}x^{n-1}) = b_{n-1} + b_0x + b_1x^2 + \cdots + b_{n-2}x^{n-2}$$

so that multiplication by x corresponds to a cyclic shift. This idea is the basis of the following algebraic characterization of cyclic codes.

Theorem 5.3.1. Let S be a cyclic code, and let S^* be the set of polynomials in K which represent code words of S. Let $g(x)$ be the monic polynomial of lowest degree in S^*. (In the degenerate case in which S consists of the zero vector alone, take $g(x) = x^n - 1$, which is identified with 0 in K). Then S^* is precisely the set of all multiples of $g(x)$ in K. Furthermore, $g(x)$ divides $x^n - 1$; that is, $x^n - 1 = h(x)g(x)$ (ordinary multiplication) for some polynomial $h(x)$. $g(x)$ is called the *generator* of S^*, or the *generator polynomial* of the code S.

Conversely, let $g_1(x)$ be any monic polynomial that is a divisor of $x^n - 1$. Then the set S^* of multiples of $g_1(x)$ in K represents a cyclic code, and furthermore, the generator of S^* is $g_1(x)$.

Proof. We prove the direct assertion first. If $b(x) \in S^*$, then since $x \cdot b(x)$ represents a cyclic shift, the hypothesis that S is cyclic implies that $x \cdot b(x) \in S^*$, and hence that $x^i \cdot b(x) \in S^*$ for all $i \geq 0$. Since S is a group code, it follows that if $c(x) = c_0 + c_1x + \cdots + c_{n-1}x^{n-1}$ is an arbitrary binary polynomial of degree $\leq n - 1$, then

$$c(x) \cdot b(x) = \left(\sum_{i=0}^{n-1} c_i x^i\right) \cdot b(x) = \sum_{i=0}^{n-1} c_i[x^i \cdot b(x)] \in S^*.$$

In particular, any multiple of $g(x)$ in K belongs to S^*.

Now let $b(x)$ be any element of S^*. Dividing $b(x)$ by $g(x)$ we have $b(x) = q(x)g(x) + r(x)$ with $\deg r(x) < \deg g(x)$. It follows that in K we have $b(x) = q(x) \cdot g(x) + r(x)$. Again using the group property, we have $r(x) = b(x) - q(x) \cdot g(x) \in S^*$. It follows by the minimality of $g(x)$ that $r(x) = 0$, so that $b(x)$ is a multiple of $g(x)$ in K. [Note that since deg $b(x) \leq n - 1$, $q(x) \cdot g(x) = q(x)g(x) = b(x)$.] To show that $g(x)$ divides $x^n - 1$, write $x^n - 1 = h(x)g(x) + r(x)$, with $\deg r(x) < \deg g(x)$. In K this equation becomes

$$0 = h(x) \cdot g(x) + r(x),$$

hence $r(x) = -h(x) \cdot g(x) \in S^*$. Again by the minimality of $g(x)$, $r(x) = 0$.

To prove the converse assertion, given the set S^* of multiples of $g_1(x)$ in K, let S be the set of binary sequences represented by the polynomials of S^*. Since S^* is closed under modulo 2 addition, it follows that S^*, hence S, is an abelian group. Thus S is a parity check code. If $b(x) \in S^*$ then $x \cdot b(x) \in S^*$, hence a cyclic shift of a sequence of S also belongs to S. Thus S is cyclic.

To show that $g_1(x)$ is the generator of S^*, we must show that $g_1(x)$ is the monic polynomial of lowest degree in S^*. Any element of S^* may be written as $b(x) \cdot g_1(x)$, where $b(x)$ is a polynomial of degree $\leq n - 1$. Dividing $b(x)g_1(x)$ by $x^n - 1$ we obtain $b(x)g_1(x) = q(x)(x^n - 1) + r(x)$, where $\deg r(x) \leq n - 1$. Since $g_1(x)$ divides $x^n - 1$ by hypothesis, it follows that $g_1(x)$ is a divisor of $r(x)$. Now in K, the equation $b(x)g_1(x) = q(x)(x^n - 1) + r(x)$ becomes $b(x) \cdot g_1(x) = r(x)$, so that the degree of $g_1(x)$ cannot exceed the degree of $b(x) \cdot g_1(x)$ [assuming $b(x) \cdot g_1(x) \neq 0$]. The result follows.

[For those familiar with algebra, Theorem 5.3.1 says that a cyclic code corresponds to an *ideal* (necessarily a principal ideal) in the algebra of polynomials modulo $x^n - 1$.]

We now examine generator and parity check matrices for a cyclic code.

Theorem 5.3.2. Let S be a cyclic code with words of length n, and let $g(x) = a_0 + a_1x + \cdots + a_{m-1}x^{m-1} + x^m$ be the generator of S^*. Then the word $\mathbf{w} = (a_0, \ldots, a_{m-1}, 1, 0, \ldots, 0)$ represented by $\mathbf{w}$, together with $k - 1 = n - m - 1$ cyclic shifts of $\mathbf{w}$, form a set of generators for S, that is, a maximal linearly independent set of code words. Furthermore, if $x^n - 1 = h(x)g(x)$ where $h(x) = h_0 + h_1x + \cdots + h_{k-1}x^{k-1} + x^k$ $(k + m = n)$, then

$$A = \overbrace{\begin{bmatrix} 0 & 0 & \ldots & 0 & 0 & 1 & h_{k-1} & \ldots & h_1 & h_0 \\ 0 & 0 & \ldots & 0 & 1 & h_{k-1} & h_{k-2} & \ldots & h_0 & 0 \\ & & & & & \vdots & & & & \\ 1 & h_{k-1} & \ldots & h_1 & h_0 & 0 & 0 & & 0 & 0 \end{bmatrix}}^{\leftarrow m \rightarrow \quad \leftarrow k \rightarrow} \updownarrow m \qquad (5.3.1)$$

is a parity check matrix for S. (In the degenerate case in which S consists of the entire space of 2^n binary n-vectors, so that $g(x)=1$, $h(x)=x^n-1=0$ in K, we may take $\mathbf{w} = (1\,0\,0 \cdots 0)$, and we may take the matrix (5.3.1) to consist of one row of zeros.)

Proof. By Theorem 5.3.1, every polynomial in the set S^* is a multiple of $g(x)$. In fact, the argument in the proof of Theorem 5.3.1 shows that each $b(x) \in S^*$ may be written as $c(x)g(x)$ where the multiplication is ordinary polynomial multiplication. If m is the degree of $g(x)$, then the degree of $c(x)$ is necessarily $\leq n - 1 - m = k - 1$. It follows that S^* consists of all polynomials of the form $(c_0 + c_1x + \cdots + c_{k-1}x^{k-1})g(x)$, that is, the set of all linear combinations of the polynomials $g(x), xg(x), \ldots,$

$x^{k-1}g(x)$. Translating this result back to S, we conclude that if $g(x) = a_0 + a_1x + \cdots + a_{m-1}x^{m-1} + x^m$, then $(a_0, a_1, \ldots, a_{m-1}, 1, 0, \ldots, 0)$, $(0, a_0, a_1, \ldots, a_{m-1}, 1, 0, \ldots, 0), \ldots, (0, \ldots, 0, a_0, a_1, \ldots, a_{m-1}, 1)$ generate S. This proves the first part of the theorem.

To prove the remaining assertion, note that a sequence $(b_0b_1 \cdots b_{n-1})$ satisfies the equation given by row t of the matrix A if and only if

$$\sum_{i=0}^{k} h_i b_{n-t-i} = 0 \qquad (\text{define } h_k = 1).$$

Since $\sum_{i=0}^{k} h_i\, b_{n-t-i}$ is the coefficient of x^{n-t} in the ordinary product of

$$h(x) = \sum_{i=0}^{k} h_i x^i \quad \text{and} \quad b(x) = \sum_{j=0}^{n-1} b_j x^j,$$

we have $A(b_0\, b_1 \cdots b_{n-1})^T = 0$ if and only if $h(x)b(x)$ has zero coefficients multiplying x^{n-t}, $t = 1, 2, \ldots, m$. Since $h(x)g(x) = x^n - 1$, it follows that $b(x) = x^i g(x)$ satisfies this condition for $i = 0, 1, \ldots, k - 1$. In other words, $\mathbf{w}$ and its $k - 1$ cyclic shifts satisfy the equations corresponding to A; and hence every code word $\mathbf{w}_1 \in S$ satisfies $A\mathbf{w}_1^T = 0$. Now the rank of A is $m = n - k$, hence $A\mathbf{w}_1^T = 0$ has 2^k solutions; since S has 2^k words, the set of solutions to $A\mathbf{w}_1^T = 0$ is precisely the set S, that is, A is a parity check matrix for S. The proof is complete.

Observe that *the rank of A = the degree of $g(x)$.*

As a first step in showing that the class of cyclic codes is identical with the class of codes generated by feedback shift registers, we examine the relationship between the minimal polynomial of a matrix and the generator polynomial of the associated cyclic code.

Lemma 5.3.3. Let T be a nonsingular binary m by m matrix whose minimal polynomial $f(\lambda)$ is of degree q. If $\mathbf{x}$ is a nonzero m-vector and $T^n\mathbf{x} = \mathbf{x}$, let A be the matrix $[\mathbf{x} \mid T\mathbf{x} \mid T^2\mathbf{x} \mid \cdots \mid T^{n-1}\mathbf{x}]$. [Notice that although T may not be of the standard form (5.1.1), the argument of Theorem 5.1.1 may be applied verbatim to show that the code S defined by A is cyclic.] To avoid degeneracy assume that $n > q$.

a. The binary sequence $\mathbf{w}$ represented by the polynomial $f(x)$ is a code word of S; hence $f(x)$ is a multiple of the generator polynomial $g(x)$ of the code.

b. If the rank of A is q, then $f(x) = g(x)$ so that we may identify the minimal polynomial of T with the generator polynomial of the associated code. In particular, if the m-vectors $\mathbf{x}_1$ and $\mathbf{x}_2$ lie on cycles of the same length t, and $n = st$ for some positive integer s, then the parity check matrices $[\mathbf{x}_1 \mid T\mathbf{x}_1 \mid \cdots \mid T^{n-1}\mathbf{x}_1]$ and $[\mathbf{x}_2 \mid T\mathbf{x}_2 \mid \cdots \mid T^{n-1}\mathbf{x}_2]$ define the same code [namely the code whose generator polynomial is $f(x)$] provided that both matrices have rank q. Conversely if $f(x) = g(x)$ then A has rank q.

c. If $f(x)$ is irreducible, then $f(x) = g(x)$ and hence the rank of A is q.

Proof. If $f(\lambda) = a_0 + a_1\lambda + \cdots + a_{q-1}\lambda^{q-1} + \lambda^q$, then

$$f(T)\mathbf{x} = a_0\mathbf{x} + a_1T\mathbf{x} + \cdots + a_{q-1}T^{q-1}\mathbf{x} + T^q\mathbf{x} = \mathbf{0};$$

that is, the binary sequence $\mathbf{w} = (a_0\, a_1 \cdots a_{q-1}\, 1\, 0 \cdots 0)$ satisfies the parity check equations of the code. This establishes (a).

Part (b) is an immediate consequence of (a) and the observation that the rank of the parity check matrix coincides with the degree of the generator polynomial.

If $f(x)$ is irreducible, then since $g(x)$ divides $f(x)$, $g(x)$ is either 1 or $f(x)$. If $g(x) = 1$, then the code consists of all 2^n binary n-vectors, an impossibility since the parity check matrix A does not consist entirely of zeros. This proves (c). Alternately, we may argue that the first q columns of A are linearly independent. For if $\sum_{i=0}^{q-1} b_iT^i\mathbf{x} = \mathbf{0}$ with not all $b_i = 0$, then Lemma 5.2.3 implies that $\sum_{i=0}^{q-1} b_iT^i = 0$, contradicting the minimality of f. On the other hand, any polynomial in T of degree $\geq q$ may be expressed as a linear combination of $I, T, T^2, \ldots, T^{q-1}$, using the fact that $f(T) = 0$. It follows that any column of A may be written as a linear combination of the first q columns, and thus the rank of A is q. The lemma is proved.

We remark that if the rank of A is $t = n - k$, the code word $\mathbf{w}^*$ corresponding to the generator polynomial $g(x)$, together with $k - 1$ cyclic shifts of $\mathbf{w}^*$, generates the code. Consequently, $g(x)$ is the nonzero polynomial of lowest degree such that $g(T)\mathbf{x} = \mathbf{0}$.

We are now ready to prove the converse of Theorem 5.1.1.

Theorem 5.3.4. Let S be a cyclic code. Then there is a feedback shift register that generates S. More precisely, if S has 2^k words of length n, then there is a nonsingular m by m matrix T of the form (5.1.1), with $m = n - k$, and an m-vector $\mathbf{x}$, such that $T^n\mathbf{x} = \mathbf{x}$ and

$$A = [\mathbf{x} \mid T\mathbf{x} \mid \cdots \mid T^{n-1}\mathbf{x}]$$

is a parity check matrix for S.

Proof. If S consists of all 2^n binary n-vectors, we may take any T with $\mathbf{x} = \mathbf{0}$. If S consists of the zero vector alone, we may take T to be an n by n matrix of the form (5.1.1) with $a_0 = 1$, $a_i = 0$, $i = 1, 2, \ldots, n - 1$, and we may take $\mathbf{x}$ to be the (column) n-vector $(0, 0, \ldots, 1)$. The corresponding parity check matrix is square and nonsingular and therefore defines the code consisting only of the zero vector. Thus assume that S has 2^k words of length n, where $0 < k < n$.

Now as in Theorems 5.3.1 and 5.3.2, $x^n - 1 = h(x)g(x)$, where $g(x) = \sum_{i=0}^{m} a_i x^i$ is the generator polynomial of the code and $h(x) = \sum_{i=0}^{k} h_i x^i$ is the polynomial whose coefficients appear in the parity check matrix A of (5.3.1). If we compare coefficients of $x^n - 1$ and $h(x)g(x)$, we obtain the identities

$$a_0 h_{n-r} + a_1 h_{n-r-1} + \cdots + a_{m-1} h_{k-r+1} + a_m h_{k-r} = 0, \qquad r = 1, 2, \ldots, n-1 \quad (5.3.2)$$

(define $a_i = 0$ if $i < 0$ or $i > m$; define $h_j = 0$ if $j < 0$ or $j > k$).
Now let

$$T = \begin{bmatrix} 0 & 1 & 0 & \cdots & 0 \\ 0 & 0 & 1 & \cdots & 0 \\ \cdot & & & & \\ \cdot & & & & \\ \cdot & & & & \\ 0 & 0 & 0 & \cdots & 1 \\ a_0 & a_1 & a_2 & \cdots & a_{m-1} \end{bmatrix}$$

and take

$$\mathbf{x} = \begin{bmatrix} 0 \\ 0 \\ \cdot \\ \cdot \\ \cdot \\ 0 \\ 1 \end{bmatrix}.$$

We claim that the parity check matrix A of (5.3.1) is

$$[\mathbf{x} \mid T\mathbf{x} \mid T^2\mathbf{x} \mid \cdots \mid T^{n-1}\mathbf{x}].$$

We may prove this by induction. The first column of A is clearly $\mathbf{x}$. Now suppose that the rth column of A is $T^{r-1}\mathbf{x}$. We note that

$$T \begin{bmatrix} y_0 \\ y_1 \\ \cdot \\ \cdot \\ \cdot \\ y_{m-2} \\ y_{m-1} \end{bmatrix} = \begin{bmatrix} y_1 \\ y_2 \\ \cdot \\ \cdot \\ \cdot \\ y_{m-1} \\ z \end{bmatrix} \quad \text{where } z = a_0 y_0 + \cdots + a_{m-1} y_{m-1}.$$

By the induction hypothesis,

$$T^r\mathbf{x} = TT^{r-1}\mathbf{x} = T(\text{column } r \text{ of } A)$$

$$= T\begin{bmatrix} h_{n-r} \\ h_{n-r-1} \\ \cdot \\ \cdot \\ \cdot \\ h_{k-r+2} \\ h_{k-r+1} \end{bmatrix} = \begin{bmatrix} h_{n-(r+1)} \\ h_{n-(r+2)} \\ \cdot \\ \cdot \\ \cdot \\ h_{k-r+1} \\ a_0h_{n-r} + a_1h_{n-r-1} + \cdots + a_{m-1}h_{k-r+1} \end{bmatrix}.$$

By (5.3.2),

$$T^r\mathbf{x} = \begin{bmatrix} h_{n-(r+1)} \\ h_{n-(r+2)} \\ \cdot \\ \cdot \\ \cdot \\ h_{k-r+1} \\ h_{k-r} \end{bmatrix} = \text{column } r + 1 \text{ of } A.$$

Finally, since $x^n - 1 = h(x)g(x)$, we have $h_0 = a_0 = 1$; thus the last column of A is

$$T^{n-1}\mathbf{x} = \begin{bmatrix} 1 \\ 0 \\ \cdot \\ \cdot \\ \cdot \\ 0 \\ 0 \end{bmatrix}.$$

Consequently,

$$T^n\mathbf{x} = \text{column 1 of } T$$

$$= \begin{bmatrix} 0 \\ 0 \\ \cdot \\ \cdot \\ \cdot \\ 0 \\ 1 \end{bmatrix} = \mathbf{x}.$$

Since a_0 is not zero, T is nonsingular and the theorem is proved.

Corollary 5.3.4.1. a. The class of cyclic codes (with the exception of the code consisting of the entire space) coincides with the class of parity check codes whose parity check matrices are of the form

$$[\mathbf{x} \mid T\mathbf{x} \mid \cdots \mid T^{n-1}\mathbf{x}],$$

where T is a nonsingular matrix of the form (5.1.1),

$$\mathbf{x} = \begin{bmatrix} 0 \\ 0 \\ \cdot \\ \cdot \\ \cdot \\ 0 \\ 1 \end{bmatrix},$$

and $T^n\mathbf{x} = \mathbf{x}$.

b. For any such code S in the latter class (excluding the degenerate case in which S consists of the zero vector alone), the minimal polynomial of T may be identified with the generator polynomial $g(x)$ of the code. If $x^n - 1 = h(x)g(x)$, then the parity check matrix $[\mathbf{x} \mid T\mathbf{x} \mid \cdots \mid T^{n-1}\mathbf{x}]$ coincides with the matrix (5.3.1).

Proof. Part (a) is an immediate consequence of the argument used in the proof of Theorem 5.3.4, along with Theorem 5.1.1. To prove (b), let $A = [\mathbf{x} \mid T_1\mathbf{x} \mid \cdots \mid T_1^{n-1}\mathbf{x}]$ be the parity check matrix of a cyclic code S, where T_1 is a nonsingular matrix of the form (5.1.1),

$$\mathbf{x} = \begin{bmatrix} 0 \\ 0 \\ \cdot \\ \cdot \\ \cdot \\ \cdot \\ 0 \\ 1 \end{bmatrix}, \quad \text{and} \quad T_1{}^n\mathbf{x} = \mathbf{x}.$$

If T_1 is m by m, it may be verified directly that the m by m submatrix formed by the first m columns of A is triangular, and therefore A has rank m. If $n = m$, then the code consists of the zero vector alone, so assume $n > m$. Since T_1 is of the form (5.1.1), Lemma 5.2.1 implies that the minimal polynomial of T_1 is of degree m. By Lemma 5.3.3b, the minimal polynomial of T_1 may be identified with the generator polynomial $g(x)$ of the code. It follows that the matrix T which corresponds to S in the argument of the proof of Theorem 5.3.4 is the same as T_1. The final assertion of (b) now follows.

As an example, consider the code associated with Fig. 5.2.1. Since the first column of the matrix A of that figure is

$$\begin{bmatrix} 0 \\ 0 \\ 0 \\ 1 \end{bmatrix},$$

A is of the form (5.3.1). The generator polynomial of the code is $g(x) = 1 + x^3 + x^4$. Also, $x^{15} - 1(= x^{15} + 1) = h(x)g(x)$, where $h(x)$, determined from the first row of A, is $1 + x^3 + x^4 + x^6 + x^8 + x^9 + x^{10} + x^{11}$. The code word $\mathbf{w} = (100110000000000)$ corresponding to $g(x)$, together with 10 cyclic shifts of $\mathbf{w}$, generate the code, that is, form a maximal linearly independent set of code words.

As another example, the parity check matrix

$$A_1 = \begin{bmatrix} 0 & 0 & 0 & 1 & 0 & 1 \\ 0 & 0 & 1 & 0 & 1 & 0 \\ 0 & 1 & 0 & 1 & 0 & 0 \\ 1 & 0 & 1 & 0 & 0 & 0 \end{bmatrix}$$

generated by the matrix

$$T = \begin{bmatrix} 0 & 1 & 0 & 0 \\ 0 & 0 & 1 & 0 \\ 0 & 0 & 0 & 1 \\ 1 & 0 & 1 & 0 \end{bmatrix}$$

(see Fig. 5.1.2) is of the form (5.3.1). The generator polynomial of the code is $g(x) = 1 + x^2 + x^4$, and $x^6 - 1 = h(x)g(x)$ where $h(x)$, determined from the first row of A_1, is $1 + x^2$. The words (101010) and (010101), corresponding to $g(x)$ and $xg(x)$, generate the code.

As a final example, let

$$A_4 = \begin{bmatrix} 0 & 1 & 1 & 0 & 1 & 1 \\ 1 & 1 & 0 & 1 & 1 & 0 \\ 1 & 0 & 1 & 1 & 0 & 1 \\ 0 & 1 & 1 & 0 & 1 & 1 \end{bmatrix}$$

be the parity check matrix whose columns consist of two repetitions of the cycle of length 3 in Fig. 5.1.2 (see p. 138). The minimal polynomial of the matrix T of Fig. 5.1.2 is $f(\lambda) = 1 + \lambda^2 + \lambda^4$, but the rank of A_4 is only 2. By Lemma 5.3.3, the generator polynomial $g(x)$ is a proper factor of $f(x)$, necessarily of degree 2 since the degree of the generator coincides with the rank of the parity check matrix. Since

$$f(x) = (1 + x + x^2)^2,$$

we have $g(x) = 1 + x + x^2$. Hence a maximal linearly independent set of code words consists of (111000), (011100), (001110), and (000111).

Since the number of check digits is 2 and the generator polynomial is $1 + x + x^2$, a parity check matrix A may be generated by the matrix $T = \begin{bmatrix} 0 & 1 \\ 1 & 1 \end{bmatrix}$ and the vector $\mathbf{x} = \begin{bmatrix} 0 \\ 1 \end{bmatrix}$. Since the code word length is 6, we have

$$A = \begin{bmatrix} 0 & 1 & 1 & 0 & 1 & 1 \\ 1 & 1 & 0 & 1 & 1 & 0 \end{bmatrix}$$

(which coincides with the first two rows of A_4). Thus $h(x) = 1 + x + x^3 + x^4$.

5.4. *Bose-Chaudhuri-Hocquenghem codes*

In this section we describe a class of multiple-error correcting codes that are very easy to construct and that perform very well at least up to word lengths of several thousand.

The procedure to be described yields, for any e and q, an e-tuple-error correcting code with word length $n = 2^q - 1$. Let T be a q by q matrix of the form (5.1.1). Assume that the minimal polynomial of T satisfies the conditions of Theorem 5.2.6 so that T has maximal period. The parity check matrix A_0 of the Bose-Chaudhuri-Hocquenghem code is shown in Table 5.4.1 ($\mathbf{x}$ is any nonzero vector).

As an example, suppose $e = 3$, $q = 4$, and the minimal polynomial of T is $f(\lambda) = \lambda^4 + \lambda + 1$. The associated Bose-Chaudhuri-Hocquenghem code is shown in Table 5.4.2.

It may be verified that the Bose-Chaudhuri-Hocquenghem codes are cyclic (see Problem 5.3). To establish that the matrix A_0 does in fact define an e-tuple-error correcting code, we prove the following theorem.

Theorem 5.4.1. Every set of $2e$ columns of the Bose-Chaudhuri-Hocquenghem matrix A_0 is linearly independent.

Table 5.4.1. Parity check matrix of the Bose-Chaudhuri-Hocquenghem code

$$A_0 = \left[\begin{array}{c|c|c|c|c|c} \mathbf{x} & T\mathbf{x} & T^2\mathbf{x} & T^3\mathbf{x} & \cdots & T^{n-1}\mathbf{x} \\ \mathbf{x} & T^3\mathbf{x} & T^6\mathbf{x} & T^9\mathbf{x} & \cdots & T^{3(n-1)}\mathbf{x} \\ \mathbf{x} & T^5\mathbf{x} & T^{10}\mathbf{x} & T^{15}\mathbf{x} & \cdots & T^{5(n-1)}\mathbf{x} \\ \cdot & & & & & \\ \cdot & & & & & \\ \cdot & & & & & \\ \mathbf{x} & T^{2e-1}\mathbf{x} & T^{4e-2}\mathbf{x} & T^{6e-3}\mathbf{x} & \cdots & T^{(2e-1)(n-1)}\mathbf{x} \end{array}\right]$$

A_0 is qe by n, $\qquad n = 2^q - 1$

Proof. Suppose the columns headed by $T^{k_1}\mathbf{x}, T^{k_2}\mathbf{x}, \ldots, T^{k_r}\mathbf{x}$ add to zero, with $r \leq 2e$. Then

$$\begin{aligned} &T^{k_1}\mathbf{x} + T^{k_2}\mathbf{x} + \cdots + T^{k_r}\mathbf{x} = \mathbf{0} \\ &(T^{k_1})^3\mathbf{x} + (T^{k_2})^3\mathbf{x} + \cdots + (T^{k_r})^3\mathbf{x} = \mathbf{0} \\ &\cdot \\ &\cdot \\ &\cdot \\ &(T^{k_1})^{2e-1}\mathbf{x} + (T^{k_2})^{2e-1}\mathbf{x} + \cdots + (T^{k_r})^{2e-1}\mathbf{x} = \mathbf{0} \end{aligned} \tag{5.4.1}$$

Table 5.4.2. Parity check matrix for a triple-error correcting Bose-Chaudhuri-Hocquenghem code

$$A_0 = \begin{bmatrix} 1&0&0&0&1&0&0&1&1&0&1&0&1&1&1 \\ 0&0&0&1&0&0&1&1&0&1&0&1&1&1&1 \\ 0&0&1&0&0&1&1&0&1&0&1&1&1&1&0 \\ 0&1&0&0&1&1&0&1&0&1&1&1&1&0&0 \\ 1&0&0&0&1&1&0&0&0&1&1&0&0&0&1 \\ 0&1&1&1&1&0&1&1&1&1&0&1&1&1&1 \\ 0&0&1&0&1&0&0&1&0&1&0&0&1&0&1 \\ 0&0&0&1&1&0&0&0&1&1&0&0&0&1&1 \\ 1&0&1&1&0&1&1&0&1&1&0&1&1&0&1 \\ 0&0&0&0&0&0&0&0&0&0&0&0&0&0&0 \\ 0&1&1&0&1&1&0&1&1&0&1&1&0&1&1 \\ 0&1&1&0&1&1&0&1&1&0&1&1&0&1&1 \end{bmatrix}$$

$$T = \begin{bmatrix} 0&1&0&0 \\ 0&0&1&0 \\ 0&0&0&1 \\ 1&1&0&0 \end{bmatrix}$$

$$f(\lambda) = \lambda^4 + \lambda + 1$$

$$\mathbf{x} = \begin{bmatrix} 1 \\ 0 \\ 0 \\ 0 \end{bmatrix}$$

Each equation of (5.4.1) is of the form $p(T)\mathbf{x} = \mathbf{0}$ where $p(T)$ is a matric polynomial. We claim that $p(T)$ must be zero, for if $p(T)$ were nonzero, then $p(T)$ would be nonsingular by Lemma 5.2.3. If the equation $p(T)\mathbf{x} = \mathbf{0}$ is premultiplied by the inverse of $p(T)$ we obtain $\mathbf{x} = \mathbf{0}$, a contradiction.

Thus we may "get rid of" the vector $\mathbf{x}$ in (5.4.1). If for the sake of abbreviation we replace T^{k_i} by y_i, we obtain

$$\begin{aligned}
&y_1 + y_2 + \cdots + y_r = 0\\
&y_1^3 + y_2^3 + \cdots + y_r^3 = 0\\
&y_1^5 + y_2^5 + \cdots + y_r^5 = 0\\
&\quad\vdots\\
&y_1^{2e-1} + y_2^{2e-1} + \cdots + y_r^{2e-1} = 0
\end{aligned}\tag{5.4.2}$$

We claim that $y_1^{2k} + y_2^{2k} + \cdots + y_r^{2k} = 0$ for $k = 1, 2, \ldots, e$. To see this, we square both sides of the first equation of (5.4.2) and note that the cross terms $y_iy_j + y_iy_j$ are zero (modulo 2). Hence

$$(y_1 + \cdots + y_r)^2 = y_1^2 + y_2^2 + \cdots + y_r^2.$$

[What we are saying here is that if $p(S)$ is any polynomial in the binary matrix S, then $[p(S)]^2 = p(S^2)$.] Similarly, to obtain $y_1^4 + \cdots + y_r^4 = 0$ we square both sides of $y_1^2 + \cdots + y_r^2 = 0$; to obtain $y_1^6 + \cdots + y_r^6 = 0$, we square both sides of $y_1^3 + \cdots + y_r^3 = 0$, etc. Thus we have

$$\begin{aligned}
&y_1 + y_2 + \cdots + y_r = 0\\
&y_1^2 + y_2^2 + \cdots + y_r^2 = 0\\
&y_1^3 + y_2^3 + \cdots + y_r^3 = 0\\
&\quad\vdots\\
&y_1^r + y_2^r + \cdots + y_r^r = 0.
\end{aligned}\tag{5.4.3}$$

We write only r equations in (5.4.3) so as to obtain a square array. The equations (5.4.3) may be written in matrix form as follows:

$$\begin{bmatrix} 1 & 1 & \cdots & 1\\ y_1 & y_2 & \cdots & y_r\\ y_1^2 & y_2^2 & \cdots & y_r^2\\ \vdots & & & \\ y_1^{r-1} & y_2^{r-1} & \cdots & y_r^{r-1} \end{bmatrix} \begin{bmatrix} y_1\\ y_2\\ y_3\\ \vdots\\ y_r \end{bmatrix} = \begin{bmatrix} 0\\ 0\\ \vdots\\ 0 \end{bmatrix}\tag{5.4.4}$$

Equation (5.4.4) is of the form $B\mathbf{y} = \mathbf{0}$, where the components of the matrix B and the vector $\mathbf{y}$ are elements in the Galois field $GF(2^q)$ (Theorem 5.2.4). The elements $y_1, \ldots, y_r$ are nonzero; hence *the equations* $B\mathbf{z} = \mathbf{0}$ *have a nontrivial solution in the Galois field* $GF(2^q)$. Consequently the determinant of B is zero. But det B is a standard determinant called the *Vandermonde* determinant; its value is $\Pi_{i>j}(y_i - y_j)$.

[To see this, note that det B is a homogeneous polynomial $g(y_1, y_2, \ldots, y_r)$ of degree $1 + 2 + \cdots + (r - 1) = r(r - 1)/2$. If $y_i = y_j$, then columns i and j of det B are identical, hence det $B = 0$. Thus for all $i \neq j$, $y_i - y_j$ is a divisor of $g(y_1, \ldots, y_r)$. Since the degree of

$$\prod_{i>j}(y_i - y_j)$$

is $r(r - 1)/2$, the result follows, up to a constant factor.]

Now in the present case the elements $y_1, y_2, \ldots y_r$, being different powers of the matrix T, are distinct. (See the remarks after Theorem 5.2.6.) Hence none of the products $y_i - y_j$, $i > j$, is zero. The conclusion is that det $B \neq 0$, a contradiction. Thus our assumption that a set of $2e$ or fewer columns is linearly dependent is untenable. Notice that if r were greater than $2e$ the above proof would break down, since we would have no way of proving that

$$y_1^{2e+1} + y_2^{2e+1} + \cdots + y_r^{2e+1} = 0.$$

The matrix A_0 of Table 5.4.2 is 12 by 15, but the rank of A_0 is less than 12 since row 10 of the matrix consists entirely of zeros, and rows 11 and 12 are identical. It is possible to give an explicit and computationally feasible procedure for obtaining the rank of A_0. First we establish a preliminary result.

Lemma 5.4.2. Let T be a binary matrix whose minimal polynomial is irreducible. If $p(T)$ is any matric polynomial in T such that $p(T)$ is not the zero matrix, then the minimal polynomial of $p(T)$ is also irreducible.

Proof. If the minimal polynomial $f(\lambda)$ of $p(T)$ is not irreducible, then there are nonzero polynomials $g(\lambda)$ and $h(\lambda)$, of lower degree than $f(\lambda)$, such that $f(\lambda) = g(\lambda)h(\lambda)$. Replacing λ by $S = p(T)$ we obtain $g(S)h(S) = f(S) = 0$. Now $g(S) = g(p(T))$ may be expressed as a matric polynomial in T; since $g(S)$ is not zero, we may (by Lemma 5.2.3) premultiply by the inverse of $g(S)$ to obtain $h(S) = 0$, which contradicts the minimality of $f(\lambda)$. This proves the lemma. [For those familiar with algebra, what we are saying here is that the minimal polynomial of any element α of a field F_1, such that α is algebraic over a field $F \subset F_1$, is irreducible. In this case, $\alpha = p(T)$, F is the modulo 2 field, and F_1 is the Galois field GF(2^q) where q is the degree of the minimal polynomial of T. (See Theorem 5.2.4.)]

The next theorem allows the precise determination of the rank.

Theorem 5.4.3. Let A_0 be the Bose-Chaudhuri-Hocquenghem parity check matrix. If $f_i(\lambda)$ is the minimal polynomial of T^i ($i = 1, 3, 5, \ldots 2e - 1$), and $f(\lambda)$ is the least common multiple of the $f_i(\lambda)$, that is, the polynomial of lowest degree which is a multiple of each $f_i(\lambda)$, then the rank of A_0 is the degree of $f(\lambda)$. Furthermore, if $i \neq j$, then $f_i(\lambda)$ and $f_j(\lambda)$ are either relatively prime or identical, so that $f(\lambda)$ is the product of the distinct members of the set $\{f_i(\lambda), i = 1, 3, \ldots, 2e - 1\}$.

Proof. Suppose that we have a finite collection of cyclic codes with parity check matrices A_i and generator polynomials $g_i(x)$, each code having words of length n. A binary sequence $\mathbf{w}$ represented by the polynomial $b(x)$ satisfies $A_i\mathbf{w}^T = \mathbf{0}$ for all i if and only if $b(x)$ is a multiple of each $g_i(x)$. Now it is immediate from the definition of a cyclic code that the intersection of the codes defined by the matrices A_i is cyclic. Furthermore, the above argument shows that the associated generator polynomial $g(x)$ is the least common multiple of the $g_i(x)$. To obtain the Bose-Chaudhuri-Hocquenghem codes we may take

$$A_i = [\mathbf{x} \mid T^i\mathbf{x} \mid T^{2i}\mathbf{x} \mid \cdots \mid T^{(n-1)i}\mathbf{x}],$$
$$i = 1, 3, \ldots, 2e - 1;$$

by Lemma 5.4.2, the minimal polynomial $f_i(\lambda)$ of T^i is irreducible; by Lemma 5.3.3c, $f_i(x) = g_i(x)$ for each i, and consequently $f(x) = g(x)$. Since the rank of the parity check matrix of a cyclic code is the degree of the generator polynomial of the code (see Theorems 5.3.1 and 5.3.2), the first part of the theorem is proved. Now if $f_i(\lambda)$ and $f_j(\lambda)$ are not identical, they must be relatively prime, for if $f_i(\lambda)$ and $f_j(\lambda)$ were to have a nontrivial common factor, the polynomials could not be irreducible, a contradiction. The theorem is proved.

For the code of Table 5.4.2, we find (see Problem 5.9) that

$$f_1(\lambda) = 1 + \lambda + \lambda^4, \qquad f_3(\lambda) = 1 + \lambda + \lambda^2 + \lambda^3 + \lambda^4,$$
$$f_5(\lambda) = 1 + \lambda + \lambda^2,$$

so that the rank of A_0 is 10. The rank of the submatrix A_i, which by the above argument is the degree of $f_i(\lambda)$, is 4 for $i = 1$ and 3, and 2 for $i = 5$. Thus two rows of the matrix A_5 must be linearly dependent on the other rows of A_5, which agrees with our previous observations that row 10 of A_0 contains only zeros, and row 11 is identical with row 12.

A procedure for obtaining the minimal polynomial of the matrix T^i without actually computing the powers of T is indicated in the solution to Problem 5.9.

If $2^q - 1$ is prime, then the rank of each matrix A_i is q. (As above, Lemma 5.4.2 implies that the minimal polynomial of T^i is irreducible; Theorem 5.2.5 then implies that T^i has maximal period, and consequently the columns of each A_i constitute a permutation of the set of all $2^q - 1$ nonzero q-vectors. Thus all the matrices A_i have the same rank. But the minimal polynomial of T is of degree q; therefore A_1 has rank q.) However, whether or not $2^q - 1$ is prime, it is possible for $f_i(\lambda)$ and $f_j(\lambda)$ to coincide when $i \neq j$. For example, if $q = 5$, then T^5 and T^9 have the same minimal polynomial. For if $p(S)$ is any polynomial in the binary matrix S, $p(S^2) = [p(S)]^2$. [See the discussion after (5.4.2).] Thus if $p(T^5) = 0$, then $p(T^{10}) = p(T^{20}) = p(T^{40}) = p(T^9) = 0$. (Note $T^{2^q-1} = T^{31} = I$). Similarly, if $p(T^9) = 0$, then $p(T^{18}) = p(T^{36}) = p(T^5) = 0$. Thus the set of polynomials annulled by T^5 coincides with the set of polynomials annulled by T^9. Similarly, if $q = 6$, we may verify that T^{13} and T^{19} have the same minimal polynomial.

5.5. *Single-error correcting cyclic codes; automatic decoding*

A shift register may be used to generate a cyclic code that corrects single errors and at the same time entirely eliminates the storage problem

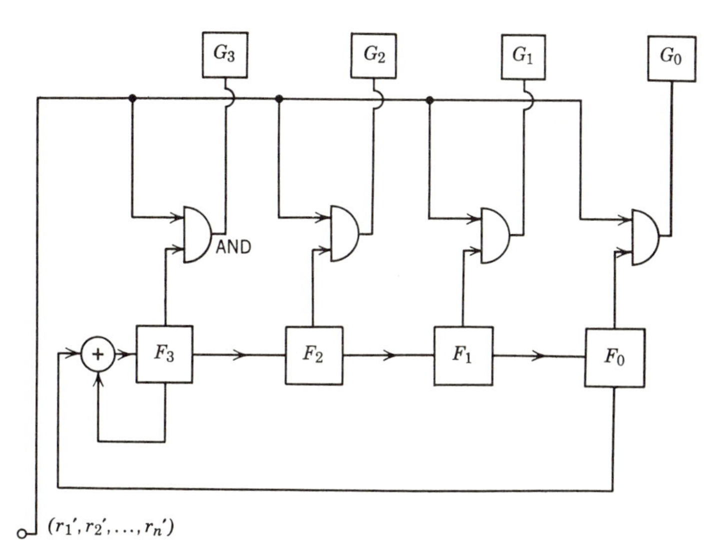

Fig. 5.5.1. Automatic decoder for the cyclic code of Fig. 5.2.1.

at the decoder. As an example, consider the shift register and associated code of Fig. 5.2.1. An automatic decoding circuit may be constructed as shown in Fig. 5.5.1. The original shift register of Fig. 5.2.1 is combined, through a number of AND gates, with another set of binary storage devices G_0, G_1, G_2, G_3. The G_i are constructed so that their state changes if a "1" input is applied. In other words the content of the G register after application of an input

$$\mathbf{y} = \begin{bmatrix} y_0 \\ y_1 \\ y_2 \\ y_3 \end{bmatrix}$$

is the modulo 2 sum of $\mathbf{y}$ and the content of G before application of the input.

In order to examine the operation of the decoding circuit, suppose that the received digits $r_1', r_2', \ldots, r_n'$ arrive at times $t = 1, 2, \ldots, n$ respectively. The columns of the parity check matrix are $\mathbf{x}$, $T\mathbf{x}$, $T^2\mathbf{x}, \ldots,$ $T^{n-1}\mathbf{x}$, where $\mathbf{x}$ is some nonzero vector.

$$\text{(In Fig. 5.2.1, } \mathbf{x} = \begin{bmatrix} 0 \\ 0 \\ 0 \\ 1 \end{bmatrix}.\text{)}$$

The F shift register is adjusted so that its content at time $t = 0$ is $\mathbf{x}$. The G register is initially set to have content zero. At time $t = 1$, digit r_1' appears on the input line. If $r_1' = 0$, the AND gates will be closed and the G register will still have content zero. If $r_1' = 1$, the AND gates will allow the content of the F register to be transferred to the G register. In short, the content of G at $t = 1$ will be $r_1'\mathbf{x}$. At $t = 2$, F has content $T\mathbf{x}$ and r_2' appears at the input line; hence the input to G is $r_2'T\mathbf{x}$. The content of G at $t = 2$ is therefore $r_1'\mathbf{x} + r_2'T\mathbf{x}$. Continuing in this fashion, the content of G at $t = n$ is

$$r_1'\mathbf{x} + r_2'T\mathbf{x} + \cdots + r_n'T^{n-1}\mathbf{x},$$

which is the corrector $\mathbf{c}$ associated with $r_1', r_2', \ldots, r_n'$. If $\mathbf{c} = \mathbf{0}$, the received sequence is accepted as the input code word. If $\mathbf{c} \neq \mathbf{0}$, we employ a comparison circuit (not shown in Fig. 5.5.1) to determine which digit is in error. The corrector is compared with the content of F as F cycles through the entire set of columns of the parity check matrix. If there is a correspondence at column j, that is, if $\mathbf{c} = T^{j-1}\mathbf{x}$, then a single error in digit j is assumed.

This brief discussion does not do justice to the subject. For further details concerning efficient encoding and decoding of cyclic codes, see Peterson (1961).

5.6. *Notes and remarks*

Cyclic codes are treated in detail by Peterson (1961), who uses the machinery of abstract algebra. Theorems 5.3.1 and 5.3.2 form the basis of Peterson's approach.

A feedback shift register is a special case of a device called a "linear sequential circuit"; see Elspas (1959), Friedland (1959), and Friedland and Stern (1959). Such devices were applied to the construction of error correcting codes by Abramson (1961), Melas (1960), and others, but Theorem 5.3.4 does not seem to have been stated explicitly anywhere.

The results of Section 5.4 are due to Bose and Chaudhuri (1960) and Hocquenghem (1959). Bose-Chaudhuri-Hocquenghem codes can be constructed in the nonbinary case, but for the proof of Theorem 5.4.1 to remain valid, additional blocks generated by $T^2\mathbf{x}$, $T^4\mathbf{x}$, . . . , $T^{2e}\mathbf{x}$ must be added to A_0. Thus in the nonbinary case the number of check digits for a given word length is greatly increased.

Very little is known about the asymptotic properties of Bose-Chaudhuri-Hocquenghem codes. According to the data of Peterson (1961), the relationship between the word length n, the number of check digits m, and the number of errors corrected e, is approximately that of the Varsharmov-Gilbert-Sacks condition for values of n up to 1000. If this behavior were to be maintained for arbitrarily large n, the asymptotic properties of the codes would be very desirable. (See the remarks at the end of Section 4.10). However, Peterson (1960) has proved that if e/n is kept constant, the transmission rate $R = k/n$ (k = number of information digits) approaches zero as n approaches infinity. But this does not necessarily imply that the asymptotic properties are bad. The difficulty is that we have proved that the errors corrected by the codes include *at least* the errors of magnitude e or smaller. It is possible that sufficiently many error patterns of weight greater than e are corrected to allow a positive transmission rate to be maintained with an arbitrarily small probability of error.

PROBLEMS

5.1 Give a parity check matrix for a cyclic code generated by a single cycle of a feedback shift register whose characteristic polynomial is

a. $\lambda^4 + \lambda + 1$

b. $\lambda^4 + \lambda^3 + \lambda^2 + \lambda + 1$

c. $\lambda^5 + \lambda^4 + 1$ (three possible codes).

5.2 Find the generator polynomial $g(x)$ of the code in Problem 5.1c which uses words of length 7, and find the polynomial $h(x)$ such that $h(x)g(x) = x^7 - 1$.

5.3 Show that the Bose-Chaudhuri-Hocquenghem codes are cyclic.

5.4 Give the parity check equations for a double-error correcting Bose-Chaudhuri-Hocquenghem code with a code-word length of 31.

5.5 The binary sequence $\mathbf{w} = (1010011)$ is a code word of a cyclic code with 3 information digits.

a. Find a parity check matrix for the code, and a feedback shift register that will generate the code.

b. Find the generator polynomial $g(x)$ of the code, and the polynomial $h(x)$ such that $h(x)g(x) = x^7 - 1$.

5.6 [The application of Cyclic Codes for Burst-Error Correction; Melas (1960).] An *error burst* of width $q + 1$ beginning at digit i is an error pattern such that r_i and r_{i+q} are in error, a digit r_j $(i < j < i + q)$ may or may not be in error, and all digits before r_i or after r_{i+q} are correct. For example, the error bursts of width 4 may be represented by the following binary sequences:

1001	r_i and r_{i+3} in error
1011	r_i, r_{i+2}, r_{i+3} in error
1101	r_i, r_{i+1}, r_{i+3} in error
1111	$r_i, r_{i+1}, r_{i+2}, r_{i+3}$ in error.

Binary sequences like these, which describe the distribution of errors within a burst without regard to the position of the burst within the code word, are called *burst-error patterns*. These are 2^{q-1} burst-error patterns of width $q + 1$.

Now consider a code generated by a feedback shift register with maximal period, as shown below.

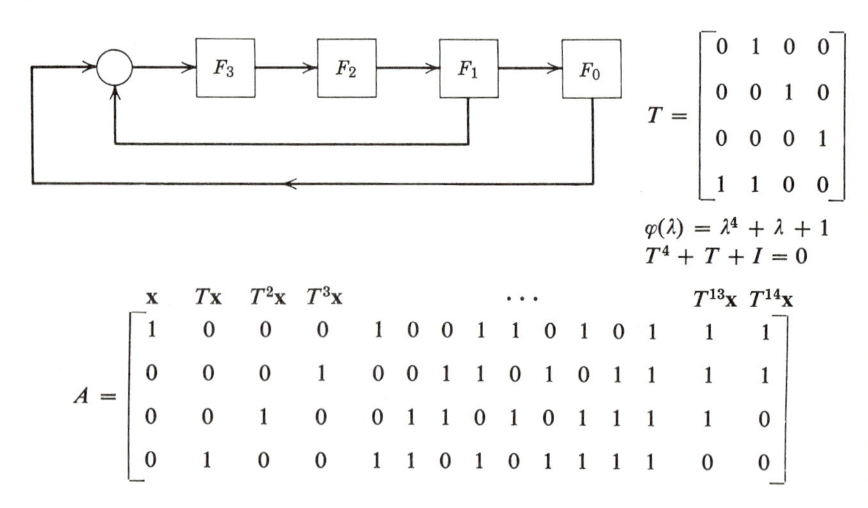

$$T = \begin{bmatrix} 0 & 1 & 0 & 0 \\ 0 & 0 & 1 & 0 \\ 0 & 0 & 0 & 1 \\ 1 & 1 & 0 & 0 \end{bmatrix}$$

$$\varphi(\lambda) = \lambda^4 + \lambda + 1$$
$$T^4 + T + I = 0$$

$$\begin{array}{c} \\ A = \end{array}\begin{array}{ccccccccccccccc} \mathbf{x} & T\mathbf{x} & T^2\mathbf{x} & T^3\mathbf{x} & & & & \cdots & & & & & & T^{13}\mathbf{x} & T^{14}\mathbf{x} \end{array}$$
$$A = \begin{bmatrix} 1 & 0 & 0 & 0 & 1 & 0 & 0 & 1 & 1 & 0 & 1 & 0 & 1 & 1 & 1 \\ 0 & 0 & 0 & 1 & 0 & 0 & 1 & 1 & 0 & 1 & 0 & 1 & 1 & 1 & 1 \\ 0 & 0 & 1 & 0 & 0 & 1 & 1 & 0 & 1 & 0 & 1 & 1 & 1 & 1 & 0 \\ 0 & 1 & 0 & 0 & 1 & 1 & 0 & 1 & 0 & 1 & 1 & 1 & 1 & 0 & 0 \end{bmatrix}$$

Consider a typical burst-error pattern, say 1101. If the burst begins at digit 1, the associated corrector is

$$(I + T + T^3)\mathbf{x} = \begin{bmatrix} 1 \\ 1 \\ 0 \\ 1 \end{bmatrix} = T^7\mathbf{x}.$$

If the burst begins at digit 2, the corrector is

$$(T + T^2 + T^4)\mathbf{x} = \begin{bmatrix} 1 \\ 0 \\ 1 \\ 0 \end{bmatrix} = T^8\mathbf{x}.$$

If the burst begins at digit 5, the corrector is $(T^4 + T^5 + T^7)\mathbf{x} = T^{11}\mathbf{x}$, etc.

a. Show that a translation of an error burst by t digits results in a translation of the associated corrector by t digits. All possible burst-error patterns of width 4 or less are listed below, along with the associated correctors (assuming that the burst begins at digit 1).

Burst-Error Pattern	*Corrector*
1	$\mathbf{x}$
11	$(I + T)\mathbf{x} = T^4\mathbf{x}$
101	$(I + T^2)\mathbf{x} = T^8\mathbf{x}$
111	$(I + T + T^2)\mathbf{x} = T^{10}\mathbf{x}$
1001	$(I + T^3)\mathbf{x} = T^{14}\mathbf{x}$
1011	$(I + T^2 + T^3)\mathbf{x} = T^{13}\mathbf{x}$
1101	$(I + T + T^3)\mathbf{x} = T^7\mathbf{x}$
1111	$(I + T + T^2 + T^3)\mathbf{x} = T^{12}\mathbf{x}$

b. Show that if a cyclic code is generated by a maximal-period feedback shift register with q stages, each error burst of width q or less results in a distinct corrector (assuming all error bursts begin at the same digit).

It follows from (a) and (b) that knowledge of the corrector and the digit at which the burst begins completely specifies the error pattern. The various burst-correcting schemes are all based on the use of additional check digits in order to determine the position of the burst. Note that a burst may begin at the end of a code word and end at the beginning of the word. In the above example, the burst-error pattern 1101 beginning at digit 14 represents an error in digits 14, 15, and 2, and results in the corrector $(I + T + T^3)T^{13}\mathbf{x} = T^5\mathbf{x}$. Because of the cyclic structure of the codes, bursts of this type will also be corrected. (Note that it is sufficient to assume that T has an irreducible minimal polynomial; however, the maximal-period case gives the longest code words.)

5.7 (Abramson, 1959). A parity check matrix A is generated by a maximal-period q-stage shift register. The matrix A is then augmented by the addition of an "all-check" digit, so that the parity check equations become

$$\begin{bmatrix} & A & \\ \hline 1 & 1 \cdots 1 & 1 \end{bmatrix} \begin{bmatrix} r_1 \\ \cdot \\ \cdot \\ \cdot \\ r_n \end{bmatrix} = 0, \qquad n = 2^q - 1$$

Show that the above code may be used to correct single errors and double adjacent errors. For example, if $q = 3$, the parity check equations might be

$$\begin{bmatrix} 1 & 0 & 0 & 1 & 0 & 1 & 1 \\ 0 & 0 & 1 & 0 & 1 & 1 & 1 \\ 0 & 1 & 0 & 1 & 1 & 1 & 0 \\ 1 & 1 & 1 & 1 & 1 & 1 & 1 \end{bmatrix} \begin{bmatrix} r_1 \\ \cdot \\ \cdot \\ \cdot \\ r_7 \end{bmatrix} = \begin{bmatrix} 0 \\ 0 \\ 0 \\ 0 \end{bmatrix}.$$

The matrix A is generated by a feedback shift register with characteristic polynomial $\lambda^3 + \lambda + 1$.

A corrector

$$\begin{bmatrix} 1 \\ 1 \\ 1 \\ 1 \end{bmatrix}$$

would be associated with a single error in r_6.

A corrector

$$\begin{bmatrix} 1 \\ 1 \\ 1 \\ 0 \end{bmatrix}$$

would be associated with a double adjacent error in r_3 and r_4. Decoding can be accomplished since the "all-check" digit distinguishes between a single error and a double error, and no two double adjacent errors result in the same corrector.

5.8 (Fire, 1959). A parity check matrix A is generated by a maximal-period q_1-stage feedback shift register. An identity matrix I_{q_2} of order $q_2 = 2q_1 - 1$ is formed. The parity check matrix for the Fire code is

$$\begin{bmatrix} I_{q_2} & | & I_{q_2} & | & I_{q_2} & | & \cdots & | & I_{q_2} \\ A & | & A & | & & \cdots & & & \end{bmatrix}$$

The length of the code words is n = least common multiple of $2^{q_1} - 1$ and q_2. The number of check digits is $q_1 + q_2$. For example, if A is the matrix of Problem 5.7 [$q_1 = 3$, $q_2 = 5$, $n = \text{lcm}\,(7, 5) = 35$], the parity check matrix of the associated Fire code is

$$\begin{bmatrix} 10000 & 10000 & 10000 & 10000 & 10000 & 10000 & 10000 \\ 01000 & 01000 & 01000 & 01000 & 01000 & 01000 & 01000 \\ 00100 & 00100 & 00100 & 00100 & 00100 & 00100 & 00100 \\ 00010 & 00010 & 00010 & 00010 & 00010 & 00010 & 00010 \\ 00001 & 00001 & 00001 & 00001 & 00001 & 00001 & 00001 \\ 10010 & 11100 & 10111 & 00101 & 11001 & 01110 & 01011 \\ 00101 & 11001 & 01110 & 01011 & 10010 & 11100 & 10111 \\ 01011 & 10010 & 11100 & 10111 & 00101 & 11001 & 01110 \end{bmatrix}$$

Show that the Fire code will correct all error bursts of width q_1 or less. For example, suppose that in the above code the corrector is

$$\begin{bmatrix} 1 \\ 0 \\ 0 \\ 1 \\ 0 \\ \hline 0 \\ 0 \\ 1 \end{bmatrix}.$$

The top five digits identify the burst-error pattern as 101, and indicate that the burst starts at digit 4, 9, 14, 19, 24, 29, or 34. Now considering only the matrix A, the corrector associated with a burst-error pattern of 101 beginning at digit 1 is

$$(I + T^2)\mathbf{x} = \begin{bmatrix} 1 \\ 1 \\ 0 \end{bmatrix} = T^6\mathbf{x}.$$

Since

$$\begin{bmatrix} 0 \\ 0 \\ 1 \end{bmatrix} = T\mathbf{x} = T^8\mathbf{x},$$

a burst-error pattern of 101 beginning at digit 3 will produce a corrector of

$$(I + T^2)T^2\mathbf{x} = T^6T^2\mathbf{x} = T^8\mathbf{x} = \begin{bmatrix} 0 \\ 0 \\ 1 \end{bmatrix}.$$

Hence we conclude from the last three digits of the corrector that the burst begins at digit 3, 10, 17, 24, or 31. Thus we have identified the 101 pattern beginning at digit 24, that is, a double error in digits 24 and 26.

5.9 If

$$T = \begin{bmatrix} 0 & 1 & 0 & 0 \\ 0 & 0 & 1 & 0 \\ 0 & 0 & 0 & 1 \\ 1 & 1 & 0 & 0 \end{bmatrix},$$

find the minimal polynomials of T^3 and T^5.

5.10 Show that the period of a nonsingular binary matrix cannot exceed $2^q - 1$, where q is the degree of the minimal polynomial of the matrix.

CHAPTER SIX

Information Sources

6.1. Introduction

We have not considered in any detail the model for a source of information. We have assumed at various times (Chapter 2) that the information to be transmitted consisted of the outcome of an experiment involving a random variable X; if successive values of X were to be transmitted, the values were assumed to be chosen independently. In other words, an information source has at times meant for us a sequence of independent, identically distributed random variables $X_1, X_2, \ldots$. At other times (Chapter 3) we have completely ignored the statistical properties of the source and focused instead on the assignment of a code word to all possible sequences the source could produce in a given interval of time; we required that the resulting code have a probability of error which was uniformly small for all possible code words so that statistical fluctuation of the source would not affect performance of the decoding scheme.

In this chapter we examine more closely the properties of information sources. We shall find that of all the sequences a source can conceivably produce in a given length of time, only a fraction are "meaningful"; the remaining sequences may be ignored when we are considering the problem of transmitting the information produced by the source. This result will allow us to take advantage of the statistical structure of a particular source to achieve a higher transmission rate for a given probability of error.

We will also try to use the concepts developed in the chapter to say something about the properties of a language as a means of communication. In addition, the results will be needed in Chapter 7 in the analysis of channels with memory. The mathematical equipment we shall use is the theory of finite Markov chains, which will be developed in this chapter.

6.2. A mathematical model for an information source

Let us begin by observing that whatever model we use for an information source, the source must produce a sequence of symbols

$$\cdots X_{t-2}X_{t-1}X_tX_{t+1}X_{t+2}\cdots$$

If we like, we may imagine the symbols being produced at a uniform rate, one per unit of time, with X_t the symbol produced at time t. The structure of the source may be described by giving the statistical dependence of X_t on the past symbols $X_{t-1}, X_{t-2}, \ldots$. As an example, suppose we have a primitive language with two symbols A and B. Suppose that in a sample of text in the language, the probability distribution of X_t is determined

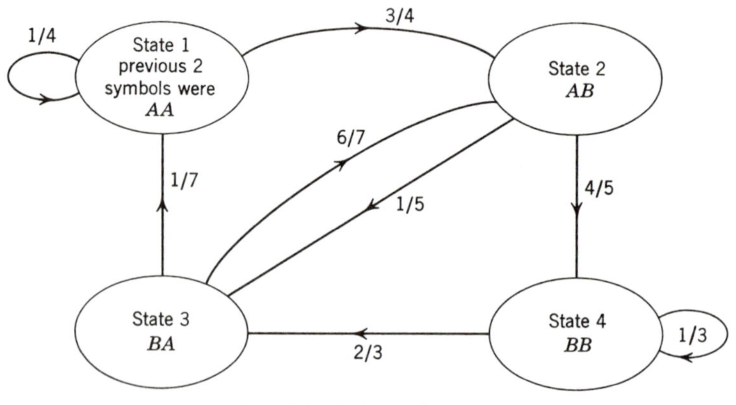

Fig. 6.2.1. Information source.

by the previous two symbols X_{t-1} and X_{t-2} but is not affected by the additional knowledge of X_{t-3}, X_{t-4}, etc. Let us assume that the conditional distribution of X_t given X_{t-1} and X_{t-2} is as follows:

$$P\{X_t = A \mid X_{t-2} = A, X_{t-1} = A\} = 1/4$$
$$P\{X_t = A \mid X_{t-2} = A, X_{t-1} = B\} = 1/5$$
$$P\{X_t = A \mid X_{t-2} = B, X_{t-1} = A\} = 1/7$$
$$P\{X_t = A \mid X_{t-2} = B, X_{t-1} = B\} = 2/3$$
$$P\{X_t = B \mid X_{t-2} = \alpha, X_{t-1} = \beta\} = 1 - P\{X_t = A \mid X_{t-2} = \alpha, X_{t-1} = \beta\} \quad (\alpha, \beta = A \text{ or } B)$$
$$P\{X_t = A \mid X_{t-3} = \alpha, X_{t-2} = \beta, X_{t-1} = \gamma\} = P\{X_t = A \mid X_{t-2} = \beta, X_{t-1} = \gamma\} \quad (\alpha, \beta, \gamma = A \text{ or } B)$$

and so on.

The source just described may be characterized by four "states" and a set of "transition probabilities." (See Fig. 6.2.1.) The source is in state 1 at time $t - 1$ if $X_{t-2} = X_{t-1} = A$, in state 2 if $X_{t-2} = A, X_{t-1} = B$, etc. If the source is in state 2 at time $t - 1$, then with probability 1/5 the source moves to state 3 at time t and $X_t = A$; with probability 4/5 the source moves to state 4 at time t and $X_t = B$.

An object of the type shown in Fig. 6.2.1 is called a *finite Markov chain*. More precisely, given a finite set $S = \{s_1, \ldots, s_r\}$ called the *set of states* and a matrix $\Pi = [p_{ij}]$, $i, j = 1, \ldots, r$, called the *transition matrix*, satisfying $p_{ij} \geq 0$ for all i, j; $\sum_{j=1}^{r} p_{ij} = 1$ for all i, we define a sequence of random variables $Z_0, Z_1, Z_2, \ldots$ as follows.

Each random variable Z_i takes on values in the set of states S. Let Z_0 have an arbitrary distribution. Having specified the joint distribution of $Z_0, Z_1, \ldots, Z_n$ we define $P\{Z_{n+1} = s_k \mid Z_0 = s_{i_0}, \ldots, Z_n = s_{i_n}\} = p_{i_n k}$; this determines the joint distribution of $Z_0, \ldots, Z_{n+1}$. The sequence of random variables $Z_0, Z_1, \ldots$ is called a *finite Markov chain*.

A square matrix whose elements are nonnegative and whose row sums are unity is called a *Markov matrix* or *stochastic matrix*; any such matrix may serve as the transition matrix of a Markov chain. Intuitively we may think of the process beginning at time $t = 0$ by choosing an initial state in accordance with a specified probability distribution. If we are in state s_i at time $t = 0$ then the process moves at $t = 1$ to a (possibly) new state, the state being chosen using the matrix Π; the number p_{ij} is the probability that the process will move to s_j at time $t = 1$ given that we are in state s_i at time $t = 0$. If we are in state s_j at $t = 1$ we move to s_k at $t = 2$ with probability p_{jk}; in this fashion we generate a sequence of states. The random variable Z_n represents the state of the process at time $t = n$; we have constructed the process so that the distribution of Z_n depends only on the state at time $n - 1$ (and on the given matrix Π) and not on the state at times $t < n - 1$. Formally, we have the *Markov* property:

$$P\{Z_n = \alpha_n \mid Z_{n-1} = \alpha_{n-1}, \ldots, Z_0 = \alpha_0\} = P\{Z_n = \alpha_n \mid Z_{n-1} = \alpha_{n-1}\}$$
$$\text{for all } \alpha_0, \alpha_1, \ldots, \alpha_n \in S. \quad (6.2.1)$$

The information source of Fig. 6.2.1 could be characterized as a finite Markov chain with transition matrix

$$\begin{bmatrix} \frac{1}{4} & \frac{3}{4} & 0 & 0 \\ 0 & 0 & \frac{1}{5} & \frac{4}{5} \\ \frac{1}{7} & \frac{6}{7} & 0 & 0 \\ 0 & 0 & \frac{2}{3} & \frac{1}{3} \end{bmatrix}$$

together with the specification that if the source enters state 1 or 3 at time t, an A is produced at that time, and similarly if the source moves to state 2 or 4 at time t, a B is produced. We may describe this situation by considering a function which assigns to the states 1 and 3 the letter A and to the states 2 and 4 the letter B. We are thus led to the concept of a *Markov information source*.

By a Markov information source we shall mean a finite Markov chain, together with a function f whose domain is the set of states S and whose range is a finite set Γ called the *alphabet* of the source. (In the example of Fig. 6.2.1, S has 4 members and $\Gamma = \{A, B\}$.) In Section 6.4 we are going to impose an additional requirement on the Markov chain, so the above description is not to be considered a formal definition of a Markov information source; such a definition will be given in that section.

In Section 6.4 we shall also discuss more general information sources.

6.3. *Introduction to the theory of finite Markov chains*

We are aiming at an eventual definition of the "uncertainty" associated with an information source. For this we shall need several concepts from Markov chain theory. Let us for the moment forget about the alphabet of the source and analyze a finite Markov chain with states $s_1, s_2, \ldots, s_r$ and transition matrix Π. As an example, consider the chain of Fig. 6.3.1. Suppose that we start at $t = 0$ by assigning an initial distribution to the states, say

$$P\{Z_0 = s_1\} = 1/3, \qquad P\{Z_0 = s_2\} = 2/3.$$
$$(Z_t = \text{state at time } t)$$

Then the distribution of states at $t = 1$ is

$$P\{Z_1 = s_1\} = \frac{1}{3}\left(\frac{1}{4}\right) + \frac{2}{3}\left(\frac{1}{2}\right) = \frac{5}{12}$$
$$P\{Z_1 = s_2\} = \frac{1}{3}\left(\frac{3}{4}\right) + \frac{2}{3}\left(\frac{1}{2}\right) = \frac{7}{12}.$$

For a general chain with r states, let

$$w_j^{(n)} = P\{Z_n = s_j\} = \text{probability of being in state } s_j \text{ at time } t = n.$$

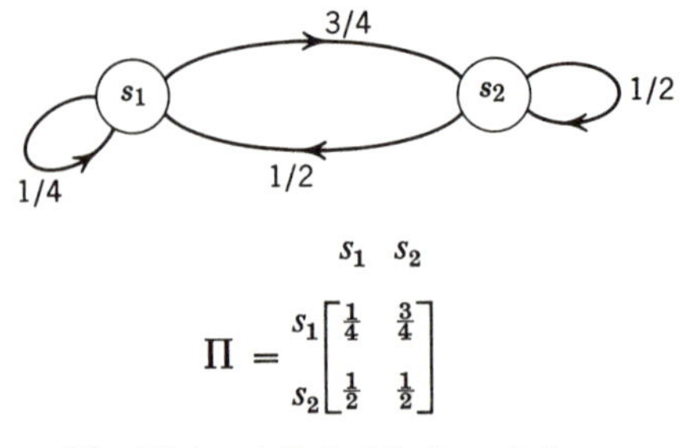

$$\Pi = \begin{array}{c} \\ s_1 \\ s_2 \end{array} \begin{array}{c} \begin{array}{cc} s_1 & s_2 \end{array} \\ \begin{bmatrix} \frac{1}{4} & \frac{3}{4} \\ \frac{1}{2} & \frac{1}{2} \end{bmatrix} \end{array}$$

Fig. 6.3.1. A finite Markov chain.

At time $t = n - 1$, the state of the chain must be $1, 2, \ldots,$ or r. Hence we may write

$$w_j^{(n)} = w_1^{(n-1)} p_{1j} + w_2^{(n-1)} p_{2j} + \cdots + w_r^{(n-1)} p_{rj}. \tag{6.3.1}$$

In matrix form, (6.3.1) is

$$[w_1^{(n)} \; w_2^{(n)} \cdots w_r^{(n)}] = [w_1^{(n-1)} \; w_2^{(n-1)} \cdots w_r^{(n-1)}] \begin{bmatrix} p_{11} & \cdots & p_{1r} \\ \cdot & & \\ \cdot & & \\ \cdot & & \\ p_{r1} & \cdots & p_{rr} \end{bmatrix}. \tag{6.3.2}$$

If we let $W^{(n)}$ be the "state distribution vector" at time n, that is, $W^{(n)} = [w_1^{(n)} \quad w_2^{(n)} \quad \cdots \quad w_r^{(n)}]$, then (6.3.2) becomes

$$W^{(n)} = W^{(n-1)}\Pi. \tag{6.3.3}$$

By iteration of (6.3.3) we obtain

$$W^{(n)} = W^{(n-1)}\Pi = W^{(n-2)}\Pi^2 = W^{(n-3)}\Pi^3 \cdots$$

and finally,

$$W^{(n)} = W^{(0)}\Pi^n. \tag{6.3.4}$$

In other words, the state distribution vector at time n is the product of the state distribution vector at the initial time $t = 0$, and the nth power of the transition matrix. In particular, suppose we start the chain in a specific state s_i at $t = 0$. Then

$$W^{(0)} = [0 \quad \cdots \quad 0 \quad 1 \quad 0 \quad \cdots \quad 0],$$

where the "1" occurs in position i.

From (6.3.4) we see that $W^{(n)}$ is the ith row of Π^n, that is,

$$W^{(n)} = [p_{i1}^{(n)} \quad p_{i2}^{(n)} \quad \cdots \quad p_{ir}^{(n)}]$$

Thus the element in row i and column j of Π^n is the probability that the chain will be in state j at time $t = n$, given that the chain is in state i at $t = 0$. In symbols,

$$p_{ij}^{(n)} = P\{Z_n = s_j \mid Z_0 = s_i\}. \tag{6.3.5}$$

Thus $p_{ij}^{(n)}$ is the "n-stage transition probability" from state i to state j. It immediately follows that Π^n is a stochastic matrix, that is, $p_{ij}^{(n)} \geq 0$ for all i, j and $\sum_{j=1}^{r} p_{ij}^{(n)} = 1$ $(i = 1, \ldots, r)$.

Let us return to the example of Fig. 6.3.1. A direct computation yields

$$\Pi^2 = \begin{bmatrix} \frac{1}{4} & \frac{3}{4} \\ \frac{1}{2} & \frac{1}{2} \end{bmatrix} \begin{bmatrix} \frac{1}{4} & \frac{3}{4} \\ \frac{1}{2} & \frac{1}{2} \end{bmatrix} = \begin{bmatrix} \frac{7}{16} & \frac{9}{16} \\ \frac{3}{8} & \frac{5}{8} \end{bmatrix}$$

$$\Pi^4 = \begin{bmatrix} \frac{7}{16} & \frac{9}{16} \\ \frac{3}{8} & \frac{5}{8} \end{bmatrix} \begin{bmatrix} \frac{7}{16} & \frac{9}{16} \\ \frac{3}{8} & \frac{5}{8} \end{bmatrix} = \begin{bmatrix} \frac{103}{256} & \frac{153}{256} \\ \frac{102}{256} & \frac{154}{256} \end{bmatrix}$$

An examination of Π^4 reveals that $p_{11}^{(4)} \approx p_{21}^{(4)}$ and $p_{12}^{(4)} \approx p_{22}^{(4)}$; in other words, the probability of being in state s_j at time $t = 4$ is almost independent of the starting state. It appears as if for large n, the state distribution at $t = n$ will approach a "steady state" condition; the probability of being in a particular state s_j at $t = n$ will be almost independent of the initial state at $t = 0$. Mathematically, we express this condition by saying that

$$\lim_{n\to\infty} p_{ij}^{(n)} = w_j \qquad (i, j = 1, \ldots, r), \tag{6.3.6}$$

where w_j is independent of i. The number w_j is called the *steady state probability* of state s_j. If there is a set of numbers $w_1, \ldots, w_r$ such that $\lim_{n\to\infty} p_{ij}^{(n)} = w_j$ $(i, j = 1, 2, \ldots, r)$ then we shall say that *steady state probabilities* exist for the chain.

We shall presently investigate the conditions under which (6.3.6) is valid; it does not hold for an arbitrary Markov chain. For the moment we concentrate on the consequences of the existence of steady state probabilities.

Theorem 6.3.1. Suppose that for a Markov chain with transition matrix $\Pi = [p_{ij}]$, $i, j = 1, \ldots, r$, we have steady state probabilities w_j $(j = 1, \ldots, r)$. Then

a. $\sum_{j=1}^{r} w_j = 1$

b. $W = [w_1, \ldots, w_r]$ is a "stationary distribution" for the chain, that is, $W\Pi = W$. Consequently, if the initial distribution is $W^{(0)} = W$ then $W^{(n)} = W$ for all n.

c. W is the unique stationary distribution for the chain, that is, if $Z = [z_1, \ldots, z_r]$, with all $z_i \geq 0$ and $\sum_{i=1}^{r} z_i = 1$, then $Z\Pi = Z$ implies $Z = W$.

Proof. To prove (a), we use the fact that $\sum_{j=1}^{r} p_{ij}^{(n)} = 1$. Allowing n to approach infinity, we obtain

$$1 = \lim_{n\to\infty} \sum_{j=1}^{r} p_{ij}^{(n)} = \sum_{j=1}^{r} \lim_{n\to\infty} p_{ij}^{(n)} = \sum_{j=1}^{r} w_j.$$

To prove (b), we note that

$$p_{ij}^{(n+1)} = p_{i1}^{(n)} p_{1j} + p_{i2}^{(n)} p_{2j} + \cdots + p_{ir}^{(n)} p_{rj} \; (i, j = 1, \ldots, r). \tag{6.3.7}$$

The above formula may be derived by probabilistic reasoning or by observing that (6.3.7) in matrix form is simply

$$\Pi^{n+1} = \Pi^n \Pi,$$

Keeping i and j fixed in (6.3.7), let $n \to \infty$. We obtain

$$w_j = w_1 p_{1j} + w_2 p_{2j} + \cdots + w_r p_{rj},$$

or in matrix form, $W = W\Pi$. Finally, if the initial distribution is $W^{(0)} = W$, then

$$\begin{aligned}
W^{(1)} &= W^{(0)}\Pi = W\Pi = W \\
W^{(2)} &= W^{(1)}\Pi = W\Pi = W \\
&\;\vdots \\
W^{(n)} &= W^{(n-1)}\Pi = W\Pi = W \quad \text{for all } n.
\end{aligned}$$

To prove (c), we note that if $Z\Pi = Z$, then

$$Z\Pi^n = Z\Pi\Pi^{n-1} = Z\Pi^{n-1} = \cdots = Z\Pi = Z.$$

Equivalently

$$z_1 p_{1j}^{(n)} + z_2 p_{2j}^{(n)} + \cdots + z_r p_{rj}^{(n)} = z_j \qquad (j = 1, \ldots, r). \tag{6.3.8}$$

Let $n \to \infty$ in (6.3.8), to obtain

$$z_1 w_j + z_2 w_j + \cdots + z_r w_j = z_j \qquad (j = 1, \ldots, r). \tag{6.3.9}$$

Since $\sum_{i=1}^{r} z_i = 1$, (6.3.9) yields $w_j = z_j$ for all j, completing the proof. (Note that the assumption that the z_i are ≥ 0 is not used.) Theorem 6.3.1c is very useful in the actual calculation of $w_1, \ldots, w_r$.

We remark that the condition $\lim_{n\to\infty} p_{ij}^{(n)} = w_j$ for $i, j = 1, 2, \ldots, r$ is equivalent to the statement that for each j, column j of Π^n approaches a vector all of whose components are w_j, that is,

$$\begin{bmatrix} p_{1j}^{(n)} \\ p_{2j}^{(n)} \\ \cdot \\ \cdot \\ \cdot \\ p_{rj}^{(n)} \end{bmatrix} \to \begin{bmatrix} w_j \\ w_j \\ \cdot \\ \cdot \\ \cdot \\ w_j \end{bmatrix};$$

alternately the matrix Π^n approaches a matrix with identical rows, that is,

$$\Pi^n \to \begin{bmatrix} w_1 & \cdots & w_r \\ & \cdot & \\ & \cdot & \\ & \cdot & \\ w_1 & \cdots & w_r \end{bmatrix}.$$

We now return to the problem of finding conditions under which steady state probabilities exist for a finite Markov chain. We have the following general result.

Theorem 6.3.2. Let Π be an r by r transition matrix associated with a finite Markov chain.

Then steady state probabilities exist if and only if there is a positive integer N such that Π^N has a positive column, that is, a column all of whose elements are >0.

The above condition is equivalent to the existence of a state s_j and a positive integer N such that starting from any initial state we can reach s_j in N steps. An immediate corollary of Theorem 6.3.2 is the result that if Π has no zeros, that is, every state may be reached in one step from any other state, then steady state probabilities exist; more generally, if for some n, Π^n has no zeros, then steady state probabilities exist. Before proving Theorem 6.3.2 let us take an example. Consider a chain with transition matrix

$$\Pi = \begin{bmatrix} 0 & 0 & 1 \\ \frac{1}{2} & \frac{1}{3} & \frac{1}{6} \\ \frac{1}{2} & \frac{1}{2} & 0 \end{bmatrix}.$$

To determine whether or not the condition of Theorem 6.3.2 is satisfied, we need not worry about the exact values of the transition probabilities

p_{ij}; we need only know which probabilities are strictly positive. Indicating positive elements by x and zero elements by 0 we compute

$$\Pi^2 = \begin{bmatrix} 0 & 0 & x \\ x & x & x \\ x & x & 0 \end{bmatrix} \cdot \begin{bmatrix} 0 & 0 & x \\ x & x & x \\ x & x & 0 \end{bmatrix} = \begin{bmatrix} x & x & 0 \\ x & x & x \\ x & x & x \end{bmatrix},$$

which has a positive column. (In fact, Π^3 has all elements >0.)

To find the steady state probabilities $w_1, \ldots, w_r$ we use the fact that $W = [w_1, \ldots, w_r]$ is the unique stationary distribution for the chain (Theorem 6.3.1c). The equations $W\Pi = W$ become

$$\begin{aligned} \tfrac{1}{2}w_2 + \tfrac{1}{2}w_3 &= w_1 \\ \tfrac{1}{3}w_2 + \tfrac{1}{2}w_3 &= w_2 \\ w_1 + \tfrac{1}{6}w_2 \qquad &= w_3. \end{aligned} \tag{6.3.10}$$

From (6.3.10) we obtain $w_3 = \frac{4}{3}w_2$ and $w_1 = \frac{7}{6}w_2$. Since $w_1 + w_2 + w_3$ must be 1 we have

$$w_1 = \tfrac{1}{3}, \qquad w_2 = \tfrac{2}{7}, \qquad w_3 = \tfrac{8}{21}.$$

We remark that if Π^N has a positive column, say column j, then column j of Π^n is positive for all $n \geq N$. (Consider $\Pi^{N+1} = \Pi\,\Pi^N$ and proceed inductively.)

Proof of Theorem 6.3.2. First suppose that Π^N has a positive column, say column j_0. If we write the equations corresponding to the matrix relation $\Pi^{n+N} = \Pi^N\Pi^n$ we obtain

$$p_{ij}^{(n+N)} = \sum_{k=1}^{r} p_{ik}^{(N)} p_{kj}^{(n)}; \quad i, j = 1, \ldots, r; \quad n = 0, 1, 2, \ldots. \tag{6.3.11}$$

Let i, m, j be any integers $\in \{1, 2, \ldots, r\}$. From (6.3.11) we have

$$p_{ij}^{(n+N)} - p_{mj}^{(n+N)} = \sum_{k=1}^{r} [p_{ik}^{(N)} - p_{mk}^{(N)}]\, p_{kj}^{(n)}. \tag{6.3.12}$$

Define sets S_1 and S_2 by

$$\begin{aligned} S_1 &= \{k\colon p_{ik}^{(N)} - p_{mk}^{(N)} \geq 0\} \\ S_2 &= \{k\colon p_{ik}^{(N)} - p_{mk}^{(N)} < 0\}. \end{aligned}$$

We may then write (6.3.12) as

$$p_{ij}^{(n+N)} - p_{mj}^{(n+N)} = \sum_{k \in S_1} [p_{ik}^{(N)} - p_{mk}^{(N)}]\, p_{kj}^{(n)} + \sum_{k \in S_2} [p_{ik}^{(N)} - p_{mk}^{(N)}]\, p_{kj}^{(n)}. \tag{6.3.13}$$

Now let

$$M_j^{(n)} = \max_{1 \le i \le r} p_{ij}^{(n)} = \text{the largest element in column } j \text{ of } \Pi^n$$

$$m_j^{(n)} = \min_{1 \le i \le r} p_{ij}^{(n)} = \text{the smallest element in column } j \text{ of } \Pi^n.$$

Using the facts that $m_j^{(n)} \le p_{kj}^{(n)} \le M_j^{(n)}$ for all k and that $p_{ik}^{(N)} - p_{mk}^{(N)}$ is ≥ 0 for $k \in S_1$ and <0 for $k \in S_2$ we obtain, from (6.3.13),

$$p_{ij}^{(n+N)} - p_{mj}^{(n+N)} \le M_j^{(n)} \sum_{k \in S_1} |p_{ik}^{(N)} - p_{mk}^{(N)}| - m_j^{(n)} \sum_{k \in S_2} |p_{ik}^{(N)} - p_{mk}^{(N)}|. \quad (6.3.14)$$

Let $q_k = |p_{ik}^{(N)} - p_{mk}^{(N)}|$; note that

$$\sum_{k \in S_1} q_k - \sum_{k \in S_2} q_k = \sum_{k=1}^{r} (p_{ik}^{(N)} - p_{mk}^{(N)}) = 1 - 1 = 0.$$

Hence we may write (6.3.14) as

$$p_{ij}^{(n+N)} - p_{mj}^{(n+N)} \le (M_j^{(n)} - m_j^{(n)}) \sum_{k \in S_1} q_k = (M_j^{(n)} - m_j^{(n)}) \sum_{k \in S_2} q_k. \quad (6.3.15)$$

Now since column j_0 of Π^N is positive, there is some $\varepsilon > 0$ for which $p_{sj_0}^{(N)} \ge \varepsilon > 0$ for $s = 1, 2, \ldots, r$ (we may take $\varepsilon = m_{j_0}^{(N)}$). If j_0 happens to belong to S_1, then

$$\sum_{k \in S_1} q_k = \sum_{k \in S_1} (p_{ik}^{(N)} - p_{mk}^{(N)})$$

$$\le \left(\sum_{k \in S_1} p_{ik}^{(N)} \right) - p_{mj_0}^{(N)} \le 1 - \varepsilon.$$

On the other hand, if $j_0 \in S_2$ then

$$\sum_{k \in S_2} q_k = \sum_{k \in S_2} (p_{mk}^{(N)} - p_{ik}^{(N)}) \le 1 - \varepsilon.$$

Thus in either case we have, from (6.3.15),

$$p_{ij}^{(n+N)} - p_{mj}^{(n+N)} \le (M_j^{(n)} - m_j^{(n)})(1 - \varepsilon). \quad (6.3.16)$$

Since (6.3.16) holds for arbitrary i and m we have

$$M_j^{(n+N)} - m_j^{(n+N)} \le (M_j^{(n)} - m_j^{(n)})(1 - \varepsilon).$$

Let $n = (l - 1)N$ to obtain

$$M_j^{(lN)} - m_j^{(lN)} \le (M_j^{([l-1]N)} - m_j^{([l-1]N)})(1 - \varepsilon). \quad (6.3.17)$$

Iterating (6.3.17) we find

$$M_j^{(lN)} - m_j^{(lN)} \le (M_j^{(N)} - m_j^{(N)})(1 - \varepsilon)^{l-1} \to 0 \quad \text{as} \quad l \to \infty. \quad (6.3.18)$$

Thus

$$\lim_{l \to \infty} [M_j^{(lN)} - m_j^{(lN)}] = 0. \quad (6.3.19)$$

Now we note that the sequences $M_j^{(n)}$ and $m_j^{(n)}$, $n = 1, 2, \ldots$ are *monotone*. To see this, suppose that $p_{i_0j}^{(n)}$ is the largest element in column j of Π^n, that is, $p_{i_0j}^{(n)} = M_j^{(n)}$. Then, using the fact that $\Pi^n = \Pi\Pi^{n-1}$ we have

$$M_j^{(n)} = p_{i_0j}^{(n)} = \sum_{k=1}^{r} p_{i_0k} p_{kj}^{(n-1)} \leq M_j^{(n-1)} \sum_{k=1}^{r} p_{i_0k} = M_j^{(n-1)}$$

so that $M_j^{(n)}$, $n = 1, 2, \ldots$, is a nonincreasing sequence.

Similarly $m_j^{(n)}$, $n = 1, 2, \ldots$, is a nondecreasing sequence. Hence $M_j^{(n)} - m_j^{(n)}$, $n = 1, 2, \ldots$, is a nonincreasing sequence; since the $M_j^{(n)}$ and $m_j^{(n)}$ are probabilities, $0 \leq m_j^{(n)} \leq M_j^{(n)} \leq 1$. But every monotone bounded sequence converges so that

$$\lim_{n\to\infty} M_j^{(n)}, \qquad \lim_{n\to\infty} m_j^{(n)}, \qquad \lim_{n\to\infty} [M_j^{(n)} - m_j^{(n)}]$$

all exist. But

$$\lim_{n\to\infty} (M_j^{(n)} - m_j^{(n)}) = \lim_{l\to\infty} (M_j^{(lN)} - m_j^{(lN)}) = 0$$

by (6.3.19). Hence

$$\lim_{n\to\infty} M_j^{(n)} = \lim_{n\to\infty} m_j^{(n)} = \text{(say) } w_j.$$

Since $m_j^{(n)} \leq p_{ij}^{(n)} \leq M_j^{(n)}$ for all i, $\lim_{n\to\infty} p_{ij}^{(n)} = w_j$ and steady state probabilities exist.

Conversely, suppose that steady state probabilities w_j exist. By Theorem 6.3.1a, $\sum_{j=1}^{r} w_j = 1$ so that at least one w_j, say w_{j_0}, is >0. Now

$$\lim_{n\to\infty} p_{ij_0}^{(n)} = w_{j_0} \qquad (i = 1, 2, \ldots, r).$$

Hence for each i there is an integer n_i such that $p_{ij_0}^{(n)} > 0$ for $n \geq n_i$. If $N = \max_{1\leq i\leq r} n_i$ then $n \geq N$ implies $p_{ij_0}^{(n)} > 0$ for all i; hence column j_0 of Π^N is >0, and the proof is complete.

Notice that if column j_0 of Π^N is positive then $m_{j_0}^{(N)} > 0$, hence $m_{j_0}^{(n)} \geq m_{j_0}^{(N)} > 0$ for $n \geq N$; thus $w_{j_0} = \lim_{n\to\infty} m_{j_0}^{(n)} > 0$. In particular, we have the following corollary to Theorem 6.3.2.

COROLLARY 6.3.2.1. If for some N, Π^N has all elements positive, then all steady state probabilities are positive.

As an example of a chain with states having a steady state probability of zero, consider the chain characterized by the transition matrix

$$\Pi = \begin{bmatrix} \frac{1}{2} & \frac{1}{2} & 0 \\ 0 & \frac{1}{3} & \frac{2}{3} \\ 0 & \frac{1}{2} & \frac{1}{2} \end{bmatrix} \qquad (6.3.20)$$

To compute w_1, w_2, w_3 we write the equations $W\Pi = W$:

$$\begin{aligned} \tfrac{1}{2}w_1 \qquad\qquad\quad &= w_1 \\ \tfrac{1}{2}w_1 + \tfrac{1}{3}w_2 + \tfrac{1}{2}w_3 &= w_2 \\ \tfrac{2}{3}w_2 + \tfrac{1}{2}w_3 &= w_3. \\ (w_1 + w_2 + w_3 &= 1) \end{aligned} \tag{6.3.21}$$

The solution to (6.3.21) is $w_1 = 0$, $w_2 = \frac{3}{7}$, $w_3 = \frac{4}{7}$. The significance of the zero value for w_1 is that s_1 is a "transient state"; the chain must eventually reach a point where a return to s_1 is impossible. We shall have more to say about transient states at the end of this section.

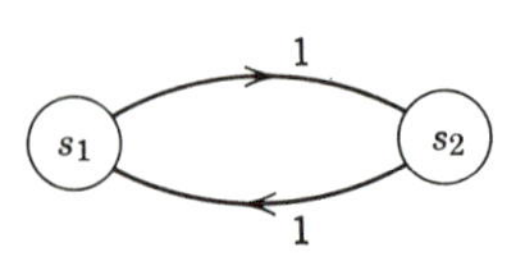

Fig. 6.3.2. A Markov chain without steady state probabilities.

A question arises as to whether the criterion of Theorem 6.3.2 is *effective*. Suppose that we wish to determine whether or not a given transition matrix has steady state probabilities. If the matrix satisfies the criterion of the theorem, then by computing the powers of the matrix we will be able to verify this fact in a finite number of steps. However, if the criterion is not satisfied, then conceivably we shall never be able to come to a decision, since the fact that none of the first n powers of the matrix have a positive column does not tell us anything about the $(n + 1)$st power. It would be desirable to have a theorem which states that if the kth power of a transition matrix Π does not have a positive column, then none of the matrices Π^n, $n \geq k$, has a positive column and hence steady state probabilities do not exist. In fact, we may take $k = 2^{r^2}$, where r is the number of states of the chain associated with Π. This will be a special case of a theorem to be proved in Chapter 7 (Theorem 7.2.2).

It is possible for a Markov chain to have a unique stationary distribution without having steady state probabilities. For example, consider the chain of Fig. 6.3.2. Since $p_{ij}^{(n)}$ oscillates back and forth from 0 to 1, steady state probabilities do not exist. There is however, a unique stationary distribution for the chain, namely the distribution that assigns probability 1/2 to each of the states s_1 and s_2.

We now try to identify the class of Markov chains which have unique stationary distributions. Consider a finite Markov chain with states $s_1, s_2, \ldots, s_r$. A set of states B is said to be *essential* if every state in B is reachable† (possibly in more than one step) from every other state in B, and it is not possible to reach a state outside of B from a state in B. It

† s_j is said to be reachable from s_i if there is a sequence of states $s_{i(1)}s_{i(2)} \cdots s_{i(t)}$, with $s_{i(1)} = s_i$, $s_{i(t)} = s_j$, such that the transition probability $p_{s_{i(n)}s_{i(n+1)}}$ is > 0 for $n = 1, 2, \ldots, t-1$.

is not hard to verify that any finite Markov chain has at least one essential set; see Problem 6.10. A chain is said to be *indecomposable* if it has exactly one essential set of states. For example, the chain characterized by the matrix (6.3.20) is indecomposable, since the states s_2 and s_3 form the unique essential class. In fact, *a chain with steady state probabilities is always indecomposable.* For suppose that a chain has at least two essential sets B_1 and B_2 (see Fig. 6.3.3).

Suppose that B_2 consists of the states $s_1, \ldots, s_m$. If the process starts in B_2 it must remain forever in B_2, and therefore $\sum_{k=1}^{m} p_{1k}^{(n)} = 1$ for all n. Allowing n to approach infinity, we conclude that $\sum_{k=1}^{m} w_k = 1$, and

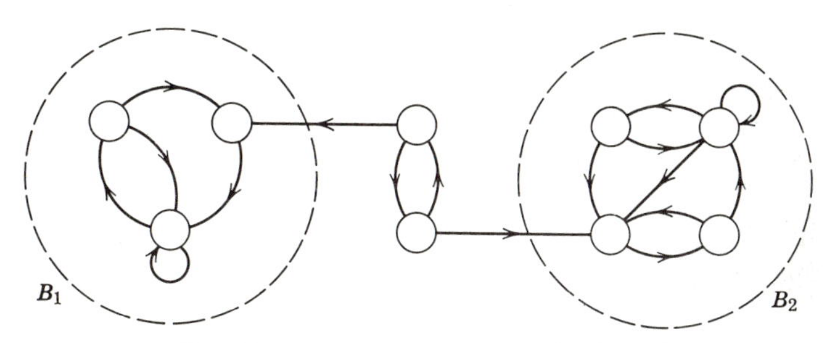

Fig. 6.3.3. A decomposable Markov chain.

consequently at least one state $s_j \in B_2$ must have a positive steady state probability w_j. But if we select any state $s_i \in B_1$, we have $p_{ij}^{(n)} = 0$ for all n; hence $w_j = 0$, a contradiction.

However, an indecomposable chain need not have steady state probabilities, as the example of Fig. 6.3.2 shows.

We now prove that an indecomposable chain always has a unique stationary distribution.

Theorem 6.3.3. Let an indecomposable Markov chain have states $s_1, \ldots, s_r$ and transition matrix Π. Then the chain has a unique stationary distribution; that is, there is a unique vector $W = [w_1 \quad w_2 \quad \cdots \quad w_r]$ such that all $w_i \geq 0$, $\sum_{i=1}^{r} w_i = 1$, and $W\Pi = W$.

Proof. Let B be the essential class of the chain. Since it is not possible to move from a state in B to a state outside of B, the submatrix of the transition matrix corresponding to the states of B is still a Markov matrix, that is, the states of B form a Markov chain. If we can find a stationary distribution for the reduced chain consisting of the states of B, we can form a stationary distribution for the original chain simply by assigning probability zero to the states that do not belong to B. Thus for the purpose of establishing the existence of a stationary distribution, we may assume without loss of generality that B is the entire set of states. We break the proof into several parts.

a. It is possible to move from any state s_i to any state s_j ($s_i = s_j$ is allowed) in at most r steps.

For if we have a sequence of states $s_{i(1)}s_{i(2)} \cdots s_{i(t)}$ where $s_{i(1)} = s_i$, $s_{i(t)} = s_j$, $s_{i(n+1)}$ is reachable in one step from $s_{i(n)}$ ($n = 1, 2, \ldots, t-1$), and $t > r$, then there is at least one state $s_{i(m)}$ ($1 < m < t$) which appears two or more times in the sequence. Thus within the sequence we have a subsequence $s_{i(m)}s_a s_b \cdots s_{i(m)}$. We may drop the states $s_a s_b \cdots s_{i(m)}$ to form a shorter path from s_i to s_j. The process may be continued until a path of length at most r results.

b. If $p_{ii} > 0$ for all i then all elements of Π^r are > 0.

For given any states s_i and s_j, we know by (a) that s_j is reachable from s_i in at most r steps. Thus $p_{ij}^{(n)} > 0$ for some $n \leq r$. Then it is possible to reach s_j from s_i in exactly r steps by moving from s_i to itself $r - n$ times, and then moving from s_i to s_j in n steps. Therefore $p_{ij}^{(r)} > 0$ for all i and j.

Now we define a matrix Π' by $\Pi' = \frac{1}{2}(\Pi + I)$, where I is an identity matrix of order r. The row sums of Π' are unity and the elements of Π' are nonnegative, so that Π' is a Markov matrix. Since $p_{ij} > 0$ implies $p_{ij}' > 0$, every state in the chain defined by Π' is reachable from every other state, so that the chain corresponding to Π' is indecomposable, with the essential class consisting of all the states. Since $p_{ii}' > 0$ for all i, it follows from (b) that all elements of $(\Pi')^r$ are positive. By Theorems 6.3.2 and 6.3.1, Π' has a stationary distribution, that is, there is a vector W whose components are nonnegative and add to unity, such that $W\Pi' = W$. But $W\Pi' = \frac{1}{2}W(\Pi + I) = \frac{1}{2}W\Pi + \frac{1}{2}W$; hence $W\Pi = W$ and W also forms a stationary distribution for Π. Thus the existence of a stationary distribution is established.

To show that the stationary distribution is unique, we proceed as follows. First assume that B does not consist of all the states in the chain. It is possible to enter the class B from any state outside of B, for otherwise the chain would contain another essential set. (If B cannot be reached from the state $s_t \notin B$, let R be the set of states which are reachable from s_t. If $s_a \in R$ and s_b is reachable from s_a, then s_b must be reachable from s_t, hence $s_b \in R$. Thus the states of R form a Markov chain, which must have at least one essential set, necessarily disjoint from B.) Thus there is a positive integer k such that $P\{Z_k \in B \mid Z_0 = s_i\} > 0$ for all $s_i \notin B$. (Of course $P\{Z_k \in B \mid Z_0 = s_i\} = 1$ for $s_i \in B$ since it is impossible to leave B.)

Let $\varepsilon = \max_{1 \le i \le r} P\{Z_k \notin B \mid Z_0 = s_i\} < 1$. Then for any state s_i,

$$\begin{aligned} P\{Z_{2k} \notin B \mid Z_0 = s_i\} &= P\{Z_k \notin B \mid Z_0 = s_i\}P\{Z_{2k} \notin B \mid Z_k \notin B, Z_0 = s_i\} \\ &\quad + P\{Z_k \in B \mid Z_0 = s_i\}P\{Z_{2k} \notin B \mid Z_k \in B, Z_0 = s_i\} \\ &= P\{Z_k \notin B \mid Z_0 = s_i\}P\{Z_{2k} \notin B \mid Z_k \notin B, Z_0 = s_i\} \\ &\le \varepsilon^2. \end{aligned}$$

By iteration, we obtain $P\{Z_{mk} \notin B \mid Z_0 = s_i\} \le \varepsilon^m \to 0$ as $m \to \infty$. Now given any positive integer n, let m be the largest integer such that $mk \le n$. Since it is impossible to leave B, $Z_n \notin B$ implies $Z_{mk} \notin B$; consequently

$$P\{Z_n \notin B \mid Z_0 = s_i\} \le P\{Z_{mk} \notin B \mid Z_0 = s_i\} \le \epsilon^m.$$

Since $m \to \infty$ as $n \to \infty$, $\lim_{n\to\infty} P\{Z_n \notin B \mid Z_0 = s_i\} = 0$. In particular if $s_j \notin B$, $\lim_{n\to\infty} p_{ij}^{(n)} = 0$ for all i. Thus all states outside of B have a steady state probability of zero. It follows that any stationary distribution must assign probability zero to these states. (If $s_j \notin B$, let $n \to \infty$ in the jth equation of $W\Pi^n = W$ to obtain $w_j = 0$.) Therefore we may ignore the states not belonging to B, in other words we may again assume without loss of generality that B consists of all states in the chain.

We now form the matrix Π' as in the proof of the first part of the theorem. We claim that a vector forms a stationary distribution for Π' if and only if it forms a stationary distribution for Π. We have already seen that $W\Pi' = W$ implies $W\Pi = W$. Conversely, if $W\Pi = W$ then $W\Pi' = \frac{1}{2}W(\Pi + I) = \frac{1}{2}W\Pi + \frac{1}{2}W = \frac{1}{2}W + \frac{1}{2}W = W$. But Π' has steady state probabilities, and therefore the stationary distribution for Π' is unique by Theorem 6.3.1. Thus the stationary distribution for Π is unique. (In addition, if the essential set consists of all the states of the chain, the fact that all elements of $(\Pi')^r$ are positive implies by Corollary 6.3.2.1 that all components of the stationary distribution are positive.)

The above discussion shows that the states in an indecomposable chain can be divided into two disjoint classes; a (never empty) class B of *essential states* and a (possibly empty) class T of *transient states*. Within the class of essential states, any state may be reached from any other state; from each state of T it is possible to reach any state of B, but it is not necessarily the case that every state of T is reachable from every other state of T. As we have seen above, the probability that the chain will be in a state of T after n steps approaches zero as n approaches infinity, hence the probability is zero that the chain will remain forever in the set T. If the chain has steady state probabilities, then it turns out (see Feller, 1950, or Chung, 1960) that each state s of B is *aperiodic* in the sense that if F is the set of positive integers n such that it is possible to move from s back to s in exactly n steps, then the greatest common divisor of the members of F is 1. On the other hand, if the chain is indecomposable but does not have steady state probabilities, then each state of B is *periodic* with the same period $d > 1$, that is, the greatest common divisor of the members of F (the so-called "period" of the chain) is d. Note that the chain of Fig. 6.3.2 has period 2. In the periodic case, the chain corresponding to the matrix Π^d has d essential sets $S_1, \ldots, S_d$; the original chain moves cyclically through the sets $S_1, \ldots, S_d$; that is, we may label the indices so that if the state at time t belongs to S_k, then the state at time $t + 1$ belongs to S_{k+1} (with the indices reduced modulo d, that is, $S_{d+1} = S_1$, $S_{d+2} = S_2$, etc.). We have a weak version of steady state probabilities in the following sense. If $s_i \in S_k$ and $s_j \in S_{k+r}$, then $p_{ij}^{(nd+r)}$ approaches a limit w_j, independent of $s_i \in S_k$, as $n \to \infty$.

In general, any finite Markov chain has a stationary distribution, but if the chain is not indecomposable, the distribution cannot be unique. For if the essential sets of a chain are $B_1, \ldots, B_k$, the chain formed by considering only the states of B_i is indecomposable, and hence has a stationary distribution. If we assign probability zero to the states outside B_i, we have a stationary distribution W_i for the original chain ($i = 1, 2, \ldots, k$). Since the W_i are distinct, we have found k different stationary distributions. In fact, if $W = \sum_{i=1}^{k} a_i W_i$ where the a_i are nonnegative numbers whose sum is unity, then

$$W\Pi = \sum_{i=1}^{k} a_i(W_i \Pi) = \sum_{i=1}^{k} a_i W_i = W,$$

so that there are uncountably many stationary distributions. Thus *the class of indecomposable finite Markov chains coincides with the class of finite Markov chains having a unique stationary distribution.* We close the discussion at this point, referring to Feller (1950), Chung (1960), or Kemeny and Snell (1960) for further details.

6.4. Information sources; uncertainty of a source

In defining what we mean by a source of information, we try to capture the physical idea of a mechanism which emits a sequence of symbols in accordance with specified statistical rules. The symbols are assumed to belong to a definite finite set, the "alphabet" of the source, and the statistical properties of the source are assumed to be time-invariant.

Definition. An *information source* is a sequence of random variables $X_0, X_1, X_2, \ldots$, such that

1. Each X_i takes values in a finite set Γ, called the *alphabet* of the source, and

2. The sequence is *stationary*, that is,

$$P\{X_{i_1} = \gamma_1, \ldots, X_{i_k} = \gamma_k\} = P\{X_{i_1+h} = \gamma_1, \ldots, X_{i_k+h} = \gamma_k\}$$

for all nonnegative integers $i_1, \ldots, i_k, h$, and all $\gamma_1, \ldots, \gamma_k \in \Gamma$.

An information source may be derived from a Markov chain as follows. Given a finite Markov chain $Z_0, Z_1, \ldots$, and a function f whose domain is the set of states of the chain and whose range is a finite set Γ, assume that the initial state Z_0 is chosen in accordance with a stationary distribution $W = [w_1, \ldots, w_r]$; that is, $P\{Z_0 = s_j\} = w_j$ for all states s_j. Then the stationary sequence $X_n = f(Z_n)$, $n = 0, 1, \ldots$, is said to be the *Markov information source* corresponding to the chain $Z_0, Z_1, \ldots$, the function f, and the distribution W. (Of course if the chain is indecomposable there is only one stationary distribution, and the source is determined by the chain together with the function f. Note also that all states s_j such that $w_j = 0$ may be removed from the chain without affecting the distribution of the sequence $\{X_n\}$.) If the chain has steady state probabilities, we call the resulting information source a *regular* Markov information source. (We may also call a Markov chain with steady state probabilities a *regular chain.*) If the chain is indecomposable, the resulting information source will be called an *indecomposable* Markov source.

Physically, we might expect that if an information source is in operation for a long period of time, a steady state condition should be reached. Thus Markov sources met in practice should have steady state probabilities. However, it is convenient mathematically to allow an information source to be derived from an arbitrary Markov chain. This will avoid annoying difficulties in Section 6.5, when we approximate an arbitrary information source by a Markov source. The approximating sources need not have steady state probabilities and in fact are not necessarily indecomposable (although if the properties of the given source are "reasonable," the approximating sources will also be "reasonable"; see Problem 6.9).

We remark that a sequence of independent, identically distributed random variables $X_0, X_1, \ldots$, taking values in a finite set $S = \{s_1, \ldots, s_r\}$, becomes a regular Markov information source if we take $p_{ij} = w_j = P\{X_k = s_j\}$, $i, j = 1, 2, \ldots, r$, $k = 0, 1, \ldots$; we may take $\Gamma = S$, and $f =$ the identity function.

We now try to arrive at a meaningful notion of "uncertainty" of an information source.

We are going to define the source uncertainty as (intuitively) the uncertainty of a particular symbol produced by the source, given that we have observed all previous symbols. Formally, we have the following.

Definition. Given an information source $X_0, X_1, \ldots$, the *uncertainty* of the source, denoted by $H\{X_n, n = 0, 1, \ldots\}$ or for brevity by $H\{\mathbf{X}\}$, is defined as $\lim_{n\to\infty} H(X_n \mid X_0, X_1, \ldots, X_{n-1})$.

Note that since

$$H(X_n \mid X_0, X_1, \ldots, X_{n-1}) \leq H(X_n \mid X_1, \ldots, X_{n-1}) \qquad \text{by Theorem 1.4.5 or Problem 1.4c}$$
$$= H(X_{n-1} \mid X_0, \ldots, X_{n-2}) \qquad \text{by stationarity,}$$

the sequence $(H(X_n \mid X_0, X_1, \ldots, X_{n-1}), n = 1, 2, \ldots)$ is nonincreasing. Since the terms of the sequence are nonnegative, the limit $H\{\mathbf{X}\}$ exists. In addition,

$$\begin{aligned} H\{\mathbf{X}\} &\leq H(X_n \mid X_0, \ldots, X_{n-1}) \\ &\leq H(X_n) \qquad \text{by Theorem 1.4.5} \\ &\leq \log |\Gamma| \qquad \text{by Theorem 1.4.2,} \end{aligned}$$

where $|\Gamma|$ is the number of letters in the alphabet Γ. We remark also that if $\{X_n\}$ is a sequence of independent, identically distributed random variables, then $H\{\mathbf{X}\} = H(X_i)$ (for any i).

It is not apparent at this point exactly what properties of an information source are revealed by knowledge of its uncertainty. This matter will be clarified in Section 6.6. First we give an alternate expression for $H\{\mathbf{X}\}$.

Theorem 6.4.1. If $X_0, X_1, \ldots$, is an information source, then

$$H\{\mathbf{X}\} = \lim_{n\to\infty} \frac{H(X_0, X_1, \ldots, X_n)}{n+1}.$$

Proof. Let $g_n = H(X_n \mid X_0, X_1, \ldots, X_{n-1}), n = 1, 2, \ldots; g_0 = H(X_0)$; then by Theorem 1.4.4,

$$\begin{aligned} \frac{H(X_0, X_1, \ldots, X_n)}{n+1} &= \frac{H(X_0) + H(X_1 \mid X_0) + \cdots + H(X_n \mid X_0, X_1, \ldots, X_{n-1})}{n+1} \\ &= \frac{g_0 + g_1 + \cdots + g_n}{n+1} \end{aligned} \tag{6.4.1}$$

But by definition of $H\{\mathbf{X}\}$, $g_n \to H\{\mathbf{X}\}$, and hence

$$\left\{\frac{H(X_0, X_1, \ldots, X_n)}{n+1}, n = 0, 1, 2, \ldots\right\},$$

being the arithmetic average of a sequence converging to $H\{\mathbf{X}\}$, also converges to $H\{\mathbf{X}\}$.

We note that the sequence $h_{n+1}/(n+1) = H(X_0, \ldots, X_n)/(n+1)$ is nonincreasing. For by (6.4.1) we have $h_{n+1} = \sum_{i=0}^{n} g_i$. Thus

$$\frac{h_{n+1}}{n+1} - \frac{h_n}{n} = \frac{\sum_{i=0}^{n} g_i}{n+1} - \frac{\sum_{i=0}^{n-1} g_i}{n}$$

$$= \frac{ng_n - \sum_{i=0}^{n-1} g_i}{n(n+1)}.$$

Since the sequence $\{g_n\}$ is nonincreasing we have $g_i \geq g_n$ for $i = 0, 1, \ldots, n-1$. Hence $ng_n - \sum_{i=0}^{n-1} g_i \leq ng_n - ng_n = 0$, and the assertion is proved.

The uncertainty of an information source is difficult to compute in most cases. However, for a certain class of Markov sources the calculation may be greatly simplified.

Definition. Consider a Markov information source with set of states $S = \{s_1, \ldots, s_r\}$, alphabet Γ, associated function $f: S \to \Gamma$, and stationary distribution $W = [w_1 \quad w_2 \quad \cdots \quad w_r]$. For each state s_k, let $s_{k1}, s_{k2}, \ldots, s_{kn_k}$ be the states that can be reached in one step from s_k, that is, the states s_j such that $p_{kj} > 0$. The source is said to be *unifilar* if for each state s_k, the letters $f(s_{k1}), \ldots, f(s_{kn_k})$ are distinct.

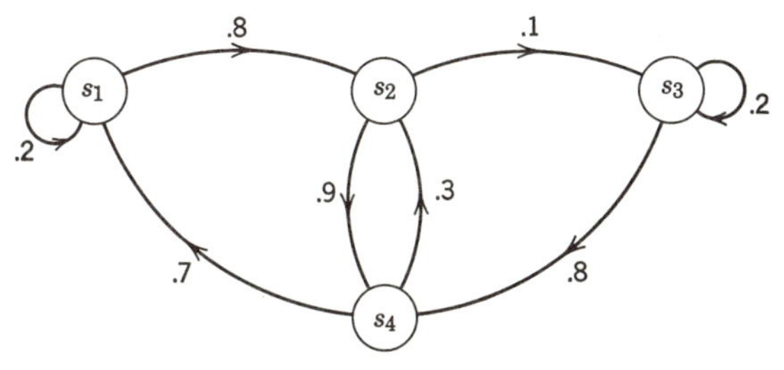

Fig. 6.4.1. (a) Unifilar Markov source. Take

$$f(s_1) = f(s_4) = A$$
$$f(s_2) = f(s_3) = B$$

(b) Nonunifilar Markov source. Take

$$f(s_1) = f(s_2) = f(s_4) = A$$
$$f(s_3) = B$$

In other words, each state reachable directly from s_k is associated with a distinct letter of the alphabet. Equivalently, the state at time t and the symbol produced at time $t + 1$ determine the state at time $t + 1$.

Examples of unifilar and nonunifilar Markov sources are shown in Fig. 6.4.1. In Fig. 6.4.1b, states s_1 and s_2 are directly reachable from s_1 but $f(s_1) = f(s_2) = A$, and therefore the source is not unifilar.

Given a unifilar Markov source, as above, let $s_{k1}, \ldots, s_{kn_k}$ be the states that are directly reachable from s_k, $k = 1, 2, \ldots, r$. We define the *uncertainty of state* s_k as $H_k = H(p_{k1}, p_{k2}, \ldots, p_{kn_k})$.

Theorem 6.4.2. The uncertainty of a unifilar Markov source is given by $H\{\mathbf{X}\} = \sum_{k=1}^{r} w_k H_k$.

Proof. The underlying Markov chain of the source is a sequence of random variables $Z_0, Z_1, \ldots$ (recall that Z_n represents the state of the process after n transitions). Now

$$\frac{H(Z_0, Z_1, \ldots, Z_n)}{n+1} = -(n+1)^{-1} \tag{6.4.2}$$
$$\times \sum_{\alpha_0, \ldots, \alpha_n \in S} P\{Z_0 = \alpha_0, \ldots, Z_n = \alpha_n\} \log P\{Z_0 = \alpha_0, \ldots, Z_n = \alpha_n\}.$$

Furthermore, we may write

$$P\{Z_0 = \alpha_0, \ldots, Z_n = \alpha_n\} = P\{Z_0 = \alpha_0\} P\{Z_1 = \alpha_1, \ldots, Z_n = \alpha_n \mid Z_0 = \alpha_0\}.$$

Now suppose that the initial state is α_0, and that the state after i transitions is α_i ($i = 1, 2, \ldots, n$); let $\gamma_i = f(\alpha_i) \in \Gamma$. The hypothesis that the source is unifilar means that no other sequences of states $\alpha_1', \ldots, \alpha_n'$ can produce the sequence of letters $\gamma_1, \ldots, \gamma_n$. In other words, *for a fixed initial state* there is a one-to-one correspondence between sequences of states and sequences of symbols from the alphabet of the source. We may therefore write (6.4.2) as

$$\begin{aligned}\frac{H(Z_0, \ldots, Z_n)}{n+1} = -(n+1)^{-1} \sum_{\substack{\gamma_1, \ldots, \gamma_n \in \Gamma \\ \alpha_0 \in S}} &(P\{Z_0 = \alpha_0\} \\ &\times P\{X_1 = \gamma_1, \ldots, X_n = \gamma_n \mid Z_0 = \alpha_0\} \\ &\times [\log P\{Z_0 = \alpha_0\} + \log P\{X_1 = \gamma_1, \ldots, X_n = \gamma_n \mid Z_0 = \alpha_0\}]) \\ &= (n+1)^{-1} [H(Z_0) + H(X_1, \ldots, X_n \mid Z_0)].\end{aligned}$$

But

$$\begin{aligned}H(X_1, \ldots, X_n \mid Z_0) &= H(Z_0, X_1, \ldots, X_n) - H(Z_0) \\ &= H(X_1, \ldots, X_n) + H(Z_0 \mid X_1, \ldots, X_n) - H(Z_0).\end{aligned}$$

Now $H(Z_0 \mid X_1, \ldots, X_n) \leq H(Z_0) \leq \log r$ and $(n+1)^{-1} \log r \to 0$ as $n \to \infty$ so that

$$\lim_{n\to\infty} \frac{H(Z_0, Z_1, \ldots, Z_n)}{n+1} = \lim_{n\to\infty} \frac{H(X_1, \ldots, X_n)}{n+1} = \lim_{n\to\infty} \frac{H(X_0, \ldots, X_{n-1})}{n+1}$$
$$= \lim_{n\to\infty} \frac{H(X_0, \ldots, X_{n-1})}{n} \cdot \frac{n}{n+1} = H\{\mathbf{X}\}$$

We conclude that $H\{\mathbf{X}\} = H\{\mathbf{Z}\}$. In this case $H\{\mathbf{Z}\}$ is easily computable. We have

$$\begin{aligned} H\{\mathbf{Z}\} &= \lim_{n\to\infty} H(Z_n \mid Z_0, Z_1, \ldots, Z_{n-1}) \\ &= H(Z_n \mid Z_{n-1}) \qquad \text{by the Markov property} \\ &= H(Z_1 \mid Z_0) = \sum_{k=1}^{r} P\{Z_0 = s_k\} H(Z_1 \mid Z_0 = s_k) \\ &= \sum_{k=1}^{r} w_k H_k. \end{aligned}$$

This completes the proof.

For the source of Fig. 6.4.1a we may compute that the steady state probabilities are $w_1 = 7/24$, $w_2 = 1/3$, $w_3 = 1/24$, $w_4 = 1/3$. The source uncertainty is therefore

$$H\{\mathbf{X}\} = \tfrac{7}{24} H(0.2, 0.8) + \tfrac{1}{3} H(0.9, 0.1) + \tfrac{1}{24} H(0.8, 0.2) + \tfrac{1}{3} H(0.7, 0.3)$$

6.5. *Order of a source; approximation of a general information source by a source of finite order*

We expect that the statistical dependence of a symbol X_n produced by an information source on the previous symbols $X_1, X_2, \ldots, X_{n-1}$ should "wear off" as $n \to \infty$. In this section we examine unifilar Markov sources in which the distribution of X_n is determined by a finite number of previous symbols $X_{n-1}, \ldots, X_{n-M}$; such sources are referred to a sources of *finite order* or *finite memory*. We then show that given an arbitrary information source we can construct in a natural way a unifilar Markov source of finite order whose uncertainty is as close as we wish to that of the original source.

Let us begin by analyzing the source shown in Fig. 6.5.1. In the diagram, if $f(s_j) = \gamma$, then all arrows entering state s_j are marked with a γ. The transition probabilities are unspecified in Fig. 6.5.1, except that we assume $p_{ij} > 0$ if states s_i and s_j are connected directly. In Table 6.5.1 we have listed all possible sequences of two and three symbols, with the corresponding final states for a given initial state. For example if $Z_{t-3} = s_2$, $X_{t-2} = B$, $X_{t-1} = A$, $X_t = A$, then $Z_t = s_1$. Blank entries correspond

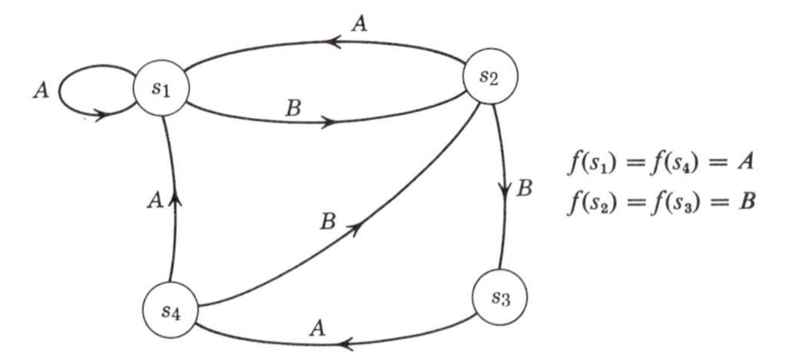

Fig. 6.5.1. Unifilar Markov information source of order three.

to impossible transitions; for example, if $Z_{t-2} = s_3$ then it is impossible that $X_{t-1} = B$, $X_t = B$. It can be seen from Table 6.5.1 that the state Z_t is determined by the three inputs X_{t-2}, X_{t-1}, X_t regardless of the initial state Z_{t-3}. For instance, if $X_{t-2} = A$, $X_{t-1} = B$, $X_t = B$, then $Z_t = s_3$ since all entries in column ABB are identical. However, the present state is not determined by the previous two inputs, since if $X_{t-1} = B$, $X_t = A$, then Z_t can be either s_1 or s_4.

Table 6.5.1. State transitions for the source of Fig. 6.5.1

	$X_{t-2}\, X_{t-1}\, X_t$							
Z_{t-3}	AAA	AAB	ABA	ABB	BAA	BAB	BBA	BBB
s_1	s_1	s_2	s_1	s_3	s_1	s_2	s_4	—
s_2	s_1	s_2	s_1	s_3	s_1	s_2	—	—
s_3	s_1	s_2	s_1	s_3	—	—	—	—
s_4	s_1	s_2	s_1	s_3	s_1	s_2	s_4	—

	$X_{t-1}\, X_t$			
Z_{t-2}	AA	AB	BA	BB
s_1	s_1	s_2	s_1	s_3
s_2	s_1	s_2	s_4	—
s_3	s_1	s_2	—	—
s_4	s_1	s_2	s_1	s_3

We define a unifilar Markov source to be of *order* M if the state Z_t is determined by the symbols $X_t, X_{t-1}, \ldots, X_{t-M+1}$, but not by the symbols $X_t, X_{t-1}, \ldots, X_{t-i}$ where $i < M - 1$. The source of Fig. 6.5.1 is therefore of order three. A unifilar Markov source is of *infinite order* if it is not of order M for any finite M. The source of Fig. 6.5.2 is of infinite order since the sequence $AAA \cdots A$ does not determine the final state; depending on the initial state, the final state could be either s_1 or s_2.

If a source is of order M, it follows that

$$P\{X_t = \gamma_t \mid X_{t-1} = \gamma_{t-1}, \ldots, X_{t-M} = \gamma_{t-M}\}$$
$$= P\{X_t = \gamma_t \mid X_{t-1} = \gamma_{t-1}, \ldots, X_{t-n} = \gamma_{t-n}\} \quad \text{for all } n \geq M.$$

Hence

$$H\{\mathbf{X}\} = H(X_t \mid X_{t-1}, \ldots, X_{t-M})$$
$$= H(X_t, X_{t-1}, \ldots, X_{t-M}) - H(X_{t-1}, \ldots, X_{t-M}).$$

A
s1
B
s3
B
s2
A
A
C
C
C
A
s4
f(s1) = f(s2) = A
f(s3) = B
f(s4) = C

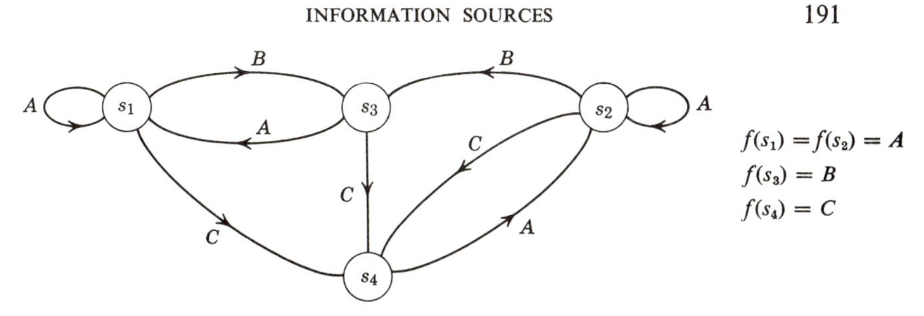

Fig. 6.5.2. Unifilar Markov source of infinite order.

Thus if a source is of order M, its uncertainty can be expressed as the difference of the uncertainty of $M + 1$ successive symbols and that of M successive symbols. In general, the uncertainty $H(X_1, \ldots, X_n)$ of n successive symbols produced by a source is sometimes referred to as the *n-gram uncertainty* of the source (For $n = 1$, the term "unigram" is used; for $n = 2$, "digram"; for $n = 3$, "trigram"). Thus the uncertainty of a source of order M is the difference of the $(M + 1)$-gram and M-gram uncertainties.

As a numerical example, we evaluate the various n-gram uncertainties for the source of Fig. 6.5.3, which may be seen to be of order 2.

$$P(A) = P\{X_t = A\} = \sum_{i=1}^{4} P\{Z_{t-1} = s_i\}P\{X_t = A \mid Z_{t-1} = s_i\}$$

$$= \tfrac{7}{24}(0.2) + \tfrac{8}{24}(0.9) + \tfrac{1}{24}(0.8) + \tfrac{8}{24}(0.7) = 5/8.$$

[Also $P(A) = P\{Z_t = s_1\} + P\{Z_t = s_4\} = w_1 + w_4$]

$$P(B) = \tfrac{7}{24}(0.8) + \tfrac{8}{24}(0.1) + \tfrac{1}{24}(0.2) + \tfrac{8}{24}(0.3) = 3/8.$$

$$P(AA) = P\{X_t = A \text{ and } X_{t+1} = A\}$$

$$= \sum_{i=1}^{4} w_i P\{X_t = A,\ X_{t+1} = A \mid Z_{t-1} = s_i\}$$

$$= \tfrac{7}{24}(0.2)(0.2) + \tfrac{8}{24}(0.9)(0.7) + \tfrac{1}{24}(0.8)(0.7) + \tfrac{8}{24}(0.7)(0.2) = 7/24.$$

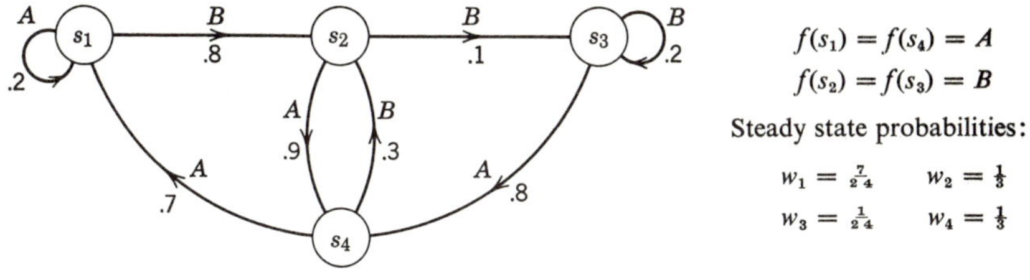

A
.2
s1
B
.8
s2
B
.1
s3
B
.2
A
.9
B
.3
A
.7
A
.8
s4
f(s1) = f(s4) = A
f(s2) = f(s3) = B
Steady state probabilities:
w1 = 7/24   w2 = 1/3
w3 = 1/24   w4 = 1/3

Fig. 6.5.3. Numerical evaluation of uncertainties.

[Also $P(AA) = P\{Z_{t+1} = s_1\} = w_1$.]

$$P(AB) = \tfrac{7}{24}(0.2)(0.8) + \tfrac{8}{24}(0.9)(0.3) + \tfrac{1}{24}(0.8)(0.3) + \tfrac{8}{24}(0.7)(0.8) = 8/24 = w_2.$$

Similarly

$$P(BA) = 8/24 = w_4, \quad \text{and} \quad P(BB) = 1/24 = w_3.$$

$$P(AAA) = P\{X_t = A,\ X_{t+1} = A,\ X_{t+2} = A\} = P\{Z_{t+1} = s_1,\ X_{t+2} = A\}.$$

$$P(ABA) = P\{X_t = A,\ X_{t+1} = B,\ X_{t+2} = A\} = P\{Z_{t+1} = s_2,\ X_{t+2} = A\},$$

and so on.

Thus

$$P(AAA) = \tfrac{7}{24}(0.2) = \frac{1.4}{24} \qquad P(BAA) = \tfrac{8}{24}(0.7) = \frac{5.6}{24}$$

$$P(AAB) = \tfrac{7}{24}(0.8) = \frac{5.6}{24} \qquad P(BAB) = \tfrac{8}{24}(0.3) = \frac{2.4}{24}$$

$$P(ABA) = \tfrac{8}{24}(0.9) = \frac{7.2}{24} \qquad P(BBA) = \tfrac{1}{24}(0.8) = \frac{0.8}{24}$$

$$P(ABB) = \tfrac{8}{24}(0.1) = \frac{0.8}{24} \qquad P(BBB) = \tfrac{1}{24}(0.2) = \frac{0.2}{24}$$

$$J_1 = \text{unigram uncertainty} = -P(A)\log P(A) - P(B)\log P(B) = -\tfrac{5}{8}\log\tfrac{5}{8} - \tfrac{3}{8}\log\tfrac{3}{8}.$$

$$\begin{aligned} J_2 &= \text{digram uncertainty} \\ &= -P(AA)\log P(AA) - P(AB)\log P(AB) \\ &\quad -P(BA)\log P(BA) - P(BB)\log P(BB) \\ &= H(7/24,\ 8/24,\ 8/24,\ 1/24). \end{aligned}$$

$$\begin{aligned} J_3 &= \text{trigram uncertainty} \\ &= -P(AAA)\log P(AAA) - P(AAB)\log P(AAB) - \cdots \\ &\quad -P(BBB)\log P(BBB) \\ &= H\Big(\underbrace{\frac{1.4}{24}, \frac{5.6}{24}}, \underbrace{\frac{7.2}{24}, \frac{0.8}{24}}, \underbrace{\frac{5.6}{24}, \frac{2.4}{24}}, \underbrace{\frac{0.8}{24}, \frac{0.2}{24}},\Big) \\ &= H\left(\frac{7}{24}, \frac{8}{24}, \frac{8}{24}, \frac{1}{24}\right) + \frac{7}{24} H\left(\frac{1.4}{1.4+5.6}, 0.8\right) + \frac{8}{24} H(0.9, 0.1) \\ &\quad + \frac{8}{24} H(0.7, 0.3) + \frac{1}{24} H(0.8, 0.2) \end{aligned}$$

(by the generalized grouping axiom; Problem 1.10).

$$\begin{aligned} H\{\mathbf{X}\} &= J_3 - J_2 \\ &= \tfrac{7}{24}H(0.2, 0.8) + \tfrac{8}{24}H(0.9, 0.1) + \tfrac{1}{24}H(0.8, 0.2) + \tfrac{8}{24}H(0.7, 0.3), \end{aligned}$$

which agrees with the expression computed at the end of Section 6.4.

To determine the order of a source, we examine the final states associated with symbol sequences of length 1, length 2, length 3, etc., until we find a number M such that for all sequences of length M, the final state is independent of the initial state; M is then the order of the source. It would be desirable to know when to stop this process and declare that the source is of infinite order. A criterion is provided by the following theorem.

Theorem 6.5.1. If a unifilar Markov information source with r states is of finite order M, then $M \leq \frac{1}{2}r(r-1)$.

Proof. Let $Q = (\alpha_1, \alpha_2, \ldots, \alpha_{\frac{1}{2}r(r-1)})$ be any sequence of $\frac{1}{2}r(r-1)$ symbols. We must show that the final state associated with Q is the same for all initial states. If u_0 and v_0 are two distinct initial states, the sequence

$$\begin{array}{l} \quad\ \ \alpha_1 \quad\ \alpha_2 \qquad \alpha_{\frac{1}{2}r(r-1)} \\ u_0 \to u_1 \to u_2 \cdots \to u_{\frac{1}{2}r(r-1)} \\ v_0 \to v_1 \to v_2 \cdots \to v_{\frac{1}{2}r(r-1)} \end{array} \qquad \begin{array}{l} \quad\ A \quad\ B \quad\ A \\ s_1 \to s_3 \to s_3 \to s_1 \\ s_2 \to s_4 \to s_7 \to s_2 \end{array}$$

a b

Fig. 6.5.4 Proof of Theorem 6.5.1

Q takes u_0 through a sequence of states $u_1, u_2, \ldots, u_{\frac{1}{2}r(r-1)}$, and v_0 through a sequence $v_1, v_2, \ldots, v_{\frac{1}{2}r(r-1)}$ (see Fig. 6.5.4a). If $u_i = v_i$ for some i $[0 < i < \frac{1}{2}r(r-1)]$ then $u_j = v_j$ for $i \leq j \leq \frac{1}{2}r(r-1)$. In particular, $u_{\frac{1}{2}r(r-1)} = v_{\frac{1}{2}r(r-1)}$ and there is nothing further to prove. Thus let us assume that $u_i \neq v_i$ $[0 \leq i < \frac{1}{2}r(r-1)]$. If for a particular i and j $[0 \leq i < j \leq \frac{1}{2}r(r-1)]$ we have $u_i = u_j$ and $v_i = v_j$, then the source must be of infinite order, which contradicts the hypothesis. To see this, suppose for example that the sequence ABA takes state s_1 into itself, and also takes s_2 into itself (see Fig. 6.5.4b). Then we cannot determine the final state at the end of the sequence $ABAABAABA \cdots$. Similarly we cannot have $u_i = v_j$ and $v_i = u_j$ for $0 \leq i < j \leq \frac{1}{2}r(r-1)$. (If the sequence ABA takes s_1 into s_2 and s_2 into s_1, then $ABAABA$ takes s_1 into s_1 and s_2 into s_2, which implies that the source must be of infinite order.)

Hence if $0 \leq i < j \leq \frac{1}{2}r(r-1)$, the (unordered) sets $\{u_i, v_i\}$ and $\{u_j, v_j\}$ are unequal, and furthermore $u_i \neq v_i$ for $0 \leq i < \frac{1}{2}r(r-1)$. However, there are only $\binom{r}{2} = \frac{1}{2}r(r-1)$ possible ways of choosing two distinct states out of r. Consequently (u_0, v_0), (u_1, v_1), $(u_2, v_2), \ldots$, $(u_{\frac{1}{2}r(r-1)-1}, v_{\frac{1}{2}r(r-1)-1})$ exhaust all possible pairs of distinct states. Thus $u_{\frac{1}{2}r(r-1)} = v_{\frac{1}{2}r(r-1)}$, and the theorem is proved.

Now given an arbitrary information source $\mathcal{S} = \{X_n, n = 0, 1, \ldots\}$, we construct a sequence of finite-order unifilar Markov sources $\mathcal{S}_n$ as follows.

Let the states of $\mathcal{S}_n$ correspond to the n-tuples $(\alpha_1, \ldots, \alpha_n)$ of symbols from the alphabet Γ of $\mathcal{S}$, such that $P\{X_1 = \alpha_1, \ldots, X_n = \alpha_n\} > 0$. Let the alphabet of $\mathcal{S}_n$ be Γ for each n. For each state $(\alpha_1, \ldots, \alpha_n)$ of $\mathcal{S}_n$ we may in one step reach states of the form $(\alpha_2, \ldots, \alpha_n, \alpha_{n+1})$. The transition probability from $(\alpha_1, \ldots, \alpha_n)$ to $(\alpha_2, \ldots, \alpha_n, \alpha_{n+1})$ is given by

$$p_{(\alpha_1,\ldots,\alpha_n)(\alpha_2,\ldots,\alpha_n,\alpha_{n+1})} = P\{X_{n+1} = \alpha_{n+1} \mid X_1 = \alpha_1, \ldots, X_n = \alpha_n\}.$$

We define the function f associated with $\mathcal{S}_n$ by $f(\alpha_1, \ldots, \alpha_n) = \alpha_n$; if the state of $\mathcal{S}_n$ at a given time is $Z_t = (\alpha_1, \ldots, \alpha_n)$, the corresponding symbol produced by $\mathcal{S}_n$ is $Y_t = f(Z_t) = \alpha_n$. Thus moving from state $(\alpha_1, \ldots, \alpha_n)$ to $(\alpha_2, \ldots, \alpha_n, \alpha_{n+1})$ corresponds to moving from a condition in which the previous n symbols were $\alpha_1, \ldots, \alpha_n$ to a condition in which the previous n symbols are $(\alpha_2, \ldots, \alpha_n, \alpha_{n+1})$.

We may verify that a stationary distribution is formed by assigning to the state $(\alpha_1, \ldots, \alpha_n)$ the probability $P\{X_1 = \alpha_1, \ldots, X_n = \alpha_n\}$ (see Problem 6.4). With this distribution, $\mathcal{S}_n$ becomes a Markov information source. Now if we know the state Z_t of $\mathcal{S}_n$ at a given time t, the symbol Y_{t+1} determines the last coordinate of the state Z_{t+1}; since the first $n - 1$ coordinates of Z_{t+1} are the same as the last $n - 1$ coordinates of Z_t, we conclude that Z_t and Y_{t+1} determine Z_{t+1} so that the source is unifilar.

Thus $\mathcal{S}_n$ is a unifilar Markov source. By construction, $\mathcal{S}_n$ has order at most n. (The order of $\mathcal{S}_n$ may be less than n; see Problem 6.11.) To find the uncertainty of $\mathcal{S}_n$, we observe that if $\{Y_n, n = 1, 2, \ldots\}$ denotes the sequence of symbols produced by $\mathcal{S}_n$, then

$$\begin{aligned}P\{Y_{n+1} = \alpha_{n+1} \mid Y_1 = \alpha_1, \ldots, Y_n = \alpha_n\} \\ &= P\{Y_{n+1} = \alpha_{n+1} \mid Z_n = (\alpha_1, \ldots, \alpha_n)\} \\ &= P\{Z_{n+1} = (\alpha_2, \ldots, \alpha_{n+1}) \mid Z_n = (\alpha_1, \ldots, \alpha_n)\} \\ &= P\{X_{n+1} = \alpha_{n+1} \mid X_1 = \alpha_1, \ldots, X_n = \alpha_n\}.\end{aligned}$$

Thus the conditional distribution of a letter, given n preceding letters, is the same for $\mathcal{S}_n$ as for $\mathcal{S}$. It follows that the uncertainty of $\mathcal{S}_n$ is

$$H_n\{\mathbf{X}\} = H(Y_{n+1} \mid Y_1, \ldots, Y_n) = H(X_{n+1} \mid X_1, \ldots, X_n).$$

Thus

$$\lim_{n\to\infty} H_n\{\mathbf{X}\} = \lim_{n\to\infty} H(X_{n+1} \mid X_1, \ldots, X_n) = H\{\mathbf{X}\},$$

so that the sources $\mathcal{S}_n$ have uncertainties which form a nonincreasing sequence converging to $H\{\mathbf{X}\}$. $\mathcal{S}_n$ is called the *nth-order approximation* to $\mathcal{S}$. If the distribution of the symbol X_t depends on a finite number M of previous symbols, then the approximation $\mathcal{S}_M$ coincides with the original source.

As an example, let us compute the first-order approximation to the source of Fig. 6.5.3. (In this case the second-order approximation coincides with the source itself.) We have (p. 191)

$$P\{X_t = A \mid X_{t-1} = A\} = \frac{P\{X_{t-1} = A,\ X_t = A\}}{P\{X_{t-1} = A\}} = \frac{7/24}{5/8} = \frac{7}{15}$$

$$P\{X_t = A \mid X_{t-1} = B\} = \frac{P\{X_{t-1} = B,\ X_t = A\}}{P\{X_{t-1} = B\}} = \frac{1/3}{3/8} = \frac{8}{9}$$

The first-order approximation is shown in Fig. 6.5.5.

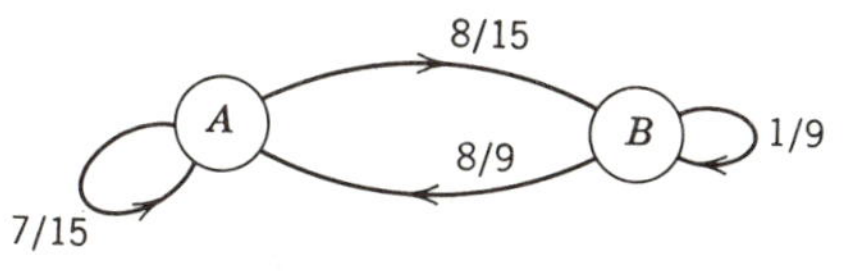

Fig. 6.5.5. First-order approximation to the source of Fig. 6.5.3.

The uncertainty of the first-order approximation is

$$\begin{aligned} H_1\{\mathbf{X}\} &= H(X_t \mid X_{t-1}) = H(X_{t-1}, X_t) - H(X_{t-1}) \\ &= J_2 - J_1 = \tfrac{5}{8}H(\tfrac{7}{15}, \tfrac{8}{15}) + \tfrac{3}{8}H(\tfrac{8}{9}, \tfrac{1}{9}). \end{aligned}$$

6.6. *The asymptotic equipartition property*

In this section we try to give a basic interpretation to the uncertainty of an information source. We shall try to make precise the notion of a "meaningful" sequence produced by a source, and we shall show that under certain conditions there are asymptotically 2^{Hn} such sequences, where H is the uncertainty of the source. First consider a sequence of independent, identically distributed random variables $X_1, X_2, \ldots$, taking values in a finite set Γ. Define another sequence of random variables $V_1, V_2, \ldots$, as follows: If $X_1 = \alpha_1, \ldots, X_n = \alpha_n$, let

$$V_n = V_n(X_1, \ldots, X_n) = -n^{-1} \log P\{X_1 = \alpha_1, \ldots, X_n = \alpha_n\}.$$

It follows from the independence of the X_i that

$$V_n(X_1, \ldots, X_n) = n^{-1} \sum_{i=1}^{n} U(X_i),$$

where $U(X_i)$ is a random variable that assumes the value $-\log P\{X_i = \alpha_i\}$ when $X_i = \alpha_i$. Thus $V_n(X_1, \ldots, X_n)$ is the arithmetic average of n independent, identically distributed random variables $U(X_i)$, with

$$E[U(X_i)] = -\sum_{\alpha} P\{X_i = \alpha\} \log P\{X_i = \alpha\} = H(X_i) = H\{\mathbf{X}\}.$$

By the weak law of large numbers,

$$V_n(X_1, \ldots, X_n) \text{ converges in probability to } H = H\{\mathbf{X}\}.$$

In other words, given $\varepsilon > 0$, $\delta > 0$, then for n sufficiently large,

$$P\{|V_n(X_1, \ldots, X_n) - H| < \delta\} \geq 1 - \varepsilon.$$

For a fixed ε and δ, let us divide the sequences $(\alpha_1, \alpha_2, \ldots, \alpha_n)$, $\alpha_i \in \Gamma$, into two sets:

$$S_1 = \{(\alpha_1, \ldots, \alpha_n)\colon |-n^{-1} \log P\{X_1 = \alpha_1, \ldots, X_n = \alpha_n\} - H| < \delta\}$$
$$S_2 = \{(\alpha_1, \ldots, \alpha_n)\colon |-n^{-1} \log P\{X_1 = \alpha_1, \ldots, X_n = \alpha_n\} - H| \geq \delta\}.$$

Then $P\{(X_1, \ldots, X_n) \in S_1\} \geq 1 - \varepsilon$. Furthermore, if $(\alpha_1, \ldots, \alpha_n) \in S_1$ then, writing $p(\alpha_1, \ldots, \alpha_n)$ for $P\{X_1 = \alpha_1, \ldots, X_n = \alpha_n\}$, we have (by definition of S_1) $|\log [p(\alpha_1, \ldots, \alpha_n)]^{-1} - nH| < n\delta$, or

$$2^{-n(H+\delta)} < p(\alpha_1, \ldots, \alpha_n) < 2^{-n(H-\delta)}. \tag{6.6.1}$$

Since

$$1 - \varepsilon \leq \sum_{(\alpha_1, \ldots, \alpha_n) \in S_1} p(\alpha_1, \ldots, \alpha_n) \leq 1, \tag{6.6.2}$$

it follows from (6.6.1) and (6.6.2), just as in the discussion of "typical sequences" in Section 1.3, that the number of sequences in S_1 is at least

$$(1 - \varepsilon)2^{n(H-\delta)} = 2^{n[H-\delta+n^{-1}\log(1-\varepsilon)]}$$

and at most $2^{n(H+\delta)}$. Thus for large n we can divide the sequences $(\alpha_1, \ldots, \alpha_n)$ into two categories: a set S_1 consisting of approximately 2^{nH} sequences, each with probability roughly 2^{-nH}, and a set S_2 whose total probability is arbitrarily small.

If the sequence $\{X_n\}$ is regarded as an information source, then the above argument indicates that when we are transmitting long sequences produced by the source by means of an encoding and decoding scheme, we can restrict our attention to a set consisting of approximately 2^{nH} "meaningful" sequences (that is, members of S_1) and neglect those remaining. The probability of correct transmission will not be unduly affected by this restriction.

We are going to prove that the above results hold for a very wide class of information sources, including all regular Markov sources.

Let $X_1, X_2, \ldots,$ be an information source with alphabet Γ, that is, a stationary sequence of random variables taking values in a finite set Γ. For each fixed sequence $\boldsymbol{\alpha} = (\alpha_1, \ldots, \alpha_m)$, $\alpha_i \in \Gamma$, define a random variable $N_{\boldsymbol{\alpha}}^n(X_1, \ldots, X_n)$ as the number of integers i $(1 \leq i \leq n - m + 1)$ such that $X_i = \alpha_1, X_{i+1} = \alpha_2, \ldots, X_{i+m-1} = \alpha_m$.

In other words, $N_{\boldsymbol{\alpha}}^{n}(X_1, \ldots, X_n)$ is the number of times $(\alpha_1, \ldots, \alpha_m)$ occurs (in sequence) in the block consisting of the first n symbols. For example, if $n = 13$, $m = 2$, $\boldsymbol{\alpha} = (AB)$, then if $(X_1, \ldots, X_{13}) = AAABBABAAAABB$ we have $N^{13}_{(AB)}(X_1, \ldots, X_{13}) = 3$; similarly if $\boldsymbol{\alpha} = AA$ and $(X_1, \ldots, X_{13})$ is as above, $N^{13}_{(AA)}(X_1, \ldots, X_{13}) = 5$.

The information source is said to be *ergodic* if, for all sequences $\boldsymbol{\alpha} = (\alpha_1, \ldots, \alpha_m)$, $N_{\boldsymbol{\alpha}}^{n}(X_1, \ldots, X_n)/n$ converges in probability to $P\{X_1 = \alpha_1, \ldots, X_m = \alpha_m\}$. In other words, the long-run relative frequency of a sequence converges stochastically to the probability assigned to the sequence.

We shall show later that any regular Markov information source is ergodic. For an example of a nonergodic source, see Problem 6.5.

The previous discussion concerning the number of meaningful sequences which a source can be reasonably expected to produce carries over to ergodic sources. Specifically, we have the following theorem.

Theorem 6.6.1. (Asymptotic Equipartition Property; also called the Shannon-McMillan Theorem.) Let $\{X_n, n = 1, 2, \ldots\}$ be an ergodic information source with alphabet Γ and uncertainty $H\{\mathbf{X}\}$. As before, define a sequence of random variables $V_n(X_1, \ldots, X_n)$ as follows: If $X_1 = \alpha_1, \ldots, X_n = \alpha_n$, let

$$V_n(X_1, \ldots, X_n) = -n^{-1} \log P\{X_1 = \alpha_1, \ldots, X_n = \alpha_n\}.$$

Then $V_n(X_1, \ldots, X_n)$ converges in probability to $H\{\mathbf{X}\}$.

Proof. We break the argument into several steps. For convenience we set $H = H\{\mathbf{X}\}$.

a. $\lim_{n\to\infty} E(V_n) = H.$

Proof of (a)

$$\begin{aligned} E(V_n) = -\sum_{\alpha_1, \ldots, \alpha_n} & P\{X_1 = \alpha_1, \ldots X_n = \alpha_n\} \\ & \times n^{-1} \log P\{X_1 = \alpha_1, \ldots, X_n = \alpha_n\} \\ = \frac{H(X_1, \ldots, X_n)}{n} & \to H \qquad \text{by Theorem 6.4.1.} \end{aligned}$$

Our ultimate objective is to show that for every $\varepsilon > 0$, $P\{H - \varepsilon \leq V_n \leq H + \varepsilon\} \to 1$ as $n \to \infty$. As a preliminary reduction, we are going to show that only the upper half of the inequality is needed, that is, if for every $\varepsilon > 0$ $P\{V_n \leq H + \varepsilon\} \to 1$ as $n \to \infty$ then the general result follows. As the first step in the reduction we prove the following result.

b. If $\varepsilon > 0, \delta > 0, \beta \geq 0$, then

$$E(V_n) - \beta \leq \delta - (\delta + \varepsilon)P\{V_n < \beta - \varepsilon\} + (\log|\Gamma|)P\{V_n > \beta + \delta\}$$
$$-n^{-1}P\{V_n > \beta + \delta\}\log P\{V_n > \beta + \delta\}. \quad (6.6.3)$$

($|\Gamma|$ is the number of elements in Γ.)

Proof of (b):

$$E(V_n) = \sum_k kP\{V_n = k\}$$
$$= \sum_{k<\beta-\varepsilon} kP\{V_n = k\} + \sum_{\beta-\varepsilon\leq k\leq\beta+\delta} kP\{V_n = k\} + \sum_{k>\beta+\delta} kP\{V_n = k\}.$$

Thus

$$E(V_n) \leq (\beta - \varepsilon)P\{V_n < \beta - \varepsilon\} + (\beta + \delta)P\{\beta - \varepsilon \leq V_n \leq \beta + \delta\}$$
$$+ \sum_{k>\beta+\delta} kP\{V_n = k\}.$$

Since $P\{\beta - \varepsilon \leq V_n \leq \beta + \delta\} \leq P\{\beta - \varepsilon \leq V_n\} = 1 - P\{V_n < \beta - \varepsilon\}$ we have

$$E(V_n) \leq (\beta + \delta) - (\delta + \varepsilon)P\{V_n < \beta - \varepsilon\} + \sum_{k>\beta+\delta} kP\{V_n = k\} \quad (6.6.4)$$

Now

$$\sum_{k>\beta+\delta} kP\{V_n = k\} = \sum_{(\alpha_1,\ldots,\alpha_n)\in S}$$
$$(-n^{-1}\log P\{X_1 = \alpha_1, \ldots, X_n = \alpha_n\})P\{X_1 = \alpha_1, \ldots, X_n = \alpha_n\}$$
$$(6.6.5)$$

where

$$S = \{(\alpha_1, \ldots, \alpha_n): -n^{-1}\log P\{X_1 = \alpha_1, \ldots, X_n = \alpha_n\} > \beta + \delta\}$$
$$= \{(\alpha_1, \ldots, \alpha_n): P\{X_1 = \alpha_1, \ldots, X_n = \alpha_n\} < 2^{-n(\beta+\delta)}\}$$

(that is, S is the set of sequences corresponding to $V_n > \beta + \delta$). The summation (6.6.5) is of the form $-n^{-1}\sum_{i=1}^m p_i \log p_i$ where $\sum_{i=1}^m p_i = P\{V_n > \beta + \delta\} = p$, and m is the number of elements in S. Now by Theorem 1.4.2.

$$-\sum_{i=1}^m \frac{p_i}{p}\log\frac{p_i}{p} \leq \log m,$$

or

$$-\sum_{i=1}^m p_i \log p_i \leq p\log m - p\log p.$$

Since the number of sequences of length n with components in Γ is $|\Gamma|^n$, we have $m \leq |\Gamma|^n$, hence

$$\sum_{k>\beta+\delta} kP\{V_n = k\} \leq n^{-1}[P\{V_n > \beta + \delta\}\log|\Gamma|^n$$
$$-P\{V_n > \beta + \delta\}\log P\{V_n > \beta + \delta\}]. \quad (6.6.6)$$

Then (6.6.3) follows from (6.6.4) and (6.6.6).

c. If $P\{V_n \leq H + \varepsilon\} \to 1$ for every $\varepsilon > 0$, then $V_n \to H$ in probability.

Proof of (c). If $P\{V_n \leq H + \varepsilon\} \to 1$ for every $\varepsilon > 0$, then $P\{V_n \leq E(V_n) + \varepsilon\} \to 1$ for every $\varepsilon > 0$ since $E(V_n) \to H$. In (6.6.3) replace β by $E(V_n)$ and let $n \to \infty$, to obtain

$$0 \leq \delta - (\delta + \varepsilon) \limsup_{n \to \infty} P\{V_n < E(V_n) - \varepsilon\}, \quad \text{or}$$

$$\limsup_{n \to \infty} P\{V_n < E(V_n) - \varepsilon\} \leq \frac{\delta}{\delta + \varepsilon}.$$

Fix ε and let $\delta \to 0$. We conclude that $P\{V_n < E(V_n) - \varepsilon\} \to 0$ for any $\varepsilon > 0$. Since $E(V_n) \to H$, we have $P\{V_n \leq H - \varepsilon\} \to 0$ for all $\varepsilon > 0$. But by hypothesis, $P\{V_n \geq H + \varepsilon\} \to 0$ as $n \to \infty$. Thus

$$P\{|V_n - H| \geq \varepsilon\} \to 0$$

for all $\varepsilon > 0$, and the result follows.

d. Let $\Gamma^{(n)}$ be the set of sequences $(\alpha_1, \ldots, \alpha_n)$, $\alpha_i \in \Gamma$. For any $\beta \geq 0$, let $D_n(\beta)$ be the largest probability assigned to any subset of $\Gamma^{(n)}$ which has at most $2^{n\beta}$ elements, that is,

$$D_n(\beta) = \max_{\substack{F \subset \Gamma^{(n)} \\ \#(F) \leq 2^{\beta n}}} P\{(X_1, \ldots, X_n) \in F\}$$

Then given $\varepsilon > 0$, $\beta \geq 0$,

$$D_n(\beta) \leq P\{V_n \leq \beta + \varepsilon\} + 2^{-n\varepsilon}.$$

Proof of (d). Let F be a subset of $\Gamma^{(n)}$ with at most $2^{n\beta}$ elements and with probability $D_n(\beta)$, that is, $P\{(X_1, \ldots, X_n) \in F\} = D_n(\beta)$. Then

$$\begin{aligned} D_n(\beta) = P(F) &= P(F \cap \{V_n \leq \beta + \varepsilon\}) + P(F \cap \{V_n > \beta + \varepsilon\}) \\ &\leq P\{V_n \leq \beta + \varepsilon\} + P(F \cap \{V_n > \beta + \varepsilon\}). \end{aligned}$$

As in (b), if $X_i = \alpha_i$, $i = 1, 2, \ldots, n$, then

$$V_n > \beta + \varepsilon \quad \text{if and only if} \quad P\{X_1 = \alpha_1, \ldots, X_n = \alpha_n\} < 2^{-n(\beta+\varepsilon)}.$$

Thus each point in $\{V_n > \beta + \varepsilon\}$ (and hence in $F \cap \{V_n > \beta + \varepsilon\}$) has probability less than $2^{-n(\beta+\varepsilon)}$. Since F has at most $2^{n\beta}$ points,

$$D_n(\beta) \leq P\{V_n \leq \beta + \varepsilon\} + 2^{n\beta}2^{-n(\beta+\varepsilon)},$$

proving (d).

e. Let $H_m = H(X_m \mid X_1, \ldots, X_{m-1})$, $m = 1, 2, \ldots$, and let M be any positive integer. If for every $\varepsilon > 0$ and every positive integer $m \geq M$, $\lim_{n \to \infty} D_n(H_m + \varepsilon) = 1$, then $V_n \to H$ in probability.

Proof of (e). By (d), $D_n(H_m + \varepsilon) \to 1$ for every $\varepsilon > 0$ implies $P\{V_n \leq H_m + \varepsilon\} \to 1$ for every $\varepsilon > 0$. By definition of the uncertainty of a source (Section 6.4), $(H_m, m = 1, 2, \ldots)$ is a nonincreasing sequence converging to H. If we choose $r \geq M$ such that $H_r \leq H + \varepsilon/2$, then $P\{V_n \leq H + \varepsilon\} \geq P\{V_n \leq H_r + \varepsilon/2\} \to 1$ for every $\varepsilon > 0$. The result now follows from (c).

Note that the results (a) through (e) hold for an arbitrary information source; the ergodic hypothesis has not yet been used.

Proof of Theorem 6.6.1. By (e) it suffices to show that $\lim_{n\to\infty} D_n(H_m + \varepsilon) = 1$ for all $\varepsilon > 0$ and all $m \geq 2$. To do this we will find (for each fixed ε and m) subsets B_n of $\Gamma^{(n)}$ such that each B_n has at most $2^{n(H_m+\varepsilon)}$ elements [hence $D_n(H_m + \varepsilon) \geq P(B_n)$] and $P(B_n) \to 1$.

Let $\delta > 0$ (to be specified later). If $\boldsymbol{\alpha} = (\alpha_1, \ldots, \alpha_n)$ is any sequence in $\Gamma^{(n)}$ and $\boldsymbol{\gamma} = (\gamma_1, \ldots, \gamma_m)$ is any sequence in $\Gamma^{(m)}$, let $z(\boldsymbol{\gamma};\ \boldsymbol{\alpha})$ be the number of times that $(\gamma_1, \ldots, \gamma_m)$ appears (in sequence) in the block consisting of the n symbols $\alpha_1, \ldots, \alpha_n$. Define B_n as the set of all $\boldsymbol{\alpha} \in \Gamma^{(n)}$ such that

$$\left|\frac{z(\boldsymbol{\gamma};\boldsymbol{\alpha})}{n} - P\{X_1 = \gamma_1, \ldots, X_m = \gamma_m\}\right| \leq \delta \quad \text{for all} \quad \boldsymbol{\gamma} \in \Gamma^{(m)}$$

The probability of B_n is then

$$P\left\{\left|\frac{N_{\boldsymbol{\gamma}}^n(X_1, \ldots, X_n)}{n} - P\{(X_1, \ldots, X_m) = \boldsymbol{\gamma}\}\right| \leq \delta \quad \text{for} \quad \text{all } \boldsymbol{\gamma} \in \Gamma^{(m)}\right\}.$$

Since $\Gamma^{(m)}$ is a finite set, $P(B_n) \to 1$ by the hypothesis of ergodicity. It remains to show that for an appropriate choice of δ, B_n has at most $2^{n(H_m+\varepsilon)}$ elements.

Now let $\mathcal{S}_{m-1}$ be the $(m-1)$th-order approximation to the original source (Section 6.5); as before we form a stationary distribution for $\mathcal{S}_{m-1}$ by assigning to the state $(\alpha_1, \ldots, \alpha_{m-1})$ the probability $w_{(\alpha_1, \ldots, \alpha_{m-1})} = P\{X_1 = \alpha_1, \ldots, X_{m-1} = \alpha_{m-1}\}$. Let $\{Y_n, n = 1, 2, \ldots\}$ be the sequence of symbols produced by $\mathcal{S}_{m-1}$ and let $\{Z_n, n = 1, 2, \ldots\}$ be the corresponding sequence of states. [Recall that if $Z_n = (\alpha_1, \ldots, \alpha_{m-1})$, then $Y_n = \alpha_{m-1}$.] If $n \geq m$ we have

$$\begin{aligned}
P\{Y_1 = \alpha_1, \ldots, Y_n = \alpha_n\}& \\
= P\{Y_1 = \alpha_1, \ldots, Y_{m-1} = \alpha_{m-1}\}& \\
\times \prod_{i=m}^{n} P\{Y_i = \alpha_i \mid Y_1 = \alpha_1, \ldots, Y_{i-1} = \alpha_{i-1}\}& \\
= P\{Z_{m-1} = (\alpha_1, \ldots, \alpha_{m-1})\}& \\
\times \prod_{i=m}^{n} P\{Y_i = \alpha_i \mid Y_{i-1} = \alpha_{i-1}, \ldots, Y_{i-m+1} = \alpha_{i-m+1}\}&
\end{aligned}$$

since the order of $\mathcal{S}_{m-1}$ is $\leq m - 1$.

Now

$$P\{Z_{m-1} = (\alpha_1, \ldots, \alpha_{m-1})\} = w_{(\alpha_1, \ldots, \alpha_{m-1})}$$
$$= P\{X_1 = \alpha_1, \ldots, X_{m-1} = \alpha_{m-1}\},$$

and

$$P\{Y_i = \alpha_i \mid Y_{i-1} = \alpha_{i-1}, \ldots, Y_{i-m+1} = \alpha_{i-m+1}\}$$
$$= P\{X_i = \alpha_i \mid X_{i-1} = \alpha_{i-1}, \ldots, X_{i-m+1} = \alpha_{i-m+1}\}$$

by definition of $\mathcal{S}_{m-1}$; thus

$$P\{Y_1 = \alpha_1, \ldots, Y_n = \alpha_n\} = P\{X_1 = \alpha_1, \ldots, X_{m-1} = \alpha_{m-1}\}$$
$$\times \prod_{\boldsymbol{\gamma} \in \Gamma^{(m)}} [P\{X_m = \gamma_m \mid X_1 = \gamma_1, \ldots, X_{m-1} = \gamma_{m-1}\}]^{z(\boldsymbol{\gamma};\boldsymbol{\alpha})} \tag{6.6.7}$$

[For example, if $\boldsymbol{\alpha} = (\alpha_1, \ldots, \alpha_{10}) = (ABAAAABABA)$ and $m = 3$, then

$$P\{Y_1 = \alpha_1, \ldots, Y_{10} = \alpha_{10}\}$$
$$= P\{X_1 = A, X_2 = B\}\,(P\{X_3 = A \mid X_1 = A, X_2 = B\})^3$$
$$\times (P\{X_3 = A \mid X_1 = A, X_2 = A\})^2\, P\{X_3 = A \mid X_1 = B, X_2 = A\}$$
$$\times P\{X_3 = B \mid X_1 = A, X_2 = A\} P\{X_3 = B \mid X_1 = B, X_2 = A\}.]$$

Now by definition of B_n, if $\boldsymbol{\alpha} = (\alpha_1, \ldots, \alpha_n) \in B_n$ then $z(\boldsymbol{\gamma};\ \boldsymbol{\alpha}) \le n(\delta + P\{X_1 = \gamma_1, \ldots, X_m = \gamma_m\})$ for all $\boldsymbol{\gamma} = (\gamma_1, \ldots, \gamma_m) \in \Gamma^{(m)}$; hence by (6.6.7),

$$P\{Y_1 = \alpha_1, \ldots, Y_n = \alpha_n\} \ge P\{X_1 = \alpha_1, \ldots, X_{m-1} = \alpha_{m-1}\}$$
$$\times \prod_{\boldsymbol{\gamma} \in \Gamma^{(m)}} [P\{X_m = \gamma_m \mid X_1 = \gamma_1, \ldots, X_{m-1} = \gamma_{m-1}\}]^{n(\delta + P\{X_1 = \gamma_1, \ldots, X_m = \gamma_m\})}$$
$$\text{for all } (\alpha_1, \ldots, \alpha_n) \in B_n. \tag{6.6.8}$$

Now

$$[P\{X_1 = \alpha_1, \ldots, X_{m-1} = \alpha_{m-1}\}]^{1/n}$$
$$\times \prod_{(\gamma_1, \ldots, \gamma_m)} [P\{X_m = \gamma_m \mid X_1 = \gamma_1, \ldots, X_{m-1} = \gamma_{m-1}\}]^{\delta} \to 1$$

as $n \to \infty$ and $\delta \to 0$. Thus we may find n_0 and δ_0 such that if $n \ge n_0$ and $0 < \delta \le \delta_0$, the above expression is $\ge 2^{-\varepsilon}$ for all possible $\alpha_1, \ldots, \alpha_{m-1}$. If δ is fixed $\in (0, \delta_0]$, then for $n \ge n_0$, (6.6.8) yields

$$P\{Y_1 = \alpha_1, \ldots, Y_n = \alpha_n\} \ge$$
$$2^{-n\varepsilon} \prod_{(\gamma_1, \ldots, \gamma_m)} [P\{X_m = \gamma_m \mid X_1 = \gamma_1, \ldots, X_{m-1} = \gamma_{m-1}\}]^{nP\{X=\gamma_1, \ldots, X_m = \gamma_m\}}$$

or

$$P\{Y_1 = \alpha_1, \ldots, Y_n = \alpha_n\} \ge 2^{-n\varepsilon} 2^{-nH_m}$$
$$\text{for } (\alpha_1, \ldots, \alpha_n) \in B_n. \tag{6.6.9}$$

Now

$$1 \geq P\{(Y_1, \ldots, Y_n) \in B_n\} = \sum_{(\alpha_1, \ldots, \alpha_n) \in B_n} P\{Y_1 = \alpha_1, \ldots, Y_n = \alpha_n\}$$
$$\geq 2^{-n\varepsilon} 2^{-nH_m} \#(B_n) \qquad \text{by (6.6.9).}$$

Thus B_n has at most $2^{n(H_m+\varepsilon)}$ elements; this completes the proof.

To show that a regular Markov source obeys the asymptotic equipartition property, we prove the following result.

Theorem 6.6.2. A regular Markov information source is ergodic.

Proof. Let $X_1, X_2, \ldots$ be a regular Markov source. Then we may assume that $X_i = f(Z_i)$ where $Z_1, Z_2, \ldots,$ is a finite Markov chain with steady state probabilities and f is a function from the set of states $S = \{s_1, \ldots, s_r\}$ of the chain to the alphabet Γ of the source.

For each sequence $\boldsymbol{\alpha} = (\alpha_1, \ldots, \alpha_m) \in \Gamma^{(m)}$, we must show that

$$\frac{N_{\boldsymbol{\alpha}}{}^n(X_1, X_2, \ldots, X_n)}{n} \to P\{X_1 = \alpha_1, \ldots, X_m = \alpha_m\} \qquad \text{in probability.}$$

Now for each $\boldsymbol{\alpha}$, let $t_1(\boldsymbol{\alpha}), \ldots, t_{k(\boldsymbol{\alpha})}(\boldsymbol{\alpha}) \in S^{(m)}$ be all possible sequences of states which produce $\boldsymbol{\alpha}$. [In other words, if

$$t_i(\boldsymbol{\alpha}) = (s_{i(1)}(\boldsymbol{\alpha}), s_{i(2)}(\boldsymbol{\alpha}), \ldots, s_{i(m)}(\boldsymbol{\alpha}))$$

then $(f(s_{i(1)}(\boldsymbol{\alpha})) \ldots, f(s_{i(m)}(\boldsymbol{\alpha})) = (\alpha_1, \ldots, \alpha_m) = \boldsymbol{\alpha}$.] It follows that

$$N_{\boldsymbol{\alpha}}{}^n(X_1, \ldots, X_n) = \sum_{i=1}^{k(\boldsymbol{\alpha})} N^n_{ti(\boldsymbol{\alpha})}(Z_1, \ldots, Z_n).$$

If we can show that for each i,

$$n^{-1} N^n_{ti(\boldsymbol{\alpha})}(Z_1, \ldots, Z_n) \to$$
$$P\{Z_1 = s_{i(1)}(\boldsymbol{\alpha}), \ldots, Z_m = s_{i(m)}(\boldsymbol{\alpha})\} \qquad \text{in probability,}$$

it will follow that

$$n^{-1} N_{\boldsymbol{\alpha}}{}^n(X_1, \ldots, X_n) \xrightarrow{\text{prob.}} \sum_{i=1}^{k(\boldsymbol{\alpha})} P\{Z_1 = s_{i(1)}(\boldsymbol{\alpha}), \ldots, Z_m = s_{i(m)}(\boldsymbol{\alpha})\}$$
$$= P\{X_1 = \alpha_1, \ldots, X_m = \alpha_m\}.$$

Thus the problem reduces to showing that the underlying Markov chain $Z_1, Z_2, \ldots,$ is ergodic.

Now construct the mth-order approximation $\mathcal{S}_m$ to the chain just as in Section 6.5; that is, construct a new Markov chain whose states are the elements of $S^{(m)}$ and whose transition probabilities are

$$p_{(s_{i(1)}\cdots s_{i(m)})(s_{i(2)}\cdots s_{i(m)}s_{i(m+1)})}$$
$$= P\{Z_{m+1} = s_{i(m+1)} \mid Z_1 = s_{i(1)}, \ldots, Z_m = s_{i(m)}\}$$
$$= P\{Z_{m+1} = s_{i(m+1)} \mid Z_m = s_{i(m)}\}$$

by the Markov property.

Since the original chain has steady state probabilities, there is a state $s_{j(1)}$ and an integer N_0 such that $s_{j(1)}$ is reachable in N_0 steps from any initial state. It follows that if $\mathbf{s} = (s_{j(1)}, \ldots, s_{j(m)})$ is any state of $\mathcal{S}_m$ having $s_{j(1)}$ as its first coordinate, then $\mathbf{s}$ is reachable from any initial state of $\mathcal{S}_m$ in $N_0 + m - 1$ steps. Thus $\mathcal{S}_m$ has steady state probabilities. (These probabilities are $w_{(s_{i(1)}, \ldots, s_{i(m)})} = P\{Z_1 = s_{i(1)}, \ldots, Z_m = s_{i(m)}\}$; see Problem 6.4.)

Now we may regard $Z_1, \ldots, Z_m$ as the components of the initial state U_1 of $\mathcal{S}_m$, and the Z_n, $n \geq m$, as the last coordinates of the succeeding states $U_2, \ldots, U_{n-m+1}$ of $\mathcal{S}_m$. Consequently, if $\mathbf{s} = (s_{i(1)}, \ldots, s_{i(m)}) \in S^{(m)}$, then

$$n^{-1}N_{\mathbf{s}}{}^n(Z_1, \ldots, Z_n) = n^{-1}N_{\mathbf{s}}^{n-m+1}(U_1, \ldots, U_{n-m+1}).$$

It follows that verifying the ergodicity condition for sequences $\mathbf{s} = (s_{i(1)}, \ldots, s_{i(m)})$ of length m in the original chain $Z_1, Z_2, \ldots,$ is equivalent to establishing the condition for sequences of length 1, that is, single states, in $\mathcal{S}_m$. Thus without loss of generality we may assume $m = 1$; it remains to show that for every state $s_j \in S$,

$$n^{-1}N_{s_j}^n(Z_1, \ldots, Z_n) \to P\{Z_1 = s_j\} = w_j \qquad \text{in probability.}$$

This is just the *weak law of large numbers* for Markov chains. We state this explicitly:

Theorem 6.6.3. Let $Z_1, Z_2, \ldots,$ be a finite Markov chain with steady state probabilities. Let the initial distribution be arbitrary. Then for each state s_j, the relative frequency $n^{-1}N_j{}^n$ of s_j converges in probability to the steady state probability w_j, that is,

$$n^{-1}N_j{}^n(Z_1, \ldots, Z_n) \xrightarrow{\text{prob.}} w_j. \tag{6.6.10}$$

Proof. It suffices to show that (6.6.10) holds when the initial distribution is such that $Z_1 = s_i$ with probability 1, where s_i is an arbitrary but fixed state. We will show that $E[(n^{-1}N_j{}^n - w_j)^2] \to 0$ as $n \to \infty$; since $P\{|n^{-1}N_j{}^n - w_j| \geq \varepsilon\} \leq \varepsilon^{-2}E[(n^{-1}N_j{}^n - w_j)^2]$ by Chebyshev's inequality (or Lemma 3.6.1) the result will follow. Define

$$\begin{aligned} U_j{}^k &= 1 \quad \text{if} \quad Z_k = s_j, \\ &= 0 \quad \text{if} \quad Z_k \neq s_j. \end{aligned} \qquad (k = 1, 2, \ldots)$$

Then

$$N_j{}^n = \sum_{k=1}^{n} U_j{}^k.$$

Now

$$E[(n^{-1}N_j{}^n - w_j)^2] = E\left[\left(n^{-1}\sum_{k=1}^{n} U_j{}^k - w_j\right)^2\right]$$

$$= n^{-2}E\left[\left(\sum_{k=1}^{n}(U_j{}^k - w_j)\right)^2\right]$$

$$= n^{-2}\sum_{k,l=1}^{n} a_{kl} \qquad (6.6.11)$$

where

$$a_{kl} = E[(U_j{}^k - w_j)(U_j{}^l - w_j)].$$

Let $m = \min(k, l)$, $d = |k - l|$. Then

$$E[U_j{}^k] = P\{U_j{}^k = 1\} = P\{Z_k = s_j\} = p_{ij}^{(k)}$$

and

$$E[U_j{}^k U_j{}^l] = P\{U_j^k = U_j^l = 1\} = P\{Z_k = Z_l = s_j\} = p_{ij}^{(m)}p_{jj}^{(d)}.$$

Thus

$$a_{kl} = p_{ij}^{(m)}p_{jj}^{(d)} - w_j p_{ij}^{(k)} - w_j p_{ij}^{(l)} + w_j{}^2. \qquad (6.6.12)$$

Let

$$M_j^{(n)} = \max_{1 \le i \le r} p_{ij}^{(n)}, \quad m_j^{(n)} = \min_{1 \le i \le r} p_{ij}^{(n)}. \qquad \text{[see after (6.3.13)]}$$

If the transition matrix of the chain is such that Π^{n_0} has a positive column, then by (6.3.18),

$$M_j{}^{(ln_0)} - m_j{}^{(ln_0)} \le (1 - \varepsilon)^{l-1}$$

where ε is the smallest element in the positive column of Π^{n_0}. Given any positive integer n, choose l so that $ln_0 \le n < (l + 1)n_0$. Then by the monotonicity of the sequences $M_j{}^{(n)}$ and $m_j{}^{(n)}$,

$$M_j^{(n)} - m_j^{(n)} \le M_j^{(ln_0)} - m_j^{(ln_0)} \le (1 - \varepsilon)^{l-1}$$

$$= \frac{(1 - \varepsilon)^{l+1}}{(1 - \varepsilon)^2} \le \frac{(1 - \varepsilon)^{n/n_0}}{(1 - \varepsilon)^2}.$$

Thus there is a positive number b and a number $t \in (0, 1)$ such that

$$M_j^{(n)} - m_j^{(n)} \le bt^n, \qquad n = 1, 2, \ldots.$$

Since

$$m_j^{(n)} \le p_{ij}^{(n)} \le M_j^{(n)} \quad \text{and} \quad m_j^{(n)} \le w_j \le M_j^{(n)},$$

we have

$$p_{ij}^{(n)} = w_j + \varepsilon_{ij}^{(n)} \quad \text{where} \quad |\varepsilon_{ij}^{(n)}| \le bt^n. \qquad (6.6.13)$$

By (6.6.12),

$$a_{kl} = w_j(\varepsilon_{ij}^{(m)} + \varepsilon_{jj}^{(d)} - \varepsilon_{ij}^{(k)} - \varepsilon_{ij}^{(l)}) + \varepsilon_{ij}^{(m)}\varepsilon_{jj}^{(d)}.$$

Thus by (6.6.13) there is a positive constant c such that

$$|a_{kl}| \leq c(t^m + t^d + t^k + t^l).$$

Now

$$\sum_{k,l=1}^{n} t^k = n\sum_{k=1}^{n} t^k \leq n\sum_{k=0}^{n} t^k \leq \frac{n}{1-t}$$

Similarly

$$\sum_{k,l=1}^{n} t^l \leq \frac{n}{1-t}$$

$$\sum_{k,l=1}^{n} t^{\min(k,l)} = \sum_{k=1}^{n} t^k + 2\sum_{k<l} t^k \leq \frac{1+2n}{1-t}$$

$$\sum_{k,l=1}^{n} t^{|k-l|} \leq 2n\sum_{k=0}^{n} t^k \leq \frac{2n}{1-t}.$$

By (6.6.11),

$$E[(n^{-1}N_j^{\,n} - w_j)^2] \leq \frac{4c(2n+1)}{n^2(1-t)} \to 0 \quad \text{as} \quad n \to \infty.$$

This proves Theorem 6.6.3 and thus finishes the proof of Theorem 6.6.2.

The asymptotic equipartition property (AEP) allows us to draw several basic conclusions about the encoding of the information produced by a source. Suppose that the information produced by an ergodic source with uncertainty H is to be transmitted through a (discrete memoryless) channel with capacity C. Suppose $H < C$, and choose R such that $H < R < C$. By the AEP, for sufficiently large n we can divide the sequences of length n produced by the source into two sets S_1 and S_2, such that S_1 has roughly 2^{nH} sequences; more precisely, for large enough n, S_1 has at least $2^{n(H-\delta)}$ and at most $2^{n(H+\delta)}$ sequences, where δ is any preassigned positive number. In particular, we may choose δ so that S_1 has fewer than 2^{nR} sequences. The total probability of the sequences in S_2 can be made $<\varepsilon/2$ for any preassigned $\varepsilon > 0$. Now we can find a code $([2^{nR}], n, \varepsilon/2)$ for the channel. Let us assign a code word of the code to each sequence in S_1; assign an arbitrary channel input n-sequence and an arbitrary decoding set to each sequence in S_2. With this coding scheme it is possible to transmit the information produced by the source with an overall probability of error $< \varepsilon$.

Thus a source with uncertainty H can be handled by a channel with capacity C provided $H < C$. The encoder assigns a code word of length n to each source sequence of length n; hence, if the channel can transmit 1 symbol per second, the source may be run at a rate not to exceed 1 symbol per second. To look at it another way, suppose that we wish to run the source at a rate of α letters per second. If the alphabet of the source has m symbols, then the equivalent binary rate of the source is $\alpha \log m = R$

binary digits per second, since in n seconds the source will theoretically produce any one of $m^{n\alpha} = 2^{nR}$ possible sequences. However, by the AEP, we may restrict our attention to $2^{Hn\alpha} = 2^{nRH/\log m}$ of these. To be able to maintain the transmission rate and transmit with high reliability through a channel with capacity C, we must have $RH/(\log m) < C$, or $R < C[(\log m)/H]$. (Note that $H \leq \log m$ so that $C[(\log m)/H] \geq C$.)

In Chapter 3 we showed that if a source produces at a rate of R binary digits per second, the rate could be maintained with an arbitrarily small error probabiity as long as $R < C$. By using the AEP, we may take advantage of the statistical structure of the source to achieve higher transmission rates; the source rate is now bounded by the larger quantity $C[(\log m)/H]$. Notice, however, that if the source is a sequence of independent random variables $X_1, X_2, \ldots$, where each X_i takes on all values in the source alphabet with equal probability, then $H = \log m$ so the upper bound remains at C. Hence if we know nothing at all about the structure of the source, the requirement $R < C$ cannot be weakened.

We may use the results of this chapter to say something about the properties of languages. For example, we may regard a portion of text in a particular language as being produced by an information source. The probabilities $P\{X_n = \alpha_n \mid X_0 = \alpha_0, \ldots, X_{n-1} = \alpha_{n-1}\}$ may be estimated from the available data about the language; in this way we can estimate the uncertainty associated with the language. A large uncertainty means, by the AEP, a large number of "meaningful" sequences. Thus given two languages with uncertainties H_1 and H_2 respectively, if $H_1 > H_2$ then in the absence of noise it is easier to communicate in the first language; more can be said in the same amount of time. On the other hand, it will be easier to reconstruct a scrambled portion of text in the second language, since fewer of the possible sequences of length n are meaningful.

6.7. *Notes and Remarks*

The notion of an information source and the idea of a Markov source are due to Shannon (1948). The present formulation of the Markov model follows Blackwell, Breiman, and Thomasian (1958). The fact that a Markov source must be unifilar for the formula of Theorem 6.4.2 to hold was pointed out by McMillan (1953).

Basic properties of Markov chains may be found in Feller (1950), Chung (1960), or Kemeny and Snell (1960).

Theorem 6.5.1 is related to a theorem of Moore (1956, Theorem 8) about sequential machines; the proof in the text was suggested to the author by D. Younger.

Theorem 6.6.1 was discovered for the case of Markov sources by Shannon (1948); the general result is due to McMillan (1953). The proof in the text follows Thomasian (1960). McMillan (1953) showed that $V_n \to H$ in L_1, that is, $E[|V_n - H|] \to 0$. Breiman (1957, 1960) showed that $V_n \to H$ with probability 1.

The criterion we have given for a process to be ergodic can be shown to be equivalent to the standard definition of ergodicity of an arbitrary stochastic process, specialized to the case of a stationary sequence of random variables with values in a finite alphabet. For a proof of equivalence see Khinchin (1957, p. 51). Further properties of ergodic processes may be found in Doob (1953).

Theorems 6.6.2 and 6.6.3 hold for indecomposable Markov chains; see Chung (1960).

PROBLEMS

6.1 Consider the unifilar Markov information source in Fig. 6.P.1.

a. Show that the source is of order 2.

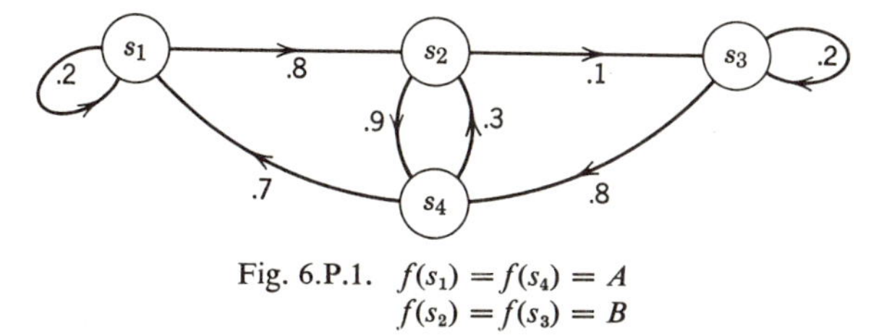

Fig. 6.P.1. $f(s_1) = f(s_4) = A$
$f(s_2) = f(s_3) = B$

b. If the system is in state s_1 at $t = 0$, find the probability that the symbol A will appear at $t = 3$.

c. Find the steady state probability of the sequence $ABBAA$.

6.2 Find the order of the unifilar Markov information source in Fig. 6.P.2 and calculate the steady state probabilities and the uncertainty of the source.

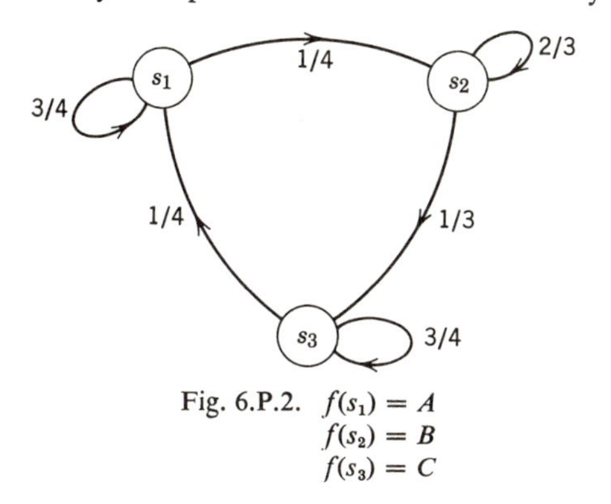

Fig. 6.P.2. $f(s_1) = A$
$f(s_2) = B$
$f(s_3) = C$

6.3 A "language" consists of four symbols: A, B, C, and the space. The symbols occur independently except for the following constraints:

i. The space cannot follow itself.

ii. If the symbol A occurs immediately after B, C, or the space, then the next symbol must be an A also, that is,

$$P\{X_t = A \mid X_{t-1} = A \quad \text{and} \quad X_{t-2} \neq A\} = 1.$$

a. Draw a unifilar Markov state diagram for the information source described above and find the order of the source.

b. Assume that if the preceding symbol is the space, then A, B, and C are equally likely; if the preceding symbol is B or C, or if the previous two symbols are both A's, then A, B, C, and the space are equally likely.

Calculate the uncertainty of the source.

c. It is desired to make a first-order approximation to the above source, that is, to consider the dependence of the present symbol on the immediately preceding symbol, but to ignore the dependence of X_t on symbols occurring at time $t - 2$ or before. Draw a state diagram corresponding to the first-order approximation and calculate the associated uncertainty.

6.4 If $\mathcal{S}_n$ is the nth-order approximation to an information source $\{\mathbf{X}\}$, show that a stationary distribution for $\mathcal{S}_n$ may be formed by taking

$$w_{(\alpha_1, \ldots, \alpha_n)} = P\{X_1 = \alpha_1, \ldots, X_n = \alpha_n\}.$$

6.5 Define a sequence of random variables as follows: Toss an unbiased coin. If the result is heads, let

$$X_n = 1 \quad \text{for} \quad \text{all } n.$$

If the result is tails, let

$$X_n = -1 \quad \text{for} \quad \text{all } n.$$

Show that the resulting sequence is stationary but not ergodic.

6.6 Consider an information source with an alphabet of two symbols A and B. Show that

$$P\{X_{t-1} = A, X_t = B\} = P\{X_{t-1} = B, X_t = A\}.$$

6.7 Consider the Markov information source of Fig. 6.P.3. Let X_0, X_1, $X_2, \ldots,$ be the associated sequence of random variables. Suppose that every third symbol of the sequence $\{X_n, n \geq 0\}$ is erased; the initial symbol to be erased is chosen with equal probability from among X_0, X_1, and X_2.

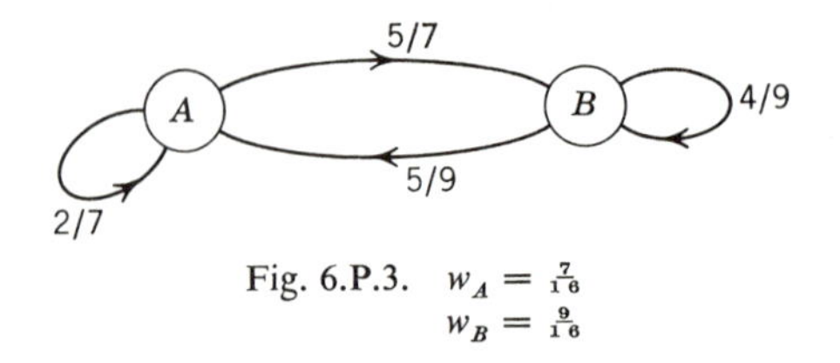

Fig. 6.P.3. $w_A = \frac{7}{16}$
$w_B = \frac{9}{16}$

The process of erasing generates a new sequence $Y_0, Y_1, Y_2, \ldots$. For example, if X_2 is the first symbol to be erased, then if

$$\{X_n, n \geq 0\} = ABAABBBAA \cdots,$$

then

$$\{Y_n, n \geq 0\} = ABe\ ABe\ BAe \cdots,$$

Find the uncertainty $H\{\mathbf{Y}\}$.

6.8 (Thomasian 1960.) a. As in the derivation of equation (6.6.3), show that if $\{X_n, n = 1, 2, \ldots\}$ is an *arbitrary* information source and V_n is defined as in Theorem 6.6.1, then

$$E(V_n) - \beta \geq -\varepsilon - (\beta - \varepsilon)P\{V_n < \beta - \varepsilon\} + (\delta + \varepsilon)P\{V_n > \beta + \delta\}$$
$$(\text{where } \varepsilon > 0,\ \delta > 0,\ \beta \geq 0).$$

b. If $V_n \rightarrow \beta$ in probability, show that necessarily $\beta = H\{\mathbf{X}\}$.

6.9 If an information source is ergodic, show that for each m, the mth-order approximation to the source is indecomposable.

6.10 Show that a finite Markov chain always has at least one essential set of states. Does this result hold for a chain with an infinite number of states?

6.11 Show that it is possible for the nth-order approximation to an information source to have order less than n.

6.12 A random variable Y takes the values $y_1, \ldots, y_M$ with probabilities $p_1, \ldots, p_M$ respectively. The symbol y_i is assigned the code word $a_{i1}a_{i2} \cdots a_{in_i}$ $(i = 1, \ldots, M)$. Assume that each code word begins with a different letter, that is, $a_{i1} \neq a_{j1}$ for $i \neq j$.

Suppose that successive values of Y are chosen independently and the resulting sequence of code characters written down. For example, if $M = 3$ and the code words assigned to y_1, y_2, and y_3 are A, BA, and CAB respectively, then the sequence $y_1y_2y_3y_1y_1y_3$ would correspond to $ABACABAACAB$.

Let $\{X_n\}$ be the sequence of code characters generated according to the above scheme. Show that $\{X_n\}$ may be regarded as a unifilar indecomposable Markov information source, and show that

$$H\{\mathbf{X}\} = \frac{H(Y)}{\bar{n}}$$

where $\bar{n} = \sum_{i=1}^{M} p_i n_i$ is the average code word length. (Intuitively,

$$\frac{H(Y)}{\bar{n}} = \frac{\text{“bits/code word”}}{\text{“symbols/code word”}} = \text{“bits/symbol.”})$$

6.13 (The Maximum Uncertainty of a Unifilar Markov Source; Shannon 1948.) Given a unifilar Markov information source with states $s_1, \ldots, s_r$ and transition probabilities p_{ij}. We define the *connection matrix* $A = [a_{ij}]$ of the source as follows. Let

$$\begin{aligned} a_{ij} &= 1 \quad \text{if} \quad p_{ij} > 0 \\ a_{ij} &= 0 \quad \text{if} \quad p_{ij} = 0. \end{aligned} \qquad (i, j = 1, \ldots, r)$$

In other words the connection matrix is formed by replacing the positive elements of the transition matrix by 1's.

For a given connection matrix, we wish to choose the probabilities p_{ij} so that the uncertainty of the source is maximized. Assume that every state may be reached from every other state, so that the source is indecomposable with the essential class consisting of all the states. It follows from the theorem of Fro-

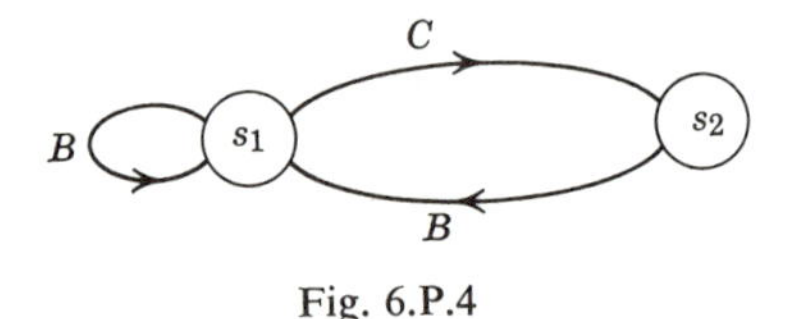

Fig. 6.P.4

benius (see Gantmacher, 1959, p. 65) that the matrix A has a positive real eigenvalue λ_0 which is greater than or equal to the magnitude of all other eigenvalues. Furthermore, the eigenvectors corresponding to λ_0 consist of all multiples of some vector $\mathbf{x}$ whose components x_i $(i = 1, \ldots, r)$ are all positive.

a. Let $N(T)$ be the total number of sequences of length T the source can produce (considering all possible initial states). Let $N_i(T)$, $i = 1, \ldots, r$, be the number of sequences of length T the source can produce when the initial state is s_i. By setting up a difference equation for the $N_i(T)$, show that there is a positive constant K such that $N(T) \leq K(\lambda_0)^T$.

b. Show that for any assignment of the probabilities p_{ij} consistent with the given connection matrix, the uncertainty of the source is $\leq \log \lambda_0$.

c. Let $p_{ij} = a_{ij}x_j/\lambda_0 x_i$, $i, j = 1, \ldots, r$. Show that under this assignment, the uncertainty $H\{\mathbf{X}\}$ attains the upper bound $\log \lambda_0$.

d. For the source of Fig. 6.P.4, find the maximum uncertainty and a set of transition probabilities for which the maximum is achieved.

CHAPTER SEVEN

Discrete Channels with Memory

7.1. Introduction

In this chapter we try to analyze channels whose behavior at a given time may depend on the "state" of the channel at that time as well as on past inputs and outputs. In the case of the discrete memoryless channel (Chapter 3), the distribution of a channel output symbol is completely determined by the corresponding input letter. In the more general situation, the distribution of the output symbol may depend on events happening arbitrarily far back in the past. We shall try to develop reasonable models for discrete channels with memory, and prove coding theorems analogous to those derived for the memoryless case. We shall then talk about the general notions of capacity, weak and strong converses for arbitrary discrete channels.

Let us begin with two examples.

Example 7.1.1. (Blackwell 1961a). Given two "trapdoors" as shown in Fig. 7.1.1. Initially (Fig. 7.1.1a), a ball labeled either 0 or 1 is placed in each of the two slots. Then (Fig. 7.1.1b) one of the trapdoors opens. (Each door has the same probability of being opened.) The ball lying above the open door then falls through and the door closes. Another ball is placed in the empty compartment (Fig. 7.1.1c) and the process is repeated.

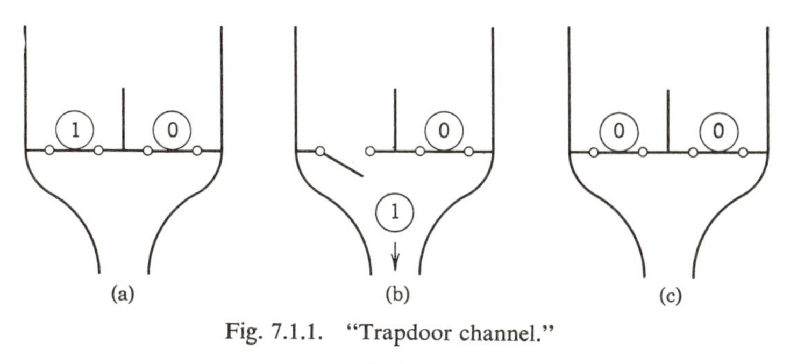

Fig. 7.1.1. "Trapdoor channel."

We may regard this description as defining a channel whose inputs correspond to the balls placed in the empty compartment and whose outputs correspond to the balls which fall through the trapdoors. Suppose that we start counting time after one of the doors has opened and a ball has dropped through. Let the symbol b_i correspond to the condition in which a ball labeled i remains in the occupied slot. We define four states

Table 7.1.1. Channel matrices and output function for the "trapdoor" channel

$$\text{Input} = 0 \qquad \begin{array}{c} \\ s_{00} \\ s_{10} \\ s_{01} \\ s_{11} \end{array} \begin{array}{c} \begin{array}{cccc} s_{00} & s_{10} & s_{01} & s_{11} \end{array} \\ \begin{bmatrix} 1 & 0 & 0 & 0 \\ 1 & 0 & 0 & 0 \\ 0 & \frac{1}{2} & 0 & \frac{1}{2} \\ 0 & \frac{1}{2} & 0 & \frac{1}{2} \end{bmatrix} \end{array} = M_0$$

$$\text{Input} = 1 \qquad \begin{array}{c} \\ s_{00} \\ s_{10} \\ s_{01} \\ s_{11} \end{array} \begin{array}{c} \begin{array}{cccc} s_{00} & s_{10} & s_{01} & s_{11} \end{array} \\ \begin{bmatrix} \frac{1}{2} & 0 & \frac{1}{2} & 0 \\ \frac{1}{2} & 0 & \frac{1}{2} & 0 \\ 0 & 0 & 0 & 1 \\ 0 & 0 & 0 & 1 \end{bmatrix} \end{array} = M_1$$

Let $g(k, s_{ij})$ = output when an input k is applied and the channel moves to state s_{ij}. Then

$g(0, s_{00}) = 0$	$g(1, s_{00}) = 1$
$g(0, s_{10}) = 1$	$g(1, s_{10}) =$ anything
$g(0, s_{01}) =$ anything (this condition will never arise)	$g(1, s_{01}) = 0$
$g(0, s_{11}) = 0$	$g(1, s_{11}) = 1$

s_{ij}, $i, j = 0$ or 1, as follows. We say that the channel is in state s_{ij} at time $t = n$ if condition b_j holds at $t = n$ and b_i held at $t = n - 1$. An input k corresponds to placing a ball labeled k in the unoccupied slot. The opening of a trapdoor then determines the corresponding output (that is, the identity of the ball which falls through) and the next state. The above information is summarized in Table 7.1.1.

The meaning of Table 7.1.1 may be clarified by means of an example. Suppose that at time $t = n$, the channel is in state s_{10}, and an input 1 is applied. Then one ball labeled 0 and one ball labeled 1 rest over the trapdoors. With probability 1/2 the 1 ball falls through, leaving the 0 ball in the occupied slot. The channel then moves to state s_{00} and emits an output $g(1, s_{00}) = 1$. With probability 1/2 the 0 ball falls through, sending the channel into state s_{01}. An output $g(1, s_{01}) = 0$ is emitted.

Thus we may describe the behavior of the "trapdoor" channel by giving matrices M_0 and M_1, called "channel matrices," whose components are the state transition probabilities under the inputs 0 and 1, respectively, and a function that associates an output with each input-state pair (x, s).

The reader may feel that this channel could be described using two states instead of four. For the purpose of analysis, however, we want to associate a unique output with each pair (x, s); this cannot be done using only two states.

Example 7.1.2. (Gilbert 1960). Consider the following model for a channel in which errors occur in bursts. We start out with two states G (for good) and B (for bad). Assume that binary digits are to be communicated at the rate of 1 per second. If we are in the good state at $t = n$,

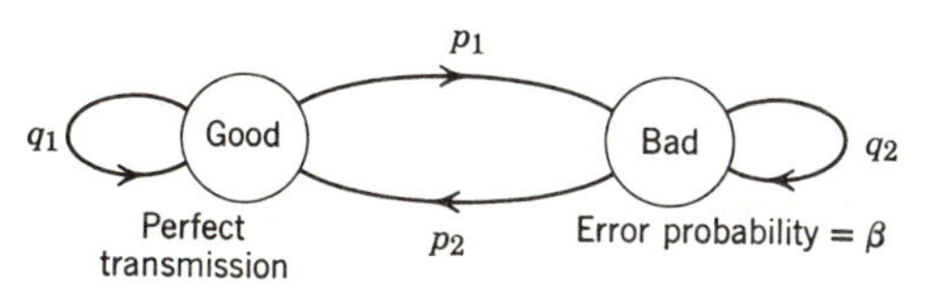

Fig. 7.1.2. Burst noise channel.

the digit transmitted at that time is received perfectly. A coin is then tossed; if the result is heads (say this happens with probability p_1), we move to the bad state at $t = n + 1$; if the result is tails (with probability $q_1 = 1 - p_1$) we stay in the good state at $t = n + 1$. In the bad state, a transmitted digit will be in error with probability β. After transmission, a coin is again tossed; with probability p_2 we move back to the good state, and with probability $q_2 = 1 - p_2$ we stay in the bad state. (See Fig. 7.1.2.) Assume $p_1, q_1, p_2, q_2, \beta \in (0, 1)$.

Let us try to analyze this example by giving state transition matrices for each input and a function defined on input-state pairs. To do this, we need four states, say $G0$, $G1$, $B0$, $B1$, where the first coordinate indicates the "good" or "bad" condition and the second coordinate indicates the most recent output. The channel matrices and output function are indicated in Table 7.1.2.

For example, if we are in state $G1$ and the input to follow will be 1, then with probability q_1 we remain in the "good" condition. Transmission is then perfect so that an output 1 is received and we remain in state $G1$. With probability p_1 we move to the "bad" condition. Having done this,

Table 7.1.2. Channel matrices and output function for the "burst noise" channel

Input = 0

	$G0$	$G1$	$B0$	$B1$
$G0$	q_1	0	$p_1(1-\beta)$	$p_1\beta$
$G1$	q_1	0	$p_1(1-\beta)$	$p_1\beta$
$B0$	p_2	0	$q_2(1-\beta)$	$q_2\beta$
$B1$	p_2	0	$q_2(1-\beta)$	$q_2\beta$

Input = 1

	$G0$	$G1$	$B0$	$B1$
$G0$	0	q_1	$p_1\beta$	$p_1(1-\beta)$
$G1$	0	q_1	$p_1\beta$	$p_1(1-\beta)$
$B0$	0	p_2	$q_2\beta$	$q_2(1-\beta)$
$B1$	0	p_2	$q_2\beta$	$q_2(1-\beta)$

The output function assigns to each state its second coordinate, regardless of the input.

---

with probability β an error is made so that the output is 0 and the new state is $B0$; with probability $1 - \beta$ the input is transmitted properly and the new state is $B1$.

Notice that in this case the output may be regarded as a function of the state alone rather than the input-state pair. It is possible to reformulate Example 7.1.1 so that the output is determined by the state; however, more than four states will be required. (See Problem 7.1.)

Now let us "connect a source" to one of the previous channels, say the channel of Example 7.1.1. Consider a Markov information source with two states a_0 and a_1, an alphabet consisting of 0 and 1 an associated function defined by $f(a_0) = 0$, $f(a_1) = 1$, and Markov transition matrix

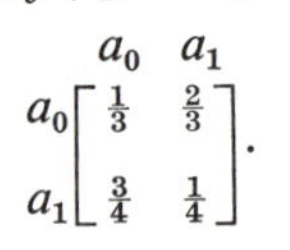

$$\begin{array}{c} \\ a_0 \\ a_1 \end{array}\begin{array}{c} \begin{array}{cc} a_0 & a_1 \end{array} \\ \begin{bmatrix} \frac{1}{3} & \frac{2}{3} \\ \frac{3}{4} & \frac{1}{4} \end{bmatrix} \end{array}.$$

Suppose that at a given moment the source is in state a_1 and the channel is in state s_{10}. With probability 3/4 the source moves to state a_0 and emits the digit 0. This digit is applied as the input to the channel; necessarily (see Table 7.1.1) the channel moves to state s_{00} and emits the output $g(0, s_{00}) = 0$. With probability 1/4 the source remains in state a_1 and emits a "1." If this digit is applied to the channel then with probability 1/2 the channel moves to state s_{00} and produces an output $g(1, s_{00}) = 1$, and with probability 1/2 the channel moves to state s_{01} and produces an output $g(1, s_{01}) = 0$. To summarize, if the source is initially in state a_1, and the channel is in state s_{10}, then with probability 3/4 the next source and channel states are (a_0, s_{00}); with probabiity $(1/4)(1/2) = 1/8$ the states are (a_1, s_{00}), and with probability 1/8 the states are (a_1, s_{01}). The corresponding inputs and outputs are determined from the states.

In general, the probability of transition from the pair of states $(a_{k_1}, s_{i_1 j_1})$ to the pair (a_k, s_{ij}) is the product of the source transition probability $p_{k_1 k}$ and the channel transition probability from $s_{i_1 j_1}$ to s_{ij} under the input k. The results are summarized in Table 7.1.3.

Table 7.1.3. A source-channel matrix for example 7.1.1

	(a_0, s_{00})	(a_0, s_{10})	(a_0, s_{01})	(a_0, s_{11})	(a_1, s_{00})	(a_1, s_{10})	(a_1, s_{01})	(a_1, s_{11})
(a_0, s_{00})	$\frac{1}{3}$	0	0	0	$\frac{1}{3}$	0	$\frac{1}{3}$	0
(a_0, s_{10})	$\frac{1}{3}$	0	0	0	$\frac{1}{3}$	0	$\frac{1}{3}$	0
(a_0, s_{01})	0	$\frac{1}{6}$	0	$\frac{1}{6}$	0	0	0	$\frac{2}{3}$
(a_0, s_{11})	0	$\frac{1}{6}$	0	$\frac{1}{6}$	0	0	0	$\frac{2}{3}$
(a_1, s_{00})	$\frac{3}{4}$	0	0	0	$\frac{1}{8}$	0	$\frac{1}{8}$	0
(a_1, s_{10})	$\frac{3}{4}$	0	0	0	$\frac{1}{8}$	0	$\frac{1}{8}$	0
(a_1, s_{01})	0	$\frac{3}{8}$	0	$\frac{3}{8}$	0	0	0	$\frac{1}{4}$
(a_1, s_{11})	0	$\frac{3}{8}$	0	$\frac{3}{8}$	0	0	0	$\frac{1}{4}$

$(a_k, s_{ij}) \rightarrow$ input $f(a_k) = k$
output $g(x_k, s_{ij})$ as given in Table 7.1.1

Thus the action of the source on the channel may be described by the matrix of Table 7.1.3, which is called the "source-channel matrix" corresponding to the given source.

In the next section we are going to use the concepts developed here to define a general model for discrete channels with memory.

7.2. *The finite-state channel*

The model we are trying to develop is that of a channel with a finite number of internal states. The present state of the channel will in a sense represent a summary of its past history. The application of an input will result in a transition to a (possibly) different state and the production of an output. Specifically, given a finite set $S = \{s_1, \ldots, s_r\}$ called the *set of states* and two finite sets $\Gamma = \{b_1, \ldots, b_t\}$ and Δ called the *input alphabet* and the *output alphabet*, respectively, we define a *finite-state channel* as a collection of r by r transition matrices (that is, matrices of nonnegative numbers with all row sums equal to unity) $M_1, \ldots, M_t$, one for each input letter, together with a function g that assigns to each pair (b, s), $b \in \Gamma$, $s \in S$, an element $g(b, s) \in \Delta$. (The matrix M_i will also be denoted by $M_{b(i)}$ and is called the *channel matrix* corresponding to the input b_i.)

The interpretation of this definition is the following (compare Examples 7.1.1 and 7.1.2). If the channel is initially in state s_j and an input b_i is applied, the channel makes a transition governed by the matrix M_i. The probability of moving from s_j to s_k is the element $M_i(s_j, s_k)$ in row j and column k of the matrix M_i. If the new state is s_k, an output $g(b_i, s_k)$ is produced.

Notice that given an initial state and a sequence of inputs, the distribution of the corresponding output sequence is determined by the channel matrices. Thus the finite-state channel may be regarded as a special case of the general discrete channel defined in Section 3.1.

Now consider a Markov information source $\mathcal{S}$ with set of states $S_0 = \{a_1, \ldots, a_q\}$, transition matrix $\Pi = [p_{ij}]$, $i, j = 1, \ldots, q$, alphabet $\Gamma = \{b_1, \ldots, b_t\}$ = the input alphabet of the channel, and function f from S_0 to Γ. As in the discussion in Section 7.1, the operation of connecting the source to the channel is described by the *source-channel matrix* Q corresponding to $\mathcal{S}$. Q is defined as a qr by qr matrix (q = number of source states, r = number of channel states) whose rows (and columns) correspond to pairs (a, s), $a \in S_0$, $s \in S$. The element in row (a_i, s_k) and column (a_j, s_l) of Q is $p_{ij}M_{f(a_j)}(s_k, s_l)$. In other words, if initially the source is in state a_i and the channel in state s_k, then with probability p_{ij} the source moves from a_i to a_j and emits an input $f(a_j)$; the channel then moves from s_k to s_l with probability $M_{f(a_j)}(s_k, s_l)$.

For purposes of analysis we must impose restrictions on the channel. We say that a finite state channel is *regular* if every finite product $M_{i(1)}M_{i(2)} \cdots M_{i(k)}$, $i(1), \ldots, i(k) \in \{1, 2, \ldots, t\}$, $k = 1, 2, \ldots$, is regular; that is, steady state probabilities exist for the Markov chain determined by the matrix $M_{i(1)}M_{i(2)} \cdots M_{i(k)}$. (The integers $i(1), \ldots, i(k)$ are not necessarily distinct.)

We remark that the product of regular matrices need not be regular. For example, take

$$M_1 = \begin{bmatrix} 0 & 1 & 0 \\ 0 & 0 & 1 \\ x & x & x \end{bmatrix} \quad \text{and} \quad M_2 = \begin{bmatrix} x & x & x \\ 1 & 0 & 0 \\ 0 & 1 & 0 \end{bmatrix}.$$

(The x's stand for positive elements.) The product M_1M_2 is

$$\begin{bmatrix} 1 & 0 & 0 \\ 0 & 1 & 0 \\ x & x & x \end{bmatrix},$$

which is not regular; in fact the chain determined by M_1M_2 has two essential sets and is therefore not indecomposable.

The following two theorems provide the justification for the restriction of regularity. Theorem 7.2.1 implies that if a regular Markov source is connected to a finite-state regular channel, the associated input sequence, output sequence, and sequence of input-output pairs are all regular Markov sources and therefore obey the asymptotic equipartition property. This will be essential in the proof of the coding theorem for the channel. Theorem 7.2.2 provides a criterion that allows us to determine in a finite number of operations whether or not a given finite state channel is regular.

Theorem 7.2.1. If a regular Markov source is connected to a finite-state regular channel, the corresponding source-channel matrix is regular.

Proof. Consider any regular Markov source with alphabet $\Gamma = \{b_1, \ldots, b_t\}$. By Theorem 6.3.2, there is a source state a^* and an integer k such that a^* is reachable from any initial state in k steps. In particular, a^* may be reached starting from a^*; hence it is possible for the source to proceed through a sequence of states of the form $a^*a_{i(1)} \cdots a_{i(k-1)}a_{i(k)}$ with $a_{i(k)} = a^*$.

Assume that the input corresponding to $a_{i(j)}$ is $b_{i(j)} \in \Gamma$. Let $M = M_{i(1)}M_{i(2)} \cdots M_{i(k)}$; by hypothesis, M is regular, and consequently there is an integer n such that M^n has a positive column. But

$$M^n = M_{i(1)} \cdots M_{i(k)}M_{i(1)} \cdots M_{i(k)} \cdots M_{i(1)} \cdots M_{i(k)};$$

hence there is some channel state s^* such that s^* is reachable in nk steps from any initial state under the sequence of source states

$$a_{i(1)} \cdots a_{i(k)}a_{i(1)} \cdots a_{i(k)} \cdots a_{i(1)} \cdots a_{i(k)}.$$

Finally, given any source-channel state pair (a, s) we may in k steps reach (a^*, s_i) for some i; but for any i, (a^*, s^*) is reachable from (a^*, s_i) in nk steps. Thus (a^*, s^*) is reachable in $k + nk$ steps from any initial state, so that if Q is the source-channel matrix, the column of $Q^{(n+1)k}$ corresponding to (a^*, s^*) is positive. The proof is complete.

The definition of regularity may be restated as follows. Given any finite input sequence $b_{i(1)}b_{i(2)} \cdots b_{i(k)}$, there must exist a channel state s^* and an integer n such that for any initial channel state, s^* is reachable under the input sequence

$$b_{i(1)}b_{i(2)} \cdots b_{i(k)}b_{i(1)}b_{i(2)} \cdots b_{i(k)} \cdots b_{i(1)}b_{i(2)} \cdots b_{i(k)} \qquad (n \text{ times}).$$

Using this criterion, we may check that the channels of Examples 7.1.1 and 7.1.2 are regular. In Example 7.1.1, if a given input sequence contains at least one "1," then it is possible to reach and remain forever in state

s_{11}; similarly, if the input sequence contains at least one "0," it is possible to reach and remain forever in state s_{00}. In Example 7.1.2, for any input sequence we may reach and remain permanently in state $B0$ (or equally well state $B1$). Thus in both cases the channels are regular.

The definition of regularity seemingly involves the checking of an infinite number of matrices. We may however, replace the definition by a criterion involving only a finite number of steps, as shown by the following theorem.

***Theorem** 7.2.2.* Let $M_1, M_2, \ldots, M_t$ be arbitrary r by r transition matrices. Define sets of matrices $A(k)$, $k = 1, 2, \ldots,$ as follows. Let

$$A(1) = \{M_1, M_2, \ldots, M_t\}.$$

If $k > 1$, let $A(k) =$ the set of all matrices that can be expressed as a product of at most k, not necessarily distinct, elements of $A(1)$. Also define

$$A = \bigcup_{k=1}^{\infty} A(k)$$

Let R be the set of all regular r by r transition matrices. Then if $A(2^{r^2}) \subset R$ then $A \subset R$.

Thus in the definition of regularity we may restrict k to be $\leq 2^{r^2}$, so that only a finite number of operations will be involved.

Proof. As noted in Section 6.3, in deciding whether or not a matrix has steady state probabilities, all we need to know about a given element of the matrix is whether it is zero or positive. In other words, without loss of generality we may assume that we are dealing with matrices with only two possible elements 0 and x (with $xx = x$, $x + x = x$). More precisely let $A'(k)$ be the set of matrices formed by replacing all positive elements of the matrices in $A(k)$ by x. Define A' and R' similarly from A and R. It follows that $A(2^{r^2}) \subset R$ if and only if $A'(2^{r^2}) \subset R'$. Since the number of r by r matrices with elements 0 and x is 2^{r^2}, $A'(k)$ must equal $A'(k+1)$ for some $k \leq 2^{r^2}$. Now by definition of the sets $A(k)$, $A(k+1) \subset A(k+2)$ for all k, and hence $A'(k+1) \subset A'(k+2)$. But if $M \in A'(k+2)$ and $M \notin A'(k+1)$ then $M = M_a M_b$ where $M_a \in A'(k+1) = A'(k)$, $M_b \in A'(1)$. Thus $M \in A'(k+1)$. Consequently $A'(k+1) = A'(k+2)$, and similarly $A'(j) = A'(k)$ for $j \geq k$, so that $A' = A'(2^{r^2})$. Consequently

$$A(2^{r^2}) \subset R \Rightarrow A' = A'(2^{r^2}) \subset R' \Rightarrow A \subset R.$$

If we take $t = 1$ in Theorem 7.2.2, we obtain as a corollary the result that in applying the criterion of Theorem 6.3.2, we need only examine the powers Π^n, $n \leq 2^{r^2}$, of a transition matrix Π in order to determine whether or not the associated Markov chain has steady state probabilities. For if S is the set of r by r transition matrices not having a positive column, an argument identical to that of Theorem 7.2.2 shows that if $A(2^{r^2}) \subset S$, then $A \subset S$. Thus if none of the matrices Π^n, $n \leq 2^{r^2}$, has a positive column, then no power of Π has a positive column so that Π is not regular.

7.3. *The coding theorem for finite-state regular channels*

Suppose that a regular Markov source is connected to a given finite-state regular channel, and Q is the resulting source-channel matrix. Q then determines a finite Markov chain with states (a_i, s_k) where the a_i are the states of the source and the s_k are the states of the channel.

By the assumption of regularity, the chain associated with Q has steady state probabilities. Given the pair (a_i, s_k) we can determine the corresponding input $f_1(a_i, s_k) = f(a_i)$ and output $g_1(a_i, s_k) = g(f(a_i), s_k)$. Thus the chain associated with Q determines a stationary sequence of input-output pairs (X_n, Y_n), $n = 1, 2, \ldots$. We may therefore define an input uncertainty $H\{\mathbf{X}\} = \lim_{n\to\infty} H(X_n \mid X_1, \ldots, X_{n-1})$, an output uncertainty $H\{\mathbf{Y}\} = \lim_{n\to\infty} H(Y_n \mid Y_1, \ldots, Y_{n-1})$, and a joint uncertainty

$$H\{\mathbf{X}, \mathbf{Y}\} = \lim_{n\to\infty} H[(X_n, Y_n) \mid (X_1, Y_1), \ldots, (X_{n-1}, Y_{n-1})].$$

We define the *information conveyed about the process* $\{\mathbf{X}\}$ *by the process* $\{\mathbf{Y}\}$ as

$$\begin{aligned} I[\{\mathbf{X}\} \mid \{\mathbf{Y}\}] &= H\{\mathbf{X}\} + H\{\mathbf{Y}\} - H\{\mathbf{X}, \mathbf{Y}\} \\ &= I[\{\mathbf{Y}\} \mid \{\mathbf{X}\}]. \end{aligned}$$

It follows from Theorem 6.4.1 that

$$I[\{\mathbf{X}\} \mid \{\mathbf{Y}\}] = \lim_{n\to\infty} n^{-1} I(X_1, \ldots, X_n \mid Y_1, \ldots, Y_n)$$

and hence that $I[\{\mathbf{X}\} \mid \{\mathbf{Y}\}]$ is nonnegative. Furthermore, $H\{\mathbf{X}, \mathbf{Y}\} - H\{\mathbf{X}\}$, which may be regarded as the conditional uncertainty of $\{\mathbf{Y}\}$ given $\{\mathbf{X}\}$, is given by $\lim_{n\to\infty} n^{-1} H(Y_1, \ldots, Y_n \mid X_1, \ldots, X_n)$ and also by

$$\lim_{n\to\infty} H(Y_n \mid X_1, \ldots, X_n, Y_1, \ldots, Y_{n-1})$$

(see Problem 7.2).

We define the *capacity* of the given channel as $C =$ the least upper bound of $I[\{\mathbf{X}\} \mid \{\mathbf{Y}\}]$, taken over all regular Markov sources. We are going to prove that it is possible to transmit information at any rate less than C with an arbitrarily small probability of error. First we define a *code* for the finite-state regular channel just as in the memoryless case; that is, a *code* (u, n, ε) is a set of input n-sequences $\mathbf{x}^{(1)}, \ldots, \mathbf{x}^{(u)}$, together with a *decision scheme*, that is, a collection of disjoint sets $B_1, \ldots, B_u$ of output n-sequences, such that

$$P\{(Y_1, \ldots, Y_n) \in B_i \mid (X_1, \ldots, X_n) = \mathbf{x}^{(i)}\} \geq 1 - \varepsilon$$

for all $i = 1, 2, \ldots, u$ and any initial state.

Thus, no matter what the past history of the channel, decoding is accomplished with a probability of error not exceeding ε.

We now prove a lemma that will be useful in the proof of the coding theorem. If Q is the source-channel matrix corresponding to a regular Markov source which is connected to a finite-state regular channel, then Q has steady state probabilities $w_{(a,s)}$, $a \in S_0 =$ set of source states, $s \in S =$ set of channel states. The steady state probability of the channel's being in state s is given by

$$w_s = \sum_a w_{(a,s)}.$$

Define a *code* $(u, n, \varepsilon;\ Q)$ as a set of input n-sequences $\mathbf{x}^{(1)}, \ldots, \mathbf{x}^{(u)}$, together with a decision scheme determined by sets $B_1, \ldots, B_u$, such that if the initial channel state is chosen at random, with state s assigned probability w_s $(s \in S)$, and the sequence $\mathbf{x}^{(i)}$ is transmitted, the probability of error, that is, the probability that the corresponding output sequence will not belong to B_i, is $\leq \varepsilon$ for all i.

Lemma 7.3.1. Let Q be a source-channel matrix for a finite-state regular channel. If for a given positive number R there is a sequence of codes $([2^{nR}], n, \varepsilon_n;\ Q)$ such that $\varepsilon_n \to 0$ as $n \to \infty$, then for any positive number $R' < R$ there is a sequence of codes $([2^{nR'}], n, \varepsilon_n')$ such that $\varepsilon_n' \to 0$ as $n \to \infty$.

Proof. Consider the following random experiment. Let $\mathbf{x}_i$ be any input n-sequence, and let B_i be any set of output n-sequences. Let (a_0, s_0) be any source-channel state pair. Allow the source and channel to make N transitions starting from (a_0, s_0) with the transition probabilities governed by the matrix Q. After N transitions, send the sequence $\mathbf{x}_i$ through the channel.

If (A_k, Z_k) is the source-channel state pair after k transitions, then the probability that the output sequence corresponding to $\mathbf{x}_i$ will not belong to

B_i is given by

$$P\{(Y_{N+1}, \ldots, Y_{N+n}) \notin B_i \mid (X_{N+1}, \ldots, X_{N+n}) = \mathbf{x}_i, (A_0, Z_0) = (a_0, s_0)\}$$
$$= \sum_{a,s} P\{(A_N, Z_N) = (a, s) \mid (A_0, Z_0) = (a_0, s_0)\}$$
$$\times P\{(Y_{N+1}, \ldots, Y_{N+n}) \notin B_i \mid (X_{N+1}, \ldots, X_{N+n}) = \mathbf{x}_i, (A_N, Z_N) = (a, s)\} \tag{7.3.1}$$
$$\xrightarrow{(N\to\infty)} \sum_{a,s} w_{(a,s)} P\{(Y_1, \ldots, Y_n) \notin B_i \mid (X_1, \ldots, X_n) = \mathbf{x}_i, (A_0, Z_0) = (a, s)\}$$
$$= \sum_s w_s P\{(Y_1, \ldots, Y_n) \notin B_i \mid (X_1, \ldots, X_n) = \mathbf{x}_i, Z_0 = s\}. \tag{7.3.2}$$

Note that the second factor of the summand of (7.3.1) is independent of N and hence coincides with the second factor of the summand of (7.3.2). Furthermore, since the number of source-channel state pairs is finite, the convergence of (7.3.1) to (7.3.2) is uniform in n, $\mathbf{x}_i$, B_i, and (a_0, s_0).

By hypothesis there are codes $([2^{nR}], n, \varepsilon_n;\ Q)$ with $\varepsilon_n \to 0$ as $n \to \infty$. But by definition of such codes, the expression (7.3.2), evaluated for a code word $\mathbf{x}_i$ and a decoding set B_i, is just the probability of error when $\mathbf{x}_i$ is sent.

For each n, consider the code $([2^{nR}], n, \varepsilon_n;\ Q)$ with code words $\mathbf{x}_i$ and decoding sets B_i. Given $\varepsilon > 0$, fix $N = N(\varepsilon)$ such that (7.3.1) and (7.3.2) differ by less than ε, uniformly in n, $\mathbf{x}_i$, B_i, and (a_0, s_0). Let n be large enough so that $\varepsilon_n < \varepsilon$ (and hence (7.3.1) is less than 2ε).

Now observe that if i and a_0 are fixed,

$$P\{(Y_{N+1}, \ldots, Y_{N+n}) \notin B_i \mid (X_{N+1}, \ldots, X_{N+n}) = \mathbf{x}_i, (A_0, Z_0) = (a_0, s_0)\} = \sum_{\mathbf{x}} p(\mathbf{x}) g_{s_0}(\mathbf{x})$$

where

$$p(\mathbf{x}) = P\{(X_1, \ldots, X_N) = \mathbf{x} \mid A_0 = a_0\}$$

and

$$g_{s_0}(\mathbf{x}) = P\{(Y_{N+1}, \ldots, Y_{N+n}) \notin B_i \mid (X_1, \ldots, X_{N+n}) = (\mathbf{x}, \mathbf{x}_i), Z_0 = s_0\},$$

and the summation is over all input N-sequences $\mathbf{x}$.

Let r be the total number of channel states. Since $\sum_{\mathbf{x}} p(\mathbf{x}) g_{s_0}(\mathbf{x}) < 2\varepsilon$, the set of all $\mathbf{x}$ such that $g_{s_0}(\mathbf{x}) \geq 4r\varepsilon$ must have a probability $< 1/2r$. For if not, we would have

$$\sum_{\mathbf{x}} p(\mathbf{x}) g_{s_0}(\mathbf{x}) \geq \sum_{\{\mathbf{x}: g_{s_0}(\mathbf{x}) \geq 4r\varepsilon\}} p(\mathbf{x}) g_{s_0}(\mathbf{x})$$
$$\geq 4r\varepsilon P\{\mathbf{x}: g_{s_0}(\mathbf{x}) \geq 4r\varepsilon\} \geq 2\varepsilon,$$

a contradiction.

If $D_{s_0} = \{\mathbf{x}\colon g_{s_0}(\mathbf{x}) \geq 4r\varepsilon\}$, then $\sum_{s_0} P(D_{s_0}) \leq r(1/2r) = \frac{1}{2}$. In particular, the sets D_{s_0} cannot exhaust the entire collection of input N-sequences, so that there is an N-sequence $\mathbf{x}_i'$ belonging to none of the D_{s_0}; for such a sequence, $g_{s_0}(\mathbf{x}_i') < 4r\varepsilon$ *for all possible initial states* s_0.

It follows that if we form codes with words $\mathbf{x}^{(i)} = (\mathbf{x}_i', \mathbf{x}_i)$, $i = 1, 2, \ldots, [2^{nR}]$, and decoding sets $B_i^* = \{(\mathbf{y}_i', \mathbf{y}_i)\colon \mathbf{y}_i \in B_i\}$ (in other words we decode by applying the original decision scheme to the last n coordinates of the received sequence), we will have constructed codes $([2^{nR}], N + n, 4r\varepsilon)$. If we let $m = N + n$ and observe that $2^{nR} = 2^{m(R - m^{-1}NR)} = 2^{mR'}$ where $R' \to R$ as $m \to \infty$, we have the desired result.

Lemma 7.3.1 is analogous to Lemma 3.5.3 in that it asserts that if we can make the "average" probability of error arbitrarily small for a given transmission rate, we can also make the maximum probability of error arbitrarily small.

We may now prove the coding theorem for the finite-state regular channel.

***Theorem** 7.3.2.* Given a finite-state regular channel with capacity $C > 0$ and a positive number $R < C$, there is a sequence of codes $([2^{nR}], n, \varepsilon_n)$ with $\varepsilon_n \to 0$ as $n \to \infty$.

Proof. By definition of C, there is a regular Markov source $\mathcal{S}$ which, when connected to the channel, yields an information $I = I[\{\mathbf{X}\} \mid \{\mathbf{Y}\}]$ which is greater than R. Let Q be the corresponding source-channel matrix. Now suppose that the initial channel state is chosen at random, with state s assigned the steady state probability w_s. Let $p(\mathbf{y} \mid \mathbf{x})$ be the probability that the n-sequence $\mathbf{y}$ is received when the n-sequence $\mathbf{x}$ is transmitted. The system of probability functions $p(\mathbf{y} \mid \mathbf{x})$ forms a discrete channel as defined in Chapter 3. Thus Lemma 3.5.2 applies (recall that we observed at the time that the assumption that the channel be memoryless is not essential to the lemma). In the lemma, we need to specify a probability function $p(\mathbf{x})$ defined on the set of all input n-sequences. Let us take $p(\mathbf{x})$ to be the probability that the source $\mathcal{S}$ produces the sequence $\mathbf{x}$, that is, $p(\mathbf{x}) = P\{(X_1, \ldots, X_n) = \mathbf{x}\}$, when the initial state of the source is chosen at random in accordance with the stationary distribution induced by the matrix Q. [Specifically, the source state a is assigned probability $w_a = \sum_s w_{(a,s)}$, where $w_{(a,s)}$ is the steady state probability of the source-channel state pair (a, s).]

We conclude that for each real number b and positive integers u and n, there is a code $(u, n, \varepsilon_n;\ Q)$ such that

$$\varepsilon_n \le u \cdot 2^{-b} + P\{(\mathbf{x}, \mathbf{y})\colon \log [p(\mathbf{y} \mid \mathbf{x})/p(\mathbf{y})] \le b\}.$$

Let $b = \frac{1}{2}n(R + I)$, $u = [2^{nR}]$. Define a sequence of random variables V_n as follows. If $(X_1, \ldots, X_n) = \boldsymbol{\alpha}$ and $(Y_1, \ldots, Y_n) = \boldsymbol{\beta}$, then

$$V_n = n^{-1} \log \frac{p(\boldsymbol{\beta} \mid \boldsymbol{\alpha})}{p(\boldsymbol{\beta})} = n^{-1} \log \frac{p(\boldsymbol{\alpha}, \boldsymbol{\beta})}{p(\boldsymbol{\alpha})p(\boldsymbol{\beta})}.$$

It follows from the asymptotic equipartition property (Theorem 6.6.1) and the fact that a regular Markov source is ergodic (Theorem 6.6.2) that V_n converges in probability to $H\{\mathbf{X}\} + H\{\mathbf{Y}\} - H\{\mathbf{X}, \mathbf{Y}\} = I$. Since

$$P\{(\mathbf{x}, \mathbf{y}) : \log [p(\mathbf{y} \mid \mathbf{x})/p(\mathbf{y})] \leq b\} = P\{V_n \leq \tfrac{1}{2}(R + I)\} \to 0 \quad \text{as} \quad n \to \infty,$$

we have $\varepsilon_n \to 0$ as $n \to \infty$. The result now follows from Lemma 7.3.1.

The weak converse to the coding theorem holds, that is, any code (u, n, ε) for the finite-state regular channel must satisfy

$$\log u \leq (nC + \log 2)/(1 - \varepsilon)$$

where C is the channel capacity (Blackwell, Breiman, and Thomasian, 1958); hence, if the transmission rate is maintained above capacity, the probability of error cannot be made to approach zero (compare Theorem 3.7.3). In fact, the probability of error must approach one with increasing code-word length (that is, the strong converse to the coding theorem holds) as shown by Wolfowitz (1963b).

7.4. *The capacity of a general discrete channel: comparison of the weak and strong converses*

It is possible to construct a definition of the capacity of a discrete channel in such a way that the definition will not depend on the properties of any specific channel. First, a code (u, n, ε) for an arbitrary discrete channel is defined in exactly the same way as a code for the finite-state channel, that is, a collection of u input n-sequences together with a decision scheme such that the maximum probability of error is $\leq \varepsilon$, regardless of the state of the channel. A nonnegative number R is called a *permissible rate of transmission* for the channel if there exist codes $([2^{nR}], n, \varepsilon_n)$ such that $\varepsilon_n \to 0$ as $n \to \infty$ (n is allowed to approach infinity through a subsequence of the positive integers).

We define the *channel capacity* C as the least upper bound of all permissible transmission rates. (Note that 0 is always a permissible rate, so that the set of permissible rates is not empty.) Thus if $R < C$, the rate R can be maintained with an arbitrarily small probability of error; this is not possible if $R > C$. In the case of the discrete memoryless or the finite-state channel, C agrees with the expressions given previously.

In general, we establish that a particular number C_0 is the capacity of a given channel by proving a *coding theorem* (such as Theorem 3.5.1 or Theorem 7.3.2) and a *weak converse* (such as Theorem 3.7.3). In other words, we prove first that for any $R < C_0$, there exist codes $([2^{nR}], n, \varepsilon_n)$ with $\varepsilon_n \to 0$ as $n \to \infty$, and hence that $C \geq C_0$. We then prove that given any sequence of codes $([2^{nR}], n, \varepsilon_n)$, $R > C_0$, there is an $\varepsilon > 0$ such that for sufficiently large n, $\varepsilon_n \geq \varepsilon$; consequently $C \leq C_0$.

We say that the *strong converse* holds for a given channel if any sequence of codes $([2^{nR}], n, \lambda_n)$, with $R > C$, must be such that $\lim_{n\to\infty} \lambda_n = 1$. To gain further insight into this concept, we introduce the notion of the *λ-capacity* of a discrete channel. Given any discrete channel, let $N(n, \lambda)$ be the maximum possible number of code words in a code that uses sequences of length n and that has maximum probability of error at most λ. If $0 < \lambda < 1$, the *λ-capacity* of the given channel is defined by

$$C(\lambda) = \limsup_{n\to\infty} n^{-1} \log N(n, \lambda) \tag{7.4.1}$$

If one is willing to use codes whose maximum probability of error is $\leq \lambda$ (instead of demanding that the maximum probability of error be made arbitrarily small), then $C(\lambda)$ may be regarded as the least upper bound of all permissible transmission rates. Specifically, we call a number R a *λ-permissible rate of transmission* if given any positive integer n_0, there exists a code $([2^{nR}], n, \lambda_n)$ with $n \geq n_0$ and $\lambda_n \leq \lambda$. It follows directly from the definition of the lim sup of a sequence of numbers that $C(\lambda)$ is the least upper bound of all λ-permissible transmission rates.

Since any code (u, n, λ_1) is automatically a code (u, n, λ_2) if $\lambda_1 < \lambda_2$, it follows that $C(\lambda)$ is nondecreasing. We have the following result.

Lemma 7.4.1. The channel capacity is given by

$$C = \lim_{\lambda\to 0} C(\lambda). \tag{7.4.2}$$

Proof. Let $C_0 = \lim_{\lambda\to 0} C(\lambda)$. If $R < C_0$ and $0 < \lambda < 1$ then $R < C(\lambda)$, hence R is λ-permissible. Thus given any n_0 we can find codes $([2^{nR}], n, \lambda_n)$ with $n \geq n_0$ and $\lambda_n \leq \lambda$. Since λ may be chosen arbitrarily small, it follows that $C \geq C_0$. On the other hand, if $R > C_0$ there is a $\lambda \in (0, 1)$ such that $R > C(\lambda)$. Since R is not λ-permissible, it follows that for sufficiently large n, no code $([2^{nR}], n, \lambda_n)$ can be such that $\lambda_n \leq \lambda$. Consequently no number greater than C_0 can be a permissible transmission rate, and hence $C \leq C_0$.

The strong converse has a striking interpretation in terms of the λ-capacity.

Lemma 7.4.2. The strong converse holds for a discrete channel if and only if

$$C(\lambda) = C \quad \text{for} \quad \text{all } \lambda \in (0, 1). \tag{7.4.3}$$

Proof. If the strong converse holds, then given any $\lambda \in (0, 1)$ and any $R > C$, for sufficiently large n no code $([2^{nR}], n, \lambda_n)$ can exist with $\lambda_n \leq \lambda$.

Therefore R is not λ-permissible for any λ; consequently $C(\lambda) \leq R$ for all λ. Since this holds for any $R > C$, we have $C(\lambda) \leq C$ for all λ. But since $C(\lambda)$ is nondecreasing, Lemma 7.4.1 yields $C(\lambda) = C$ for all $\lambda \in (0, 1)$. Conversely, suppose that $C(\lambda) = C$ for all $\lambda \in (0, 1)$. Given $R > C$ and $\lambda \in (0, 1)$. Since $R > C(\lambda)$, R is not λ-permissible and therefore if we have any sequence of codes $([2^{nR}], n, \lambda_n)$, we must have $\lambda_n > \lambda$ eventually. In other words $\lambda_n \to 1$ as $n \to \infty$, and the strong converse holds.

We are going to give an example of a channel for which the strong converse fails to hold. Our example will be a binary channel with the curious feature that $\lim_{\lambda \to 0} C(\lambda) = 0$, $\lim_{\lambda \to 1} C(\lambda) = 1$ [thus violating (7.4.3)]. Hence if we desire to transmit information with an arbitrarily small probability of error, the only permissible transmission rate is zero, while if we are willing to put up with an arbitrarily high probability of error, we can approach the maximum transmission rate of a noiseless binary channel.

Consider a channel whose inputs and outputs are binary sequences of length $n (n = 1, 2, \ldots)$. The action of the channel is described by specifying that if the output sequence $\mathbf{y}$ differs from the input sequence $\mathbf{x}$ in exactly i places $(i = 0, 1, \ldots, n)$ then

$$p(\mathbf{y} \mid \mathbf{x}) = \left[(n + 1)\binom{n}{i}\right]^{-1}.$$

Thus the number of transmission errors is uniformly distributed between 0 and n, and all error patterns of the same weight have the same probability. (A formal definition of the channel may be readily supplied.)

We shall establish the following result.

Theorem 7.4.3. For $\lambda > \frac{3}{4}$, $C(\lambda) \geq 1 - H(2\lambda - 1)$, where for brevity we write $H(x)$ for the uncertainty $-x \log x - (1 - x) \log (1 - x)$; consequently $\lim_{\lambda \to 1} C(\lambda) = 1$. Furthermore, for all λ, $C(\lambda) \leq 1 - H(\frac{1}{2}(1 - \lambda))$; consequently $\lim_{\lambda \to 0} C(\lambda) = 0$.

Proof. The Varsharmov-Gilbert-Sacks condition (Theorem 4.5.3) states that if

$$2^{n-k} > \sum_{i=0}^{2t-1} \binom{n-1}{i},$$

it is possible to construct a binary code with 2^k words of length n which will correct t-tuple and all smaller errors. Now

$$\sum_{i=0}^{an} \binom{n}{i} \leq 2^{nH(a)} \quad \text{if} \quad 0 \leq a < \tfrac{1}{2} \qquad \text{(Lemma 4.7.2)}$$

Thus if $2^{n-k} > 2^{(n-1)H((2t-1)/(n-1))}$, in particular if $2^{n-k} > 2^{nH(2t/n)}$ (and $2t/n < \frac{1}{2}$), we can find a binary code with 2^k words of length n which corrects all errors of weight t or less. Set $R = k/n$ and $\alpha = 2t/n$. Then the above condition implies that if $R < 1 - H(\alpha)$, for sufficiently large n we can construct a code with at least 2^{nR} words of length n which corrects all errors of magnitude $\leq n\alpha/2$. Let us use this code in the given channel. If $p(e')$ is the probability of correct decoding for any specific word of the code, and Z is the number of transmission errors, then

$$\begin{aligned} p(e') &= P\{Z \leq \tfrac{1}{2}n\alpha\}p(e' \,|\, Z \leq \tfrac{1}{2}n\alpha) + P\{Z > \tfrac{1}{2}n\alpha\}p(e' \,|\, Z > \tfrac{1}{2}n\alpha) \\ &\geq P\{Z \leq \tfrac{1}{2}n\alpha\}p(e' \,|\, Z \leq \tfrac{1}{2}n\alpha) = \frac{\tfrac{1}{2}n\alpha + 1}{n+1} \cdot 1 \\ &\geq \tfrac{1}{2}\alpha. \end{aligned}$$

It follows that R is $(1 - \frac{1}{2}\alpha)$-permissible, where R may be chosen arbitrarily close to $1 - H(\alpha)$. Hence $C(1 - \frac{1}{2}\alpha) \geq 1 - H(\alpha) = 1 - H(1 - \alpha)$ $(\alpha < \frac{1}{2})$, and the first part of the theorem is proved.

To prove the second part of the theorem, consider any code for the given channel, say a code with N words of length n and maximum probability of error λ. Since each error pattern of weight i or $n - i$ has probability

$$\left[(n+1)\binom{n}{i}\right]^{-1},$$

which decreases as i varies from 0 to $\frac{1}{2}n$, it follows that if r is the smallest integer such that $2r/(n + 1) \geq 1 - \lambda$, each decoding set of the code has at least $2\sum_{j=0}^{r-2}\binom{n}{j}$ sequences. [The way to achieve a probability of correct transmission of $1 - \lambda$ with the smallest possible number of sequences in a decoding set is to correct the errors of weight 0 and n, 1 and $n - 1, \ldots,$ $r - 2$ and $n - (r - 2)$, and some or all of the errors of weight $r - 1$ and $n - (r - 1)$.]

Hence, as in Theorem 4.2.2,

$$N \leq \frac{2^n}{2\sum_{j=0}^{r-2}\binom{n}{j}} \leq \frac{2^{n-1}}{\sum_{j=0}^{[\frac{1}{2}(n+1)(1-\lambda)-2]}\binom{n}{j}}$$

Using the fact that

$$n^{-1}\log\left[\sum_{i=0}^{an}\binom{n}{i}\right] \to H(a) \quad \text{as} \quad n \to \infty \quad \text{if} \quad 0 \leq a < \tfrac{1}{2}$$

(Lemma 4.7.2), we obtain

$$C(\lambda) = \limsup_{n\to\infty} n^{-1} \log N(n, \lambda) \leq 1 - H(\tfrac{1}{2}(1 - \lambda)),$$

and the theorem follows.

In the example just given, the variance of the number of transmission errors is of order greater than n (actually of order n^2), as opposed to the case of a binary symmetric channel, where the variance is of the order of n. Wolfowitz (1963a) suggests that this fact is essentially responsible for the nonconstant behavior of $C(\lambda)$.

7.5. *Notes and remarks*

The finite-state channel was introduced and analyzed by Blackwell, Breiman, and Thomasian (1958). These authors achieved greater generality than we have in this chapter by allowing the Markov chains involved in the definition of the channel to be indecomposable rather than simply regular. However, they used several ideas from martingale theory in their presentation, and we have avoided this here.

Theorem 7.2.2 is due to Thomasian (1963).

Just as in the case of the discrete memoryless channel (Theorem 3.6.4), there is an exponential bound on the rate of approach toward zero of the probability of error of codes with a fixed transmission rate; see Blackwell (1961b).

A channel that fits naturally into the finite-state model is the so-called *discrete finite-memory channel.* Intuitively, a channel with memory m should have the property that the distribution of the state of the channel at the end of an input sequence of length n, $n \geq m$, depends only on the last m inputs, and not on any previous inputs or states. Now if an input sequence $x_{i(1)} \cdots x_{i(m)}$ is applied, the distribution of the state at the end of the sequence may be found from the matrix $M = M_{i(1)}M_{i(2)} \cdots M_{i(m)}$, where $M_{i(k)}$ is the channel transition matrix corresponding to the input $x_{i(k)}$. To say that the state distribution at the end of the sequence is independent of the initial state is to say that the matrix M has identical rows. Thus a finite-memory channel may be regarded as a finite-state channel with the property that for some positive integer m, each matrix $M_{i(1)}M_{i(2)} \cdots M_{i(m)}$ (and hence each matrix $M_{i(1)}M_{i(2)} \cdots M_{i(n)}$, $n \geq m$) has identical rows. A transition matrix with identical rows has at least one positive column, and thus a finite-memory channel is regular.

Our approach to the discrete finite-memory channel is that of Blackwell, Breiman, and Thomasian (1958). Alternate treatments may be found in Khinchin (1957), Feinstein (1959), and Wolfowitz (1961).

The results of this chapter have given no clue as to how to compute the capacity of a given finite-state channel. Very little is known in this area. An important initial step has been taken by Wolfowitz (1963b), who showed the existence of a procedure whereby we can achieve any preassigned approximation to the capacity in a finite number of steps. However, practical algorithms do not exist at present. One of the few explicit results is that of Gilbert (1960), who found the capacity of the channel of Example 7.1.2. The capacity of the channel of Example 7.1.1 is apparently unknown.

There is a channel that in a sense combines some of the features of the discrete memoryless channel and the discrete channel with memory. Suppose we have a collection of discrete memoryless channels, one for each element s in some arbitrary set S. A code word is to be transmitted through one of the channels, but the identity of the channel to be used is unknown to the sender and may be regarded as being chosen by "nature." We wish to construct codes that maintain a given transmission rate and at the same time have a low probability of error *regardless of which channel is used.* Formally, we define *a compound channel* (the terminology is that of Wolfowitz, 1961) as a system of probability functions $p_n(\beta_1, \ldots, \beta_n \mid \alpha_1, \ldots, \alpha_n;\ s)$, where $\alpha_1, \ldots, \alpha_n$ belong to a specified finite set Γ called the *input alphabet*, $\beta_1, \ldots, \beta_n$ belong to a specified finite set Γ' called the *output alphabet*, s belongs to an arbitrary set S called the *set of states*, n ranges over the positive integers, and finally for each s and n we have

$$p_n(\beta_1, \ldots, \beta_n \mid \alpha_1, \ldots, \alpha_n;\ s) = p_1(\beta_1 \mid \alpha_1;\ s)p_1(\beta_2 \mid \alpha_2;\ s) \cdots p_1(\beta_n \mid \alpha_n;\ s)$$

The compound channel is thus a special case of the general discrete channel, and a code for the compound channel is therefore defined in the usual way. Notice that we require that the input alphabets be the same for all the individual memoryless channels in the collection indexed by the set S, and similarly for the output alphabets.

Now if $p(x)$ is any probability distribution on the input alphabet, then for each $s \in S$ we may calculate the information $I(s;\ p(x))$ processed by the discrete memoryless channel corresponding to s. The following theorem was proved by Wolfowitz (1960, 1961) and independently by Blackwell, Breiman, and Thomasian (1959).

The capacity of the compound channel is

$$C = \sup_{p(x)} \inf_{s \in S} I(s;\ p(x)).$$

In accordance with the discussion in Section 7.4, the proof may be accomplished by proving a coding theorem and a weak converse. The strong converse to the coding theorem also holds, as was shown by Wolfowitz (1960, 1961).

The notion of the λ-capacity of a channel and the discovery that there are channels for which the strong converse does not hold are due to Wolfowitz (1963a). The example discussed in Theorem 7.4.3 is the author's (1965).

In certain physical situations we may have a discrete channel whose state at the beginning of an input sequence is known to the sender or to the receiver, or to both. For a discussion of this type of problem, see Wolfowitz (1961).

PROBLEMS

7.1 a. Reformulate Example 7.1.1 so that the output is a function of the channel state alone rather than the input-state pair.

b. Show that any finite-state channel model may be reformulated so that the output is a function of the channel state alone, and furthermore, if the original channel is regular so is the new channel, and the capacity of the new channel is the same as that of the original.

7.2 Let $\{X_n, n = 1, 2, \ldots\}$ be a sequence of inputs produced by a regular Markov source that is connected to a finite-state regular channel, and let $\{Y_n, n = 1, 2, \ldots\}$ be the corresponding output sequence. Define the *conditional uncertainty of the sequence* $\{Y_n\}$ *given the sequence* $\{X_n\}$ as

$$H(\{\mathbf{Y}\} \mid \{\mathbf{X}\}) = \lim_{n\to\infty} H(Y_n \mid X_1, \ldots, X_n, Y_1, \ldots, Y_{n-1}).$$

Show that

$$H(\{\mathbf{Y}\} \mid \{\mathbf{X}\}) = \lim_{n\to\infty} n^{-1} H(Y_1, \ldots, Y_n \mid X_1, \ldots, X_n).$$

Hence show that $H\{\mathbf{X}, \mathbf{Y}\} = H\{\mathbf{X}\} + H(\{\mathbf{Y}\} \mid \{\mathbf{X}\})$.

7.3 Let $X_n, n = 1, 2, \ldots$, and $Y_n, n = 1, 2, \ldots$, be stationary sequences of random variables defined on the same sample space, such that the sequence of pairs (X_n, Y_n) is also stationary. (Assume that the X_n and Y_n can take on only finitely many possible values.) Define the conditional uncertainty of the sequence $\{Y_n\}$ given the sequence $\{X_n\}$ as in Problem 7.2. Do the results of that problem still hold?

CHAPTER EIGHT

Continuous Channels

8.1. *Introduction*

In this chapter we consider communication channels that are "continuous" in two senses. First we allow the input and output alphabets to contain an uncountably infinite number of elements, although we require that the material to be transmitted through the channel be in the form of a discrete sequence of symbols. In other words, once every second (say) we transmit a symbol from a possibly uncountable alphabet. This is the so-called *time-discrete, amplitude-continuous* channel. Second, we allow the transmission of information to be continuous in time. In other words, we transmit a "time function" $f(t)$ defined on an interval of real numbers, whose range is also an interval of real numbers. This is the so-called *time-continuous, amplitude-continuous* channel.

The first channel mentioned above is somewhat easier to formalize. Given an arbitrary set S called the *set of states*, we define a time-discrete, amplitude-continuous channel as a system of probability density functions $p_n(y_1, \ldots, y_n \mid x_1, \ldots, x_n; \ s), s \in S, n = 1, 2, \ldots,$ the x_i and y_j real; that is, a system of nonnegative (Borel measurable) functions such that

$$\int_{-\infty}^{\infty} \cdots \int_{-\infty}^{\infty} p_n(y_1, \ldots, y_n \mid x_1, \ldots, x_n; s)\, dy_1 \cdots dy_n = 1 \quad \text{for all } n, s, x_1, \ldots, x_n.$$

Thus for each input sequence and each state, we have a probability measure on the set of all output sequences (in this case, Euclidean n-space). (We could have achieved greater generality by allowing arbitrary probability measures on the set of all output sequences for each input sequence and state.) A *code* (u, n) for such a channel is defined in the now familiar way as a collection of sequences $\mathbf{x}^{(1)}, \ldots . \mathbf{x}^{(u)}$ belonging to Euclidean n-space E^n, together with a *decision scheme*, that is, a partition of E^n into disjoint (Borel) sets $B_1, \ldots, B_u$. A code (u, n, λ) is a code (u, n) with maximum probability of error $\leq \lambda$, that is, a code (u, n) such that

$$P\{(Y_1, \ldots, Y_n) \in B_i \mid (X_1, \ldots, X_n) = \mathbf{x}^{(i)}\} \geq 1 - \lambda$$

for all possible code words $\mathbf{x}^{(i)}$ and all possible states s.

In a given problem, physical considerations may require that some restriction be placed on the set of allowable inputs. For example, if each component of a code word is represented by the amplitude of a pulse, we may wish to require that $|x_i| \leq$ some constant M for all i, or possibly we may have an "average power" limitation of the form $n^{-1} \sum_{i=1}^{n} x_i^2 \leq M$. Given (Borel) subsets F_n of Euclidean n-space ($n = 1\ 2, \ldots$) we may define a *restricted* time-discrete, amplitude-continuous channel corresponding to the F_n as a system of probability density functions

$$p_n(y_1, \ldots, y_n \mid x_1, \ldots, x_n;\ s),$$

defined only for $(x_1, \ldots, x_n) \in F_n$. A code (u, n) and a code (u, n, λ) are defined in exactly the same way as before, except that the code words are now required to belong to the set F_n.

8.2. *The time-discrete Gaussian channel*

In many physical communication situations, disturbances enter in the form of an unwanted "noise" superimposed on the signal to be transmitted, thus producing a result in the form of signal plus noise, from which the desired signal is to be extracted. The assumption that the noise is Gaussian seems to be reasonable in a wide variety of physical settings; the assumption is definitely desirable from the mathematical point of view, as we shall see in the course of the chapter.

A *time-discrete Gaussian channel* is a time-discrete, amplitude-continuous channel whose probability density functions exhibit no dependence on the state s, and are given explicitly by

$$p_n(y_1, \ldots, y_n \mid x_1, \ldots, x_n) = (2\pi N)^{-\frac{1}{2}n} \exp\left[-\sum_{j=1}^{n} \frac{(y_j - x_j)^2}{2N}\right]. \tag{8.2.1}$$

Thus the output sequence is obtained from the input sequence by the addition of a sequence of independent Gaussian random variables with mean 0 and variance N. We shall see that if no restriction is placed on the choice of code words, arbitrarily high transmission rates can be maintained with an arbitrarily small probability of error. We are going to make a restriction on the "average power" of the input sequence; in other words, we introduce a restricted channel determined by sets $F_n = \{(x_1, \ldots, x_n)\colon n^{-1} \sum_{i=1}^{n} x_i^2 \leq M\}$. The resulting channel is called a *time-discrete Gaussian channel with average power limitation M*.

We may establish a relationship between the number of words in a code and the probability of error by proving a result which is analogous to the Fundamental Lemma 3.5.2 and its extension in Problem 3.11. The result is not restricted to the Gaussian case.

Lemma 8.2.1. Given an arbitrary restricted time-discrete, amplitude-continuous channel whose restrictions are determined by sets F_n and whose density functions exhibit no dependence on the state s, let n be a fixed positive integer, and $p(\mathbf{x})$ an arbitrary probability density function on Euclidean n-space. Write $p(\mathbf{y} \mid \mathbf{x})$ for the density $p_n(y_1, \ldots, y_n \mid x_1, \ldots, x_n)$, and write F for F_n.

For any real number a, let

$$A = \left\{(\mathbf{x}, \mathbf{y})\colon \log \frac{p(\mathbf{y} \mid \mathbf{x})}{p(\mathbf{y})} > a\right\}.$$

Then for each positive integer u, there is a code (u, n, λ) (with all code words belonging to F), such that

$$\lambda \le ue^{-a} + P\{(\mathbf{X}, \mathbf{Y}) \notin A\} + P\{\mathbf{X} \notin F\} \tag{8.2.2}$$

where

$$P\{(\mathbf{X}, \mathbf{Y}) \in A\} = \int \cdots \int_A p(\mathbf{x}, \mathbf{y})\, d\mathbf{x}\, d\mathbf{y}, \qquad p(\mathbf{x}, \mathbf{y}) = p(\mathbf{x})p(\mathbf{y} \mid \mathbf{x}),$$

and

$$P\{\mathbf{X} \in F\} = \int \cdots \int_F p(\mathbf{x})\, d\mathbf{x}.$$

[For convenience we assume that *all logarithms in this chapter are to the base e:* thus we have a term ue^{-a} in (8.2.2) instead of $u \cdot 2^{-a}$.]

Proof. The argument proceeds exactly as in Lemma 3.5.2 (and Problem 3.11), with sums replaced by integrals, but we will go through the details for convenience. Let ε be the expression on the right side of (8.2.2). If $\varepsilon \ge 1$, there is nothing to prove, so assume $0 < \varepsilon < 1$. Pick (if possible) a sequence $\mathbf{x}^{(1)} \in F$ such that

$$P\{\mathbf{Y} \in A_{\mathbf{x}^{(1)}} \mid \mathbf{X} = \mathbf{x}^{(1)}\} \ge 1 - \varepsilon \quad \text{where} \quad A_{\mathbf{x}} = \{\mathbf{y} : (\mathbf{x}, \mathbf{y}) \in A\};$$

choose the decoding set B_1 to be $A_{\mathbf{x}^{(1)}}$. Having chosen $\mathbf{x}^{(1)}, \ldots, \mathbf{x}^{(k-1)}$ and $B_1, \ldots, B_{k-1}$, select $\mathbf{x}^{(k)} \in F$ such that

$$P\left\{\mathbf{Y} \in A_{\mathbf{x}^{(k)}} - \bigcup_{i=1}^{k-1} B_i \mid \mathbf{X} = \mathbf{x}^{(k)}\right\} \ge 1 - \varepsilon;$$

set $B_k = A_{\mathbf{x}^{(k)}} - \bigcup_{i=1}^{k-1} B_i$. If the process does not terminate in a finite number of steps, then the sequences $\mathbf{x}^{(i)}$ and decoding sets B_i, $i = 1, 2, \ldots, u$, form the desired code. Thus assume that the process terminates after t steps. (Conceivably $t = 0$.) We will show $t \ge u$ by showing that

$\varepsilon \le te^{-a} + P\{(\mathbf{X}, \mathbf{Y}) \notin A\} + P\{\mathbf{X} \notin F\}$. We proceed as follows. Let $B = \bigcup_{j=1}^{t} B_j$. (If $t = 0$, take $B = \emptyset$.) Then

$$
\begin{aligned}
P\{(\mathbf{X}, \mathbf{Y}) \in A\} &= \int_{(\mathbf{x},\mathbf{y})\in A} p(\mathbf{x}, \mathbf{y})\, d\mathbf{x}\, d\mathbf{y} = \int_{\mathbf{x}} p(\mathbf{x}) \int_{\mathbf{y}\in A_{\mathbf{x}}} p(\mathbf{y} \mid \mathbf{x})\, d\mathbf{y}\, d\mathbf{x} \\
&= \int_{\mathbf{x}} p(\mathbf{x}) \int_{\mathbf{y}\in B\cap A_{\mathbf{x}}} p(\mathbf{y} \mid \mathbf{x})\, d\mathbf{y}\, d\mathbf{x} \\
&\quad + \int_{\mathbf{x}} p(\mathbf{x}) \int_{\mathbf{y}\in \bar{B}\cap A_{\mathbf{x}}} p(\mathbf{y} \mid \mathbf{x})\, d\mathbf{y}\, d\mathbf{x} \\
&= \int_{\mathbf{x}} p(\mathbf{x}) \int_{\mathbf{y}\in B\cap A_{\mathbf{x}}} p(\mathbf{y} \mid \mathbf{x})\, d\mathbf{y}\, d\mathbf{x} \\
&\quad + \int_{\mathbf{x}\in F} p(\mathbf{x}) \int_{\mathbf{y}\in \bar{B}\cap A_{\mathbf{x}}} p(\mathbf{y} \mid \mathbf{x})\, d\mathbf{y}\, d\mathbf{x} \\
&\quad + \int_{\mathbf{x}\notin F} p(\mathbf{x}) \int_{\mathbf{y}\in \bar{B}\cap A_{\mathbf{x}}} p(\mathbf{y} \mid \mathbf{x})\, d\mathbf{y}\, d\mathbf{x}.
\end{aligned}
\tag{8.2.3}
$$

Since $B \cap A_{\mathbf{x}} \subset B$, the first term of (8.2.3) cannot exceed

$$\int_{\mathbf{x}} p(\mathbf{x}) \int_{\mathbf{y}\in B} p(\mathbf{y} \mid \mathbf{x})\, d\mathbf{y}\, d\mathbf{x} = P\{\mathbf{Y} \in B\}.$$

But since $B_i \subset A_{\mathbf{x}^{(i)}}$,

$$P\{\mathbf{Y} \in B\} = \sum_{i=1}^{t} P\{\mathbf{Y} \in B_i\} \le \sum_{i=1}^{t} P\{\mathbf{Y} \in A_{\mathbf{x}^{(i)}}\}.$$

Now if $\mathbf{y} \in A_{\mathbf{x}^{(i)}}$, then $(\mathbf{x}^{(i)}, \mathbf{y}) \in A$; hence $\log [p(\mathbf{y} \mid \mathbf{x}^{(i)})/p(\mathbf{y})] > a$, or $p(\mathbf{y}) < p(\mathbf{y} \mid \mathbf{x}^{(i)})e^{-a}$. Consequently

$$P\{\mathbf{Y} \in A_{\mathbf{x}^{(i)}}\} = \int_{\mathbf{y}\in A_{\mathbf{x}^{(i)}}} p(\mathbf{y})\, d\mathbf{y} \le e^{-a} \int_{\mathbf{y}\in A_{\mathbf{x}^{(i)}}} p(\mathbf{y} \mid \mathbf{x}^{(i)})\, d\mathbf{y} \le e^{-a}.$$

Therefore the first term of (8.2.3) is $\le te^{-a}$.

To estimate the second term, we claim that

$$\int_{\mathbf{y}\in \bar{B}\cap A_{\mathbf{x}}} p(\mathbf{y} \mid \mathbf{x})\, d\mathbf{y} < 1 - \varepsilon \quad \text{for all } \mathbf{x} \in F. \tag{8.2.4}$$

First assume $\mathbf{x}$ = one of the code words $\mathbf{x}^{(k)}$. Then we have

$$\bar{B} \cap A_{\mathbf{x}} = \emptyset. \tag{8.2.5}$$

To establish (8.2.5), we show (as in Lemma 3.5.2) that if $\mathbf{y} \in A_{\mathbf{x}^{(k)}}$ then $\mathbf{y}$ necessarily belongs to B. Since $B_k = A_{\mathbf{x}^{(k)}} - \bigcup_{i=1}^{k-1} B_i$, if $\mathbf{y} \in A_{\mathbf{x}^{(k)}}$ and $\mathbf{y} \notin \bigcup_{i=1}^{k-1} B_i$, then $\mathbf{y} \in B_k \subset B$; on the other hand, if $\mathbf{y} \in A_{\mathbf{x}^{(k)}}$ and $\mathbf{y} \in \bigcup_{i=1}^{k-1} B_i$, then, since B is the union of all the decoding sets, we have $\mathbf{y} \in B$. Thus (8.2.4) is proved when $\mathbf{x}$ is a code word. Now assume that $\mathbf{x} \neq$ any of the $\mathbf{x}^{(i)}$. If $\int_{\mathbf{y} \in \bar{B} \cap A_{\mathbf{x}}} p(\mathbf{y} \mid \mathbf{x})\, d\mathbf{y} \geq 1 - \varepsilon$, then, since $\mathbf{x}$ is assumed $\in F$ and $\bar{B} \cap A_{\mathbf{x}} = A_{\mathbf{x}} - \bigcup_{i=1}^{t} B_i$, we could enlarge the code by taking $\mathbf{x}^{(t+1)} = \mathbf{x}$, $B_{t+1} = \bar{B} \cap A_{\mathbf{x}}$, contradicting the assumption that the process terminates after t steps. Thus the second term of (8.2.3) is $\leq 1 - \varepsilon$.

We estimate the third term of (8.2.3) by observing that $\bar{B} \cap A_{\mathbf{x}}$ is a subset of the entire space E^n, and hence the third term cannot exceed

$$\int_{\mathbf{x} \notin F} p(\mathbf{x}) \int_{\mathbf{y}} p(\mathbf{y} \mid \mathbf{x})\, d\mathbf{y}\, d\mathbf{x} = P\{\mathbf{X} \notin F\}.$$

We conclude that $P\{(\mathbf{X}, \mathbf{Y}) \in A\} \leq te^{-a} + 1 - \varepsilon + P\{\mathbf{X} \notin F\}$, and the result follows.

As in the discussion of Section 7.4, we say that a number R is a *permissible rate of transmission* for a given time-discrete, amplitude-continuous channel (restricted or not) if there exist codes $([e^{nR}], n, \lambda_n)$ with $\lambda_n \to 0$ as $n \to \infty$ (where n is allowed to approach infinity through a subsequence of the positive integers).

The *channel capacity* is defined as the least upper bound of all permissible transmission rates.

Let C be the capacity of the time-discrete, amplitude-continuous Gaussian channel with average power limitation M and "noise" variance N [see (8.2.1)]. We are going to show that

$$C = \tfrac{1}{2} \log \left(1 + \frac{M}{N}\right). \tag{8.2.6}$$

As indicated in Section 7.4, in order to establish (8.2.6) we shall prove a *coding theorem* which states that $C \geq \frac{1}{2} \log(1 + M/N)$ [or equivalently, if $R < \frac{1}{2} \log(1 + M/N)$, we can find codes $([e^{nR}], n, \lambda_n)$ with $\lambda_n \to 0$ as $n \to \infty$], and a *weak converse* which states that $C \leq \frac{1}{2} \log(1 + M/N)$. We first prove the coding theorem.

Theorem 8.2.2. $C \geq \frac{1}{2} \log(1 + M/N)$.

Proof. In Lemma 8.2.1, take

$$p(\mathbf{x}) = (2\pi M_0)^{-\frac{1}{2}n} \exp\left(-\sum_{j=1}^{n} \frac{x_j^{\,2}}{2M_0}\right) \quad \text{where} \quad M_0 < M.$$

In other words, choose the components $X_1, \ldots, X_n$ of the input independently, each component having a Gaussian distribution with mean 0 and variance $M_0 < M$. The third term of (8.2.2) then becomes

$$P\{\mathbf{X} \notin F\} = P\left\{n^{-1}\sum_{j=1}^{n} X_j^{\,2} > M\right\}.$$

But $n^{-1}\sum_{j=1}^{n} X_j^{\,2}$ is the arithmetic average of n independent, identically distributed random variables with expected value $E(X_j^{\,2}) = M_0 < M$. By the weak law of large numbers, the arithmetic average converges in probability to M_0, hence $P\{\mathbf{X} \notin F\} \to 0$ as $n \to \infty$.

Now we note that the output sequence $\mathbf{Y}$ is given by $\mathbf{Y} = \mathbf{X} + \mathbf{Z}$, where the components of $\mathbf{X}$ are independent Gaussian random variables with mean 0 and variance M_0, and the components of $\mathbf{Z} = \mathbf{Y} - \mathbf{X}$ are independent Gaussian random variables with mean 0 and variance N [see (8.2.1)]. By the form of (8.2.1), the conditional distribution of $\mathbf{Z}$ given $\mathbf{X}$ is the same as the (unconditional) distribution of $\mathbf{Z}$, and hence $\mathbf{X}$ and $\mathbf{Z}$ are independent. Since sums of independent normally distributed random variables are also normally distributed, the components of $\mathbf{Y}$ are independent Gaussian random variables with mean 0 and variance $N + M_0$. With this in mind, let us examine the second term of (8.2.2). We have

$$\log \frac{p(\mathbf{y} \mid \mathbf{x})}{p(\mathbf{y})} \le a$$

$$\Leftrightarrow \log \left[\frac{(2\pi N)^{-\frac{1}{2}n} \exp\left(-\sum_{j=1}^{n}(y_j - x_j)^2/2N\right)}{[2\pi(N + M_0)]^{-\frac{1}{2}n} \exp\left(-\sum_{j=1}^{n} y_j^{\,2}/2(N + M_0)\right)}\right] \le a$$

$$\Leftrightarrow \tfrac{1}{2}n \log\left(1 + \frac{M_0}{N}\right) + \frac{1}{2}\sum_{j=1}^{n}\left[\frac{y_j^{\,2}}{(N + M_0)} - \frac{(y_j - x_j)^2}{N}\right] \le a.$$

Let $W_j = Y_j^{\,2}/2(N + M_0) - Z_j^{\,2}/2N$. Since $E(Y_j^{\,2}) = N + M_0$ and $E(Z_j^{\,2}) = N$, W_j has mean 0. We may compute the variance of W_j as follows:

$$\operatorname{Var} W_j = E(W_j^{\,2}) = \frac{1}{4}\left\{E\left[\left(\frac{Y_j}{\sqrt{N + M_0}}\right)^4\right] + E\left[\left(\frac{Z_j}{\sqrt{N}}\right)^4\right] - \frac{2E(Y_j^{\,2}Z_j^{\,2})}{N(N + M_0)}\right\}.$$

Since $Y_j/\sqrt{N + M_0}$ and $Z_j/\sqrt{N}$ are each normal (0, 1),

$$E\left[\left(\frac{Y_j}{\sqrt{N + M_0}}\right)^4\right] = E\left[\left(\frac{Z_j}{\sqrt{N}}\right)^4\right] = 3.$$

Also $E(Y_j^2 Z_j^2) = E[(X_j^2 + 2X_jZ_j + Z_j^2)Z_j^2]$; since X_j and Z_j are independent,

$$E(Y_j^2 Z_j^2) = E(X_j^2)E(Z_j^2) + 2E(X_j)E(Z_j^3) + E(Z_j^4) = M_0N + 0 + 3N^2.$$

Thus

$$\operatorname{Var} W_j = \frac{1}{4}\left[6 - \frac{2N(M_0 + 3N)}{N(M_0 + N)}\right] = \frac{M_0}{N + M_0}.$$

Now let $V_n = \sum_{j=1}^n W_j$. Then $E(V_n) = 0$, $\operatorname{Var} V_n = \sum_{j=1}^n \operatorname{Var} W_j = nM_0/(N + M_0)$ by independence of the W_j; furthermore,

$$\log \frac{p(\mathbf{y} \mid \mathbf{x})}{p(\mathbf{y})} \leq a \Leftrightarrow \tfrac{1}{2}n \log \left(1 + \frac{M_0}{N}\right) + V_n \leq a.$$

Let us choose $a = \frac{1}{2}n \log (1 + M_0/N) - n\delta$, where $\delta > 0$. With this choice, and with the aid of Chebyshev's inequality, the second term of (8.2.2) becomes

$$P\{(\mathbf{X}, \mathbf{Y}) \notin A\} = P\left\{(\mathbf{x}, \mathbf{y})\colon \log \frac{p(\mathbf{y} \mid \mathbf{x})}{p(\mathbf{y})} \leq a\right\}$$
$$= P\{V_n \leq -n\delta\} \leq \frac{E(V_n^2)}{n^2\delta^2} = \frac{M_0}{(N + M_0)\delta^2} \frac{1}{n}.$$

Thus $P\{(\mathbf{X}, \mathbf{Y}) \notin A\} \to 0$ as $n \to \infty$.

Finally, in Lemma 8.2.1 let $u = [e^{nR}]$ where $R < \frac{1}{2} \log (1 + M/N)$. The first term of (8.2.2) becomes

$$u \cdot e^{-a} \leq e^{nR} e^{-\frac{1}{2}n \log (1 + M_0/N)} e^{n\delta}.$$

If M_0 is chosen sufficiently close to M, and δ sufficiently close to 0, we will have $R - \frac{1}{2} \log (1 + M_0/N) + \delta < 0$, and $u \cdot e^{-a}$ will approach 0 as $n \to \infty$. The theorem follows.

8.3. Uncertainty in the continuous case

Before proving the converse to the coding theorem, we must extend the notion of uncertainty to the continuous case, and examine some of the properties of the resulting uncertainty function. If X is an absolutely continuous random variable, that is, a random variable having a probability density function $p(x)$, we define the *uncertainty of X* as

$$H(X) = -\int_{-\infty}^{\infty} p(x) \log p(x)\, dx,$$

provided the integral exists.

In contrast to the situation in the discrete case, $H(X)$ may be arbitrarily large, positive or negative. To see this, let X be uniformly distributed between 0 and b. We then have $H(X) = \log b$, which takes on all real values as b ranges over $(0, \infty)$. In fact $H(X)$ may be $+\infty$ or $-\infty$. For example, let $p(x) = x^{-1}(\log x)^{-2}$, $x \geq e$; $p(x) = 0$, $x < e$. Note that

$$\int_{-\infty}^{\infty} p(x)\, dx = \int_{e}^{\infty} \frac{d(\log x)}{(\log x)^2} = 1.$$

Then

$$H(X) = -\int_{e}^{\infty} x^{-1}(\log x)^{-2}(-\log x - 2\log\log x)\, dx$$

$$\geq \int_{e}^{\infty} \frac{dx}{x \log x} = \int_{e}^{\infty} \frac{d(\log x)}{\log x} = +\infty.$$

As another example, let I_n $(n = 2, 3, 4, \ldots)$ be any system of disjoint intervals of real numbers such that the length of I_n is $(n \log n)^{-2}$. Let $K = \sum_{n=2}^{\infty} n^{-1}(\log n)^{-2}$; note that $K < \infty$ since $\int_{2}^{\infty} x^{-1}(\log x)^{-2}\, dx < \infty$.

Define $p(x) = n/K$ if $x \in I_n$ $(n = 2, 3, 4, \ldots)$; $p(x) = 0$ otherwise. Then

$$\int_{-\infty}^{\infty} p(x)\, dx = K^{-1} \sum_{n=2}^{\infty} n^{-1}(\log n)^{-2} = 1,$$

so that $p(x)$ is a probability density function. Furthermore,

$$H(X) = -\sum_{n=2}^{\infty} \frac{(n/K)\log(n/K)}{(n \log n)^2} = \log K - K^{-1} \sum_{n=2}^{\infty} (n \log n)^{-1}$$

But $\sum_{n=2}^{\infty} (n \log n)^{-1} = \infty$ since $\int_{2}^{\infty} (x \log x)^{-1}\, dx = \infty$, so that

$$H(X) = -\infty.$$

Note that in the above example, $\bigcup_{n=2}^{\infty} I_n$ may be taken to be a bounded set; if this is done, then all the moments of X are finite, and yet $H(X) = -\infty$. We shall see later in this section that the finiteness of the variance of X implies that $H(X) < \infty$. If $p(x) \leq L < \infty$ for all x then $H(X) \geq -\log L > -\infty$, so that boundedness of the density is sufficient to eliminate the case $H(X) = -\infty$.

Now if X and Y are random variables which are defined on the same sample space and which have a joint density $p(x, y)$, we may define the *joint uncertainty* of X and Y as

$$H(X, Y) = -\int_{-\infty}^{\infty}\int_{-\infty}^{\infty} p(x, y) \log p(x, y)\, dx\, dy$$

and the *conditional uncertainty of Y given X* as

$$H(Y \mid X) = -\int_{-\infty}^{\infty}\int_{-\infty}^{\infty} p(x, y) \log p(y \mid x)\, dx\, dy$$

where $p(y \mid x) = p(x, y)/p(x)$ is the conditional density of Y given X.

We may prove results that are the continuous analog of the results of Section 1.4. The following lemma is useful.

Lemma 8.3.1. Let $p(x)$ and $q(x)$ be arbitrary probability density functions.

a. If $-\int_{-\infty}^{\infty} p(x) \log q(x)\, dx$ is finite, then $-\int_{-\infty}^{\infty} p(x) \log p(x)\, dx$ exists, and furthermore

$$-\int_{-\infty}^{\infty} p(x) \log p(x)\, dx \le -\int_{-\infty}^{\infty} p(x) \log q(x)\, dx, \tag{8.3.1}$$

with equality if and only if $p(x) = q(x)$ for almost all x (with respect to Lebesgue measure).

b. If $-\int_{-\infty}^{\infty} p(x) \log p(x)\, dx$ is finite, then $-\int_{-\infty}^{\infty} p(x) \log q(x)\, dx$ exists, and (8.3.1) holds.

Proof. We will prove (a); the argument of (b) is quite similar. Since $\log b \le b - 1$ with equality if and only if $b = 1$ (see Fig. 1.4.1) we have $p(x) \log [q(x)/p(x)] \le q(x) - p(x)$, with equality if and only if $p(x) = q(x)$. (Define $q(x)/p(x) = 0$ when $p(x) = q(x) = 0$, and take $0 \cdot \infty = 0$.) Thus

$$\int_{-\infty}^{\infty} p(x) \log \frac{q(x)}{p(x)}\, dx \le \int_{-\infty}^{\infty} q(x)\, dx - \int_{-\infty}^{\infty} p(x)\, dx = 1 - 1 = 0,$$

with equality if and only if $p(x) = q(x)$ for almost all x. Now

$$-p(x) \log p(x) = p(x) \log \frac{q(x)}{p(x)} - p(x) \log q(x). \tag{8.3.2}$$

(Note that there can be no set of positive measure on which $q(x) = 0$ and $p(x) > 0$, for if there were, $-\int_{-\infty}^{\infty} p(x) \log q(x)\, dx$ would be $+\infty$. Thus the right side of the previous equation will not be of the form $-\infty + \infty$ except possibly on a set of measure 0.)

It follows that

$$-\int_{-\infty}^{\infty} p(x) \log p(x)\, dx \leq -\int_{-\infty}^{\infty} p(x) \log q(x)\, dx.$$

Finally, if equality holds in (8.3.1), the finiteness of $-\int_{-\infty}^{\infty} p(x) \log q(x)\, dx$ allows us to conclude from (8.3.2) that

$$\int_{-\infty}^{\infty} p(x) \log \frac{q(x)}{p(x)}\, dx = 0,$$

and hence that $p(x) = q(x)$ for almost all x. The lemma is proved.

Lemma 8.3.1 has the following immediate consequences.

Lemma 8.3.2. Let X and Y be random variables with a joint density $p(x, y)$. Assume that $H(X)$ and $H(Y)$ are both finite. Then:

a. $H(X, Y)$ exists, and $H(X, Y) \leq H(X) + H(Y)$, with equality if and only if X and Y are independent.

b. $H(Y \mid X)$ and $H(X \mid Y)$ exist, and $H(X, Y) = H(X) + H(Y \mid X) = H(Y) + H(X \mid Y)$.

c. $H(Y \mid X) \leq H(Y)$ [also $H(X \mid Y) \leq H(X)$], with equality if and only if X and Y are independent.

Thus if we define the *information conveyed about X by Y* by $I(X \mid Y) = H(X) - H(X \mid Y)$, we have $I(X \mid Y) = I(Y \mid X)$, and $I(X \mid Y) \geq 0$, with equality if and only if X and Y are independent.

Proof. We note that by Lemma 8.3.1(a),

$$-\int_{-\infty}^{\infty}\int_{-\infty}^{\infty} p(x, y) \log p(x, y)\, dx\, dy \leq -\int_{-\infty}^{\infty}\int_{-\infty}^{\infty} p(x, y) \log p(x)p(y)\, dx\, dy,$$

with equality if and only if $p(x, y) = p(x)p(y)$ for almost all (x, y). The hypothesis that $H(X)$ and $H(Y)$ are finite implies that the integral on the right is finite, so that the application of Lemma 8.3.1 is legitimate. This proves (a).

To prove (b), we note that

$$-\int_{-\infty}^{\infty}\int_{-\infty}^{\infty} p(x, y) \log p(y \mid x)\, dx\, dy = -\int_{-\infty}^{\infty}\int_{-\infty}^{\infty} p(x, y) \log p(x, y)\, dx\, dy + \int_{-\infty}^{\infty}\int_{-\infty}^{\infty} p(x, y) \log p(x)\, dx\, dy.$$

[The first integral on the right exists and is $<\infty$ by (a); the second integral on the right is finite by hypothesis. Thus the integral on the left exists, and we have $H(Y \mid X) = H(X, Y) - H(X)$.]

Finally, (c) follows immediately from (a) and (b) since $H(X, Y) = H(X) + H(Y \mid X) \leq H(X) + H(Y)$, with equality if and only if X and Y are independent.

Just as in Section 1.4, we may extend the notion of uncertainty to random vectors. For example, if $X = (X_1, \ldots, X_n)$, and $X_1, \ldots, X_n$ have a joint density $p(x_1, \ldots, x_n)$, we define

$$H(X) = -\int_{-\infty}^{\infty} \cdots \int_{-\infty}^{\infty} p(x_1, \ldots, x_n) \log p(x_1, \ldots, x_n)\, dx_1 \cdots dx_n.$$

The above results will remain valid in the more general context.

We shall now prove a series of results which will be needed in the proof of the converse to the coding theorem for the time-discrete Gaussian channel. First, we show that the Gaussian distribution has maximum uncertainty among all distributions with a given variance.

Theorem 8.3.3. Let X be an absolutely continuous random variable with density $p(x)$. If X has finite variance σ^2, then $H(X)$ exists, and $H(X) \leq \frac{1}{2} \log (2\pi e \sigma^2)$, with equality if and only if X is Gaussian (with variance σ^2).

Proof. Let X be an arbitrary random variable with density $p(x)$, mean μ, and variance σ^2. Let $q(x) = (2\pi\sigma^2)^{-1/2} e^{-(x-\mu)^2/2\sigma^2}$. But

$$-\int_{-\infty}^{\infty} p(x) \log q(x)\, dx = \int_{-\infty}^{\infty} p(x)\left[\frac{1}{2} \log (2\pi\sigma^2) + \frac{(x-\mu)^2}{2\sigma^2}\right] dx$$

$$= \tfrac{1}{2} \log (2\pi\sigma^2) + \frac{\sigma^2}{2\sigma^2} = \tfrac{1}{2} \log (2\pi e \sigma^2).$$

The result now follows from Lemma 8.3.1a.

We are going to be interested in situations in which a code word is chosen at random and then transmitted through a Gaussian channel.

Since the number of words in a code is finite, we are applying a discrete input to a channel and receiving a continuous output. We must examine the interplay between the uncertainty functions in the discrete and continuous cases.

Let X be a discrete random variable that takes on the values $x_1, \ldots, x_M$. Suppose that Y is a random variable such that for each i, Y has a conditional density given $X = x_i$, say $p(y \mid x_i)$. (In the case that will be of most interest to us, $Y = X + Z$, where Z is a Gaussian random variable.) Then Y is absolutely continuous, with density

$$p(y) = \sum_{i=1}^{M} p(x_i)p(y \mid x_i).$$

We may define the *uncertainty of* Y *given that* $X = x_i$ as

$$H(Y \mid X = x_i) = -\int_{-\infty}^{\infty} p(y \mid x_i) \log p(y \mid x_i)\, dy,$$

and the *uncertainty of* Y *given* X as

$$H(Y \mid X) = \sum_{i=1}^{M} p(x_i) H(Y \mid X = x_i).$$

Furthermore, we may define the *uncertainty of* X *given that* $Y = y$ as

$$H(X \mid Y = y) = -\sum_{i=1}^{M} p(x_i \mid y) \log p(x_i \mid y),$$

where $p(x_i \mid y) = p(x_i)p(y \mid x_i)/p(y)$ is the conditional probability that $X = x_i$ given $Y = y$. The *uncertainty of* X *given* Y is defined as

$$H(X \mid Y) = \int_{-\infty}^{\infty} p(y) H(X \mid Y = y)\, dy.$$

[Notice that the above definitions closely parallel the corresponding definitions in the discrete case; furthermore, $H(X \mid Y)$ is always finite and in fact is bounded by $\log M$.]

Now assume for simplicity that $H(Y \mid X = x_i)$ is finite for each i, and that $H(Y)$ is finite. It follows that

$$H(Y \mid X) \leq H(Y), \tag{8.3.3}$$

with equality if and only if X and Y are independent. For

$$H(Y \mid X) = -\sum_{i=1}^{M} p(x_i) \int_{-\infty}^{\infty} p(y \mid x_i) \log p(y \mid x_i)\, dy.$$

As in the proof of Lemma 8.3.1,

$$\int_{-\infty}^{\infty} p(y \mid x_i) \log \frac{p(y)}{p(y \mid x_i)}\, dy \le \int_{-\infty}^{\infty} p(y \mid x_i)\left[\frac{p(y)}{p(y \mid x_i)} - 1\right] dy = 0,$$

with equality if and only if $p(y) = p(y \mid x_i)$ for almost all y. Thus

$$\sum_{i=1}^{M} p(x_i) \int_{-\infty}^{\infty} p(y \mid x_i) \log \frac{p(y)}{p(y \mid x_i)}\, dy \le 0,$$

with equality if and only if $p(y) = p(y \mid x_i)$ for almost all (x_i, y); in other words, $H(Y \mid X) \le H(Y)$, with equality if and only if X and Y are independent. Furthermore,

$$H(Y) + H(X \mid Y) = H(X) + H(Y \mid X). \tag{8.3.4}$$

For

$$H(X \mid Y) = \int_{-\infty}^{\infty} p(y)\left[-\sum_{i=1}^{M} p(x_i \mid y) \log p(x_i \mid y)\right] dy$$

$$= -\sum_{i=1}^{M} p(x_i) \int_{-\infty}^{\infty} p(y \mid x_i) \log \left[\frac{p(x_i)p(y \mid x_i)}{p(y)}\right] dy$$

$$= H(X) + H(Y \mid X) - H(Y).$$

We define the *information conveyed about X by Y*, just as in the discrete case, by

$$I(X \mid Y) = H(X) - H(X \mid Y).$$

By (8.3.4),

$$I(X \mid Y) = H(Y) - H(Y \mid X);$$

furthermore, by (8.3.3), $I(X \mid Y) \ge 0$, that is,

$$H(X \mid Y) \le H(X), \tag{8.3.5}$$

with equality if and only if X and Y are independent.

We now prove a convexity result that is analogous to Theorem 3.3.1.

Theorem 8.3.4. Let X be a discrete random variable with values $x_1, \ldots, x_M$; let Y be a random variable such that for each i, Y has a conditional density $p(y \mid x_i)$ given that $X = x_i$. Assume $H(Y \mid X = x_i)$ is finite for $i = 1, 2, \ldots, M$. Suppose that for each $k = 1, 2 \ldots, n$, $p_k(\alpha)$, $\alpha = x_1, \ldots, x_M$, is an assignment of probabilities to the values of X.

Let $p_0(\alpha)$ be a convex linear combination of the $p_k(\alpha)$, that is, $p_0(\alpha) = \sum_{k=1}^{n} a_k p_k(\alpha)$, where all $a_k \ge 0$ and $\sum_{k=1}^{n} a_k = 1$.

Then, assuming $H_k(Y)$ is finite for each $k = 1, 2, \ldots, n$,

$$I_0(X \mid Y) \geq \sum_{k=1}^{n} a_k I_k(X \mid Y),$$

where the index k indicates that the quantity in question is computed under the assignment $p_k(\alpha)$.

Proof. We first compute $H_0(Y \mid X)$:

$$H_0(Y \mid X) = -\sum_{i=1}^{M} p_0(x_i) \int_{-\infty}^{\infty} p_0(y \mid x_i) \log p_0(y \mid x_i)\, dy.$$

Since $p_0(y \mid x_i)$ is just the conditional density $p(y \mid x_i)$,

$$\begin{aligned} H_0(Y \mid X) &= -\sum_{k=1}^{n} \sum_{i=1}^{M} a_k p_k(x_i) \int_{-\infty}^{\infty} p(y \mid x_i) \log p(y \mid x_i)\, dy \\ &= \sum_{k=1}^{n} a_k H_k(Y \mid X). \end{aligned} \tag{8.3.6}$$

We must show that $\sum_{k=1}^{n} a_k I_k(X \mid Y) - I_0(X \mid Y) \leq 0$. But

$$\sum_{k=1}^{n} a_k I_k(X \mid Y) - I_0(X \mid Y) = \sum_{k=1}^{n} a_k H_k(Y) - H_0(Y) \qquad \text{[by (8.3.6)]}$$

By hypothesis, $-\int_{-\infty}^{\infty} p_k(y) \log p_k(y)\, dy = H_k(Y)$ is finite, hence by Lemma 8.3.1b, $-\int_{-\infty}^{\infty} p_k(y) \log p_0(y)\, dy$ exists and

$$-\int_{-\infty}^{\infty} p_k(y) \log p_0(y)\, dy \geq -\int_{-\infty}^{\infty} p_k(y) \log p_k(y)\, dy.$$

If we multiply the above inequality by a_k and then sum over $k = 1, 2, \ldots, n$, we see that $H_0(Y)$ exists and $H_0(Y) \geq \sum_{k=1}^{n} a_k H_k(Y)$. The theorem is proved.

Note that Theorem 8.3.4 and the discussion preceding it extend immediately to the case in which X and Y are replaced by random vectors.

8.4. *The converse to the coding theorem for the time-discrete Gaussian channel*

We now apply the results of Section 8.3 to prove the converse to the coding theorem. Just as in the case of the discrete memoryless channel, the key step is *Fano's inequality* (compare Theorem 3.7.1).

Theorem 8.4.1. Given an arbitrary code (s, n) for a time-discrete Gaussian channel, consisting of words $\mathbf{x}^{(1)}, \ldots, \mathbf{x}^{(s)}$. Let $\mathbf{X} = (X_1, \ldots, X_n)$ be a random vector which takes on the value $\mathbf{x}^{(i)}$ with probability $p(\mathbf{x}^{(i)})$, $i = 1, 2, \ldots, s$, where $\sum_{i=1}^{s} p(\mathbf{x}^{(i)}) = 1$. Let $\mathbf{Y} = (Y_1, \ldots, Y_n)$ be the corresponding output sequence.

If $p(e)$ is the overall probability of error of the code, computed for the given input distribution, then

$$H(\mathbf{X} \mid \mathbf{Y}) \leq H(p(e), 1 - p(e)) + p(e) \log (s - 1).$$

Proof. The first part of the proof proceeds exactly as in Theorem 3.7.1. If $g(\mathbf{y})$ is the input selected by the decoder when the sequence $\mathbf{y}$ is received, then by the grouping axiom (Section 1.2, Axiom 3),

$$H(\mathbf{X} \mid \mathbf{Y} = \mathbf{y}) = H(q, 1 - q) + qH(1) + (1 - q)H(q_1, \ldots, q_{s-1})$$

where $q = P\{\mathbf{X} = g(\mathbf{y}) \mid \mathbf{Y} = \mathbf{y}\} = 1 - p(e \mid \mathbf{y})$ and the q_i are of the form

$$\frac{p(\mathbf{x} \mid \mathbf{y})}{\sum_{\mathbf{x} \neq g(\mathbf{y})} p(\mathbf{x} \mid \mathbf{y})}$$

Since $H(q_1, \ldots, q_{s-1}) \leq \log (s - 1)$,

$$H(\mathbf{X} \mid \mathbf{Y} = \mathbf{y}) \leq H(p(e \mid \mathbf{y}), 1 - p(e \mid \mathbf{y})) + p(e \mid \mathbf{y}) \log (s - 1). \quad (8.4.1)$$

We now must deviate slightly from the proof of Theorem 3.7.1. As in the discrete case, we must show that the uncertainty of a convex linear combination of probabilities is greater than or equal to the convex linear combination of the uncertainties, but in this case the convex linear combination is an integral rather than a summation. However, we may reason as follows.

Let V be a random variable that equals 1 if a decoding error occurs, and 0 otherwise. Then

$$\int_{-\infty}^{\infty} p(\mathbf{y}) H(p(e \mid \mathbf{y}), 1 - p(e \mid \mathbf{y}))\, d\mathbf{y} = \int_{-\infty}^{\infty} p(\mathbf{y}) H(V \mid \mathbf{Y} = \mathbf{y})\, d\mathbf{y} = H(V \mid \mathbf{Y}).$$

By (8.3.5),

$$H(V \mid \mathbf{Y}) \leq H(V) = H(p(e), 1 - p(e)). \quad (8.4.2)$$

Thus if we multiply (8.4.1) by $p(\mathbf{y})$ and integrate over all $\mathbf{y}$, we find using (8.4.2) that $H(\mathbf{X} \mid \mathbf{Y}) \leq H(p(e), 1 - p(e)) + p(e) \log (s - 1)$, as required. (The fact that the channel is Gaussian implies that all uncertainties appearing in the above proof are finite, and thus there is no difficulty in applying the results of Section 8.3.)

We shall need one more preliminary result.

Lemma 8.4.2. Let $\mathbf{X} = (X_1, \ldots, X_n)$ be a random vector with a finite number of possible values. Suppose that $\mathbf{X}$ is applied as the input to a time-discrete Gaussian channel, and that $\mathbf{Y}$ is the corresponding output. Then

$$I(\mathbf{X} \mid \mathbf{Y}) \leq \sum_{i=1}^{n} I(X_i \mid Y_i)$$

(compare Lemma 3.7.2).

Proof. We have $I(\mathbf{X} \mid \mathbf{Y}) = H(\mathbf{Y}) - H(\mathbf{Y} \mid \mathbf{X})$; note that since the channel is Gaussian, $H(\mathbf{Y})$ and $H(\mathbf{Y} \mid \mathbf{X})$ are automatically finite. We compute

$$H(\mathbf{Y} \mid \mathbf{X}) = - \sum_{x_1, \ldots, x_n} p(x_1, \ldots, x_n) \int_{-\infty}^{\infty} \cdots \int_{-\infty}^{\infty} p(y_1, \ldots, y_n \mid x_1, \ldots, x_n) \times \log p(y_1, \ldots, y_n \mid x_1, \ldots, x_n)\, dy_1 \cdots dy_n$$

where $p(y_1, \ldots, y_n \mid x_1, \ldots, x_n)$ is given by (8.2.1).

Since $p(y_1, \ldots, y_n \mid x_1, \ldots, x_n) = p(y_1 \mid x_1) \cdots p(y_n \mid x_n)$, it follows that $H(\mathbf{Y} \mid \mathbf{X}) = \sum_{i=1}^{n} H(Y_i \mid X_i)$. Since $H(\mathbf{Y}) = H(Y_1, \ldots, Y_n) \leq \sum_{i=1}^{n} H(Y_i)$ by Lemma 8.3.2a, the result is established.

We now prove the weak converse.

Theorem 8.4.3. Given a time-discrete Gaussian channel with average power limitation M and noise variance N. Any code (s, n) for such a channel must satisfy

$$\log s \leq \frac{nC_0 + \log 2}{1 - \overline{p(e)}}$$

where $C_0 = \frac{1}{2} \log (1 + M/N)$ and $\overline{p(e)}$ is the average probability of error of the code.

Proof. Let the code words be $\mathbf{x}^{(1)} = (x_{11}, x_{12}, \ldots, x_{1n})$, $\mathbf{x}^{(2)} = (x_{21}, x_{22}, \ldots, x_{2n}), \ldots, \mathbf{x}^{(s)} = (x_{s1}, x_{s2}, \ldots, x_{sn})$. Let $\mathbf{X} = (X_1, \ldots, X_n)$ be a code word chosen at random, with all words equally likely, and let $\mathbf{Y} = (Y_1, \ldots, Y_n)$ be the output sequence when $\mathbf{X}$ is transmitted. Now the overall probability of a decoding error associated with the transmission of a code word chosen according to the uniform distribution is $\overline{p(e)}$ (see Section 3.5). Thus by Theorem 8.4.1,

$$\begin{aligned} H(\mathbf{X} \mid \mathbf{Y}) &\leq H(\overline{p(e)}, 1 - \overline{p(e)}) + \overline{p(e)} \log (s - 1) \\ &\leq \log 2 + \overline{p(e)} \log s. \end{aligned} \tag{8.4.3}$$

Again using the fact that all words have the same probability, we have

$$I(\mathbf{X} \mid \mathbf{Y}) = H(\mathbf{X}) - H(\mathbf{X} \mid \mathbf{Y}) = \log s - H(\mathbf{X} \mid \mathbf{Y}). \tag{8.4.4}$$

By Lemma 8.4.2,

$$I(\mathbf{X} \mid \mathbf{Y}) \leq \sum_{i=1}^{n} I(X_i \mid Y_i). \tag{8.4.5}$$

Thus far the proof is an exact duplicate of the argument of Theorem 3.7.3, but we must now deviate. We cannot conclude that $I(X_i \mid Y_i) \leq C_0$ for all i, but we shall show that

$$\sum_{i=1}^{n} I(X_i \mid Y_i) \leq nC_0. \tag{8.4.6}$$

It will then follow from (8.4.3)–(8.4.6) that $nC_0 \geq \log s - \log 2 - \overline{p(e)} \log s$, as asserted.

To prove (8.4.6), we proceed as in Problem 3.11b. Let X be chosen with uniform probability from the sn digits x_{ij}, $i = 1, 2, \ldots, s$, $j = 1, 2, \ldots, n$, of the code vocabulary. Since X may be selected by first choosing a digit j at random and then selecting a code word at random and examining the jth digit of the word, we have

$$P\{X = \alpha\} = n^{-1} \sum_{j=1}^{n} P\{X_j = \alpha\}.$$

Let Y be the output when X is transmitted. It follows from Theorem 8.3.4 that

$$I(X \mid Y) \geq n^{-1} \sum_{j=1}^{n} I(X_j \mid Y_j). \tag{8.4.7}$$

Since the average power limitation of the channel is M, we have $n^{-1} \sum_{j=1}^{n} x_{ij}^2 \leq M$, $i = 1, \ldots, s$. It follows that the variance of X is $\leq M$, for

$$E(X^2) = (sn)^{-1} \sum_{i=1}^{s} \sum_{j=1}^{n} x_{ij}^2 \leq s^{-1} \sum_{i=1}^{s} M = M,$$

$$\text{and} \quad \operatorname{Var} X = E(X^2) - (EX)^2 \leq E(X^2) \leq M.$$

Now we recall that $Y = X + Z$, where Z is Gaussian with mean 0 and variance N, and X and Z are independent. (See the proof of Theorem 8.2.2.) Thus $\operatorname{Var} Y = \operatorname{Var} X + \operatorname{Var} Z \leq M + N$. By Theorem 8.3.3,

$$H(Y) \leq \tfrac{1}{2} \log [2\pi e(M + N)].$$

But $H(Y \mid X) = H(Z)$ (compare Problem 1.9); by Theorem 8.3.3, $H(Z) = \frac{1}{2} \log (2\pi eN)$, so that $I(X \mid Y) = H(Y) - H(Y \mid X) \leq \frac{1}{2} \log (1 + M/N) = C_0$. By (8.4.7), $\sum_{j=1}^{n} I(X_j \mid Y_j) \leq nC_0$, and the proof is complete.

We now prove the strong converse to the coding theorem, that is, the result that if the transmission rate is maintained above C_0, the probability of error must approach 1 with increasing code-word length. Specifically, we have the following theorem.

Theorem 8.4.4. If $\varepsilon > 0$ and $0 \leq \lambda < 1$, then for sufficiently large n, any code (s, n, λ) for the time-discrete Gaussian channel with average power limitation must satisfy

$$\log s < n(C_0 + \varepsilon)$$

where $C_0 = \frac{1}{2}\log(1 + M/N)$.

Proof. The proof is a "sphere-packing" argument similar to the proof of Theorem 4.8.1. In that proof we estimated the number of sequences in the decoding sets; here we estimate the volume of the sets.

Let the code words be $\mathbf{x}^{(1)}, \ldots, \mathbf{x}^{(s)}$ and let the corresponding decoding sets be $B_1, \ldots, B_s$. First we approximate each B_i by a bounded set. Let δ be a fixed positive number, and define $B_i^* = B_i \cap D_n$, where D_n is the sphere in Euclidean n-space with center at the origin and radius $[n(M + N + \delta)]^{1/2}$. It follows that if $\mathbf{X} = \mathbf{x}^{(i)}$ is transmitted, the probability that the output sequence $\mathbf{Y}$ will fall outside of B_i^* is

$$\begin{aligned} P\{\mathbf{Y} \notin B_i^* \mid \mathbf{X} = \mathbf{x}^{(i)}\} &\leq P\{\mathbf{Y} \notin B_i \mid \mathbf{X} = \mathbf{x}^{(i)}\} + P\{\mathbf{Y} \notin D_n \mid \mathbf{X} = \mathbf{x}^{(i)}\} \\ &\leq \lambda + P\{\mathbf{Y} \notin D_n \mid \mathbf{X} = \mathbf{x}^{(i)}\}. \end{aligned} \tag{8.4.8}$$

We shall show that

$$P\{\mathbf{Y} \notin D_n \mid \mathbf{X} = \mathbf{x}^{(i)}\} \to 0 \quad \text{as} \quad n \to \infty. \tag{8.4.9}$$

Let $\mathbf{x}^{(i)} = (x_{i1}, \ldots, x_{in})$, $\mathbf{Y} = (Y_1, \ldots, Y_n)$. Then $\mathbf{Y} = \mathbf{x}^{(i)} + \mathbf{Z}$, where the components $Z_1, \ldots, Z_n$ of $\mathbf{Z}$ are independent Gaussian random variables with mean 0 and variance N. It follows that

$$n^{-1}\sum_{k=1}^{n} Y_k^2 = n^{-1}\sum_{k=1}^{n} x_{ik}^2 + n^{-1}\sum_{k=1}^{n} Z_k^2 + 2n^{-1}\sum_{k=1}^{n} x_{ik}Z_k. \tag{8.4.10}$$

The first term of (8.4.10) is $\leq M$ by the average power limitation; the second term converges in probability to N by the weak law of large numbers. The variance of the third term is

$$\frac{4}{n^2}\sum_{k=1}^{n} x_{ik}^2 N \leq \frac{4NM}{n} \to 0 \quad \text{as} \quad n \to \infty.$$

By Chebyshev's inequality, the third term converges in probability to zero. Thus $P\{n^{-1}\sum_{k=1}^{n} Y_k^2 > M + N + \delta\} \to 0$ as $n \to \infty$. But since this probability is $P\{\mathbf{Y} \notin D_n \mid \mathbf{X} = \mathbf{x}^{(i)}\}$, (8.4.9) is established.

Thus we have a new system of decoding sets† B_i^* such that $B_i^* \subset D_n$ for $i = 1, 2, \ldots, s$. For n sufficiently large, the new code (s, n, λ^*) has, by (8.4.8) and (8.4.9), λ^* arbitrarily close to λ, say $\lambda^* \leq (1 + \lambda)/2 < 1$.

† Strictly speaking, the sets B_i^* do not form a decision scheme as defined in Chapter 3, since the sets do not exhaust the space E^n. However, the B_i^* may be extended to form a partition of E^n. The nature of the extension has no effect on the proof.

Now we shall find a lower bound on the volume of the decoding sets. The lower bound arises from the fact that the decoding sets must be large enough to make the probability of correct transmission at least $1 - \lambda^*$ for each code word. This result, plus the upper bound on the volume resulting from the requirement that $B_i^* \subset D_n$, will yield the strong converse.

Let ε_n be the smallest positive number such that

$$P\left\{n^{-1}\sum_{k=1}^{n} Z_k^2 > N(1 - \varepsilon_n)\right\} \geq \frac{1 + \lambda}{2}.$$

Note that $\varepsilon_n \to 0$ as $n \to \infty$ since $n^{-1}\sum_{k=1}^{n} Z_k^2$ converges in probability to N.

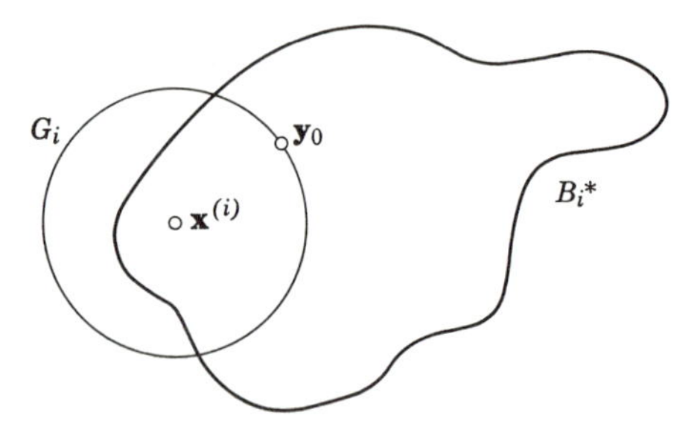

Fig. 8.4.1. Proof of Theorem 8.4.4.

Let G_i be a sphere of radius $[nN(1 - \varepsilon_n)]^{1/2}$ and center at $\mathbf{x}^{(i)}$. We claim that

$$\text{volume } (B_i^*) \geq \text{volume } (G_i), \qquad i = 1, 2, \ldots, s. \tag{8.4.11}$$

To establish (8.4.11), suppose that the volume of B_i^* is less than the volume of G_i for some i. We shall then prove that

$$P\{\mathbf{Y} \in B_i^* \mid \mathbf{X} = \mathbf{x}^{(i)}\} < P\{\mathbf{Y} \in G_i \mid \mathbf{X} = \mathbf{x}^{(i)}\}. \tag{8.4.12}$$

But

$$P\{\mathbf{Y} \in B_i^* \mid \mathbf{X} = \mathbf{x}^{(i)}\} \geq 1 - (1 + \lambda)/2$$

and

$$P\{\mathbf{Y} \in G_i \mid \mathbf{X} = \mathbf{x}^{(i)}\} = P\left\{n^{-1}\sum_{k=1}^{n} Z_k^2 \leq N(1 - \varepsilon_n)\right\} \leq 1 - \frac{1 + \lambda}{2}.$$

This contradicts (8.4.12), and therefore establishes (8.4.11).

To verify (8.4.12), we examine Fig. 8.4.1. Let $\mathbf{y}_0$ be any point on the surface of G_i. Since the Gaussian density $p_n(\mathbf{y} \mid \mathbf{x})$ of (8.2.1) increases as the distance from $\mathbf{x}$ to $\mathbf{y}$ decreases, it follows that

$$P\{\mathbf{Y} \in B_i^* - G_i \mid \mathbf{X} = \mathbf{x}^{(i)}\} = \int_{B_i^*-G_i} p_n(\mathbf{y} \mid \mathbf{x}^{(i)})\, d\mathbf{y}$$
$$\leq \int_{B_i^*-G_i} p_n(\mathbf{y}_0 \mid \mathbf{x}^{(i)})\, d\mathbf{y} = p_n(\mathbf{y}_0 \mid \mathbf{x}^{(i)})[\text{volume}\,(B_i^* - G_i)]. \tag{8.4.13}$$

($B_i^* - G_i$ is the set of points belonging to B_i^* but not to G_i.)

Similarly,

$$P\{\mathbf{Y} \in G_i - B_i^* \mid \mathbf{X} = \mathbf{x}^{(i)}\} = \int_{G_i-B_i^*} p_n(\mathbf{y} \mid \mathbf{x}^{(i)})\, d\mathbf{y}$$
$$\geq \int_{G_i-B_i^*} p_n(\mathbf{y}_0 \mid \mathbf{x}^{(i)})\, d\mathbf{y} = p_n(\mathbf{y}_0 \mid \mathbf{x}^{(i)})[\text{volume}\,(G_i - B_i^*)] \tag{8.4.14}$$

Subtracting (8.4.13) from (8.4.14) we obtain

$$P\{\mathbf{Y} \in G_i - B_i^* \mid \mathbf{X} = \mathbf{x}^{(i)}\} - P\{\mathbf{Y} \in B_i^* - G_i \mid \mathbf{X} = \mathbf{x}^{(i)}\}$$
$$\geq p_n(\mathbf{y}_0 \mid \mathbf{x}^{(i)})[\text{volume}\,(G_i - B_i^*) - \text{volume}\,(B_i^* - G_i)],$$

or

$$P\{\mathbf{Y} \in G_i \mid \mathbf{X} = \mathbf{x}^{(i)}\} - P\{\mathbf{Y} \in B_i^* \mid \mathbf{X} = \mathbf{x}^{(i)}\}$$
$$\geq p_n(\mathbf{y}_0 \mid \mathbf{x}^{(i)})[\text{volume}\,(G_i) - \text{volume}\,(B_i^*)]. \tag{8.4.15}$$

Thus if volume $(B_i^*) <$ volume (G_i), then (8.4.12) follows from (8.4.15).

Finally, the bound (8.4.11) and the requirement that each B_i^* be a subset of D_n imply that

$$\sum_{i=1}^{s} \text{volume}\,(G_i) \leq \text{volume}\left(\bigcup_{j=1}^{s} B_j^*\right) \leq \text{volume}\,(D_n). \tag{8.4.16}$$

If V_n is the volume of an n-dimensional sphere of radius 1, then (8.4.16) yields

$$sV_n[nN(1 - \varepsilon_n)]^{n/2} \leq V_n[n(M + N + \delta)]^{n/2},$$

or

$$\log s \leq \frac{n}{2} \log \frac{M + N + \delta}{N(1 - \varepsilon_n)}$$

If δ is chosen sufficiently small and n is sufficiently large,

$$\log s < n[\tfrac{1}{2} \log\,(1 + M/N) + \varepsilon].$$

The proof is complete.

8.5. *The time-continuous Gaussian channel*

In this section we shall try to construct a reasonable model for a channel that has as its input a function $s(t)$ defined on the interval $[-T/2, T/2]$, and whose output is obtained from the input by the addition of a sample function $n(t)$, $-T/2 \leq t \leq T/2$, of a stationary Gaussian random process. We shall try to show that such a channel may be regarded as a natural generalization of the time-discrete Gaussian channel. In the discussion to follow, we must use certain results about Hilbert space and second-order random processes. The relevant material is summarized in the Appendix.

Let $n(t)$, $-\infty < t < \infty$, be a stationary Gaussian process with zero mean, continuous covariance function $R(\tau)$, and spectral density $N(\omega)$. We assume that for each T, 0 is not an eigenvalue of the integral operator on $L_2[-T/2, T/2]$ associated with R, that is, the operator defined by

$$y(t) = \int_{-T/2}^{T/2} R(t - \tau)x(\tau)\, d\tau, \qquad -\frac{T}{2} \leq t \leq \frac{T}{2}.$$

By the Karhunen-Loève theorem (Appendix, Theorems 3.4, 3.5), the restriction $n_T(t)$ of $n(t)$ to the interval $[-T/2, T/2]$ may be represented by the equation

$$n_T(t) = \sum_{k=1}^{\infty} Z_k e_k(t;\, T),$$

where the e_k are an orthonormal basis for $L_2[-T/2, T/2]$ consisting of eigenfunctions of the integral operator mentioned above, and the Z_k are orthogonal (hence independent) Gaussian random variables with mean 0 and variance $\lambda_k(T) =$ the eigenvalue of e_k. Specifically, $Z_k = \int_{-T/2}^{T/2} n(t)e_k(t;\, T)\, dt$, where the integral is defined in a mean square sense (Appendix, Section 3). The Karhunen-Loève series converges in mean square (and since the process is Gaussian, the series also converges with probability 1) for each $t \in [-T/2, T/2]$. Now (since 0 is not an eigenvalue), the functions e_k span $L_2[-T/2, T/2]$ (Appendix, Theorem 1.14). Thus each function $s_T \in L_2[-T/2, T/2]$ has an expansion of the form

$$s_T(t) = \sum_{k=1}^{\infty} s_k(T)e_k(t;\, T)$$

where the series converges in $L_2[-T/2, T/2]$.

Thus the function s_T is determined by the sequence of numbers $s_k(T)$, and the process n_T is essentially determined by the sequence of random variables Z_k. The above discussion suggests that in effect the channel adds to the sequence of numbers $s_k(T)$ the sequence of random variables Z_k, or equivalently adds to the sequence of numbers $s_k(T)[\lambda_k(T)]^{-1/2}$ the sequence of random variables $Z_k^* = Z_k[\lambda_k(T)]^{-1/2}$.

Just as in the time-discrete case, it will turn out that if no constraint at all is put on the channel inputs, it will be possible to transmit at arbitrarily high rates with arbitrarily small error probabilities. Let us examine the form an "average power" limitation might take. The "energy" in a signal of the form $x(t) = \sum_{k=1}^{\infty} x_k e_k(t;\ T)$ is

$$\int_{-T/2}^{T/2} x^2(t)\,dt = \sum_{k=1}^{\infty} x_k^2,$$

and hence the "average power" in the signal is

$$T^{-1}\int_{-T/2}^{T/2} x^2(t)\,dt = T^{-1}\sum_{k=1}^{\infty} x_k^2.$$

We are led to the following definition.

Let K be a positive real number, fixed for the remainder of the discussion. Any sequence $\{Z_k^*, k = 1, 2, \ldots\}$ of independent Gaussian random variables with mean 0 and variance 1 is said to define a *time-continuous Gaussian channel.* A *code* (u, T) for such a channel is a set of real sequences $\mathbf{x}^{(i)} = (x_{i1}, x_{i2}, \ldots)$, $i = 1, 2, \ldots, u$, called *code words*, such that

$$\sum_{k=1}^{\infty} x_{ik}^2 \leq KT, \qquad i = 1, 2, \ldots, u,$$

together with a *decision scheme*, that is, a partition of the set of all sequences of real numbers into disjoint Borel sets $B_1, \ldots, B_u$. The *probability of error* when $\mathbf{x}^{(i)}$ is transmitted is $p(e \mid \mathbf{x}^{(i)}) = P\{\mathbf{x}^{(i)} + \mathbf{Z} \notin B_i\}$, where $\mathbf{Z} = (Z_1^*, Z_2^*, \ldots)$. A *code* (u, T, β) is a code (u, T) with maximum probability of error $\leq \beta$, that is, $\max_{1 \leq i \leq u} p(e \mid \mathbf{x}^{(i)}) \leq \beta$. A number R is called a *permissible rate of transmission* for the channel if there exist codes $([e^{RT}], T, \beta_T)$ such that $\beta_T \rightarrow 0$ as $T \rightarrow \infty$. (T is allowed to approach infinity through a subsequence of the positive reals.) The *channel capacity* C is the supremum of all permissible transmission rates.

It is shown in the Appendix (Theorem 4.5) that a way of generating inputs which satisfy the "average power" constraint for the time-continuous Gaussian channel is the following. Let s be any function in $L_2(-\infty, \infty)$ whose Fourier transform S vanishes whenever the spectral density N of the process $n(t)$ vanishes. Assume in addition that

$$(2\pi)^{-1}\int_{-\infty}^{\infty}\frac{|S(\omega)|^2}{N(\omega)}\,d\omega \le KT.$$

Then for any T, the restriction $s_T(t) = \sum_{k=1}^{\infty} s_k(T)e_k(t;\ T)$ of s to the interval $[-T/2, T/2]$ satisfies

$$\sum_{k=1}^{\infty}\frac{s_k^{\,2}(T)}{\lambda_k(T)} \le (2\pi)^{-1}\int_{-\infty}^{\infty}\frac{|S(\omega)|^2}{N(\omega)}\,d\omega \le KT,$$

and thus corresponds to an allowable channel input.

The capacity is established by the following theorem.

Theorem 8.5.1. $C = K/2$.

Proof. We first prove a *coding theorem* which states that $C \ge K/2$, or equivalently if $R < K/2$ there exist codes $([e^{RT}], T, \beta_T)$ with $\beta_T \to 0$ as $T \to \infty$. Let T be a positive integer, and let $n = \alpha T$, where α is a fixed positive integer. Consider the time-discrete Gaussian channel with noise variance 1 and average power limitation $KT/n = K/\alpha$. By Theorem 8.2.2, if $R' < \frac{1}{2}\log(1 + K/\alpha)$, there exist codes $([e^{nR'}], n, \varepsilon_n)$ for this channel with $\varepsilon_n \to 0$ as $n \to \infty$. If $(x_1, \ldots, x_n)$ is a code word for the time-discrete channel, we may form a code word for the time-continuous channel by taking $\mathbf{x} = (x_1, x_2, \ldots, x_n, 0, 0, \ldots)$. Since

$$n^{-1}\sum_{i=1}^{n} x_i^{\,2} \le \frac{K}{\alpha} \quad \text{or} \quad \sum_{i=1}^{n} x_i^{\,2} \le KT,$$

such a procedure is allowable. If B_n is the decoding set of $(x_1, \ldots, x_n)$ in the time-discrete channel, we take the decoding set B of $\mathbf{x}$ to be the *cylinder* with base B_n, that is, $B = \{\mathbf{y}: (y_1, \ldots, y_n) \in B_n\}$. In this way we construct a code for the time-continuous channel with precisely the same error probabilities as the corresponding code for the time-discrete channel. Since $nR' = (\alpha R')T$, it follows that if $R = \alpha R' < (\alpha/2)\log(1 + K/\alpha)$, there exist codes $([e^{RT}], T, \beta_T)$ with $\beta_T \to 0$ as $T \to \infty$. Hence $C \ge (\alpha/2)\log(1 + K/\alpha)$. But this holds *for arbitrary* α; therefore we may let $\alpha \to \infty$ to obtain $C \ge K/2$.

We now prove the *weak converse* to the coding theorem, that is, $C \le K/2$. To do this we shall show that any code (u, T, β) with $0 < \beta < \frac{1}{2}$ must satisfy

$$\log u < \frac{\frac{1}{2}KT + \log 2}{1 - 2\beta}.$$

Let $\mathbf{x}^{(1)}, \ldots, \mathbf{x}^{(u)}$ be the code words and let $B_1, \ldots, B_u$ be the corresponding decoding sets. The idea is to approximate the given code by a code for the time-discrete Gaussian channel. To do this, we first approximate the sets B_i by "finite-dimensional" sets, that is, cylinders. We appeal to a standard approximation theorem [for example, Halmos (1950), p. 56, Theorem D] which implies† that for each B_i there is a cylinder B_i^*, whose base $B_{in_i}^*$ is a Borel set in Euclidean n_i-space, such that

$$P\{\mathbf{x}^{(j)} + \mathbf{Z} \in B_i \triangle B_i^*\} \leq \frac{\beta}{2u}, \quad i, j = 1, \ldots, u$$

where $B_i \triangle B_i^* = (B_i - B_i^*) \cup (B_i^* - B_i)$ is the symmetric difference between B_i and B_i^*. (The notation $B_i - B_i^*$ signifies, as before, the set of elements which belong to B_i but not to B_i^*.) The sets B_i^* may not be disjoint, but if we define sets A_i by

$$A_i = B_i^* - \bigcup_{j \neq i} B_j^*,$$

the A_i are a disjoint collection of cylinders. If we set $\mathbf{Y}_i = \mathbf{x}^{(i)} + \mathbf{Z}$, we have

$$\begin{aligned} P\{\mathbf{Y}_i \in A_i\} &= P\{\mathbf{Y}_i \in B_i^*\} - P\left\{\mathbf{Y}_i \in B_i^* \cap \left(\bigcup_{j \neq i} B_j^*\right)\right\} \\ &\geq P\{\mathbf{Y}_i \in B_i^*\} - \sum_{j \neq i} P\{\mathbf{Y}_i \in B_i^* \cap B_j^*\} \\ &\geq P\{\mathbf{Y}_i \in B_i\} - \frac{\beta}{2u} - \sum_{j \neq i} P\{\mathbf{Y}_i \in B_i^* \cap B_j^*\}. \end{aligned}$$

We claim that

$$B_i^* \cap B_j^* \subset (B_i \triangle B_i^*) \cup (B_j \triangle B_j^*) \quad \text{for} \quad i \neq j.$$

To see this, write $B_i^* \cap B_j^* = [(B_i^* \cap B_i) \cup (B_i^* \cap \bar{B}_i)] \cap B_j^*$ (the bar denotes complementation). Since the B_i are disjoint,

$$B_i^* \cap B_i \cap B_j^* \subset B_i^* \cap \bar{B}_j \cap B_j^* \subset B_j^* - B_j.$$

Furthermore, $B_i^* \cap \bar{B}_i \cap B_j^* \subset B_i^* - B_i$. The assertion follows. Thus

$$P\{\mathbf{Y}_i \in A_i\} \geq P\{\mathbf{Y}_i \in B_i\} - \frac{\beta}{2u} - (u-1)\frac{\beta}{u} \geq P\{\mathbf{Y}_i \in B_i\} - \beta \geq 1 - 2\beta.$$

Since the number of code words is finite, there is an integer n such that the base of each cylinder A_i is n-dimensional, that is, for each A_i there is a Borel set A_{in} in Euclidean n-space such that $(y_1, y_2, \ldots) \in A_i$ if and only if $(y_1, y_2, \ldots, y_n) \in A_{in}$. We may now form a code $(u, n, 2\beta)$ for a time-discrete Gaussian channel with noise variance 1, using the code words

† If μ_i is the measure induced by $\mathbf{x}^{(i)} + \mathbf{Z}$, and $\mu = \sum_{k=1}^{u} \mu_k$, then there is a cylinder B_i^* such that $\mu(B_i \triangle B_i^*) \leq \beta/2u$.

$\mathbf{w}^{(i)} = (x_{i1}, \ldots, x_{in})$ = the first n coordinates of $\mathbf{x}^{(i)}$, and the decoding sets† A_i. Since

$$n^{-1}\sum_{k=1}^{n} x_{ik}^{\ 2} \le n^{-1}\sum_{k=1}^{\infty} x_{ik}^{\ 2} \le \frac{KT}{n},$$

Theorem 8.4.3 yields

$$\log u < \frac{(n/2)\log(1 + KT/n) + \log 2}{1 - 2\beta}.$$

But $f(x) = x \log(1 + b/x)$ is an increasing function for $x > 0$ if b is positive [the derivative $f'(x)$ is decreasing for $x > 0$ and $f'(x) \to 0$ as $x \to \infty$, so that $f'(x) > 0$ for $x > 0$]. Now $\lim_{x \to \infty} x \log(1 + b/x) = b$, hence

$$\log u < \frac{KT/2 + \log 2}{1 - 2\beta}.$$

The proof is complete.

It is possible to give an *explicit* procedure for constructing codes for the time-continuous Gaussian channel which maintain any transmission rate up to half the channel capacity with an arbitrarily small probability of error. Before giving the procedure, we need some preliminary results.

Lemma 8.5.2. Given any code (u, n) for the time-discrete Gaussian channel, the ideal observer for the code, calculated for all code words equally likely, is a minimum-distance decoder; that is, for each received sequence $\mathbf{y} = (y_1, \ldots, y_n)$ the decoder selects the code word $\mathbf{x} = (x_1, \ldots, x_n)$ which minimizes the Euclidean distance $[\sum_{i=1}^{n} (x_i - y_i)^2]^{1/2}$ between $\mathbf{x}$ and $\mathbf{y}$.

Proof. The discussion in Section 3.4 may be taken over essentially *in toto;* the ideal observer for a given input distribution selects the input $\mathbf{x}$ which maximizes the conditional probability $p(\mathbf{x} \mid \mathbf{y})$ that $\mathbf{x}$ was transmitted given that $\mathbf{y}$ is received. If all inputs are equally likely, then the decoder selects the input which maximizes $p(\mathbf{y} \mid \mathbf{x})$, where $p(\mathbf{y} \mid \mathbf{x})$ is given by (8.2.1). The result then follows from the form of the Gaussian density.

It follows from Lemma 8.5.2 that if all code words are equidistant from the origin, the ideal observer with equally likely inputs selects the code word $\mathbf{x}$ which maximizes $\sum_{i=1}^{n} x_i y_i$. Let us call this procedure the *correlation detector*. We have the following result.

Lemma 8.5.3. Let $\mathbf{Y} = (Y_1, \ldots, Y_n)$ be the received sequence when the input $\mathbf{x}^{(1)} = (x_{11}, \ldots, x_{1n})$ is transmitted through a time-discrete

† The footnote to Theorem 8.4.4 is applicable here also.

Gaussian channel, that is, $\mathbf{Y} = \mathbf{x}^{(1)} + \mathbf{Z}$, where $\mathbf{Z} = (Z_1, \ldots, Z_n)$ is a sequence of independent Gaussian random variables with mean 0 and variance N. If $\mathbf{x}^{(2)} = (x_{21}, \ldots, x_{2n})$ is any point in Euclidean n-space, define $\gamma = \gamma(\mathbf{x}^{(1)}, \mathbf{x}^{(2)}) = \sum_{k=1}^{n} x_{1k}x_{2k}$. Also define $G(\mathbf{x}^{(i)}) = \sum_{k=1}^{n} x_{ik} Y_k$, $i = 1, 2$.

Assume that $\sum_{k=1}^{n} x_{ik}^2 = L$, $i = 1, 2$. Then $P\{G(\mathbf{x}^{(2)}) \geq G(\mathbf{x}^{(1)})\}$, which is the probability that the correlation detector will favor $\mathbf{x}^{(2)}$ over $\mathbf{x}^{(1)}$, is given by

$$F^*\left[-\left(\frac{L-\gamma}{2N}\right)^{1/2}\right]$$

where $F^*(t) = \int_{-\infty}^{t} (2\pi)^{-1/2} e^{-s^2/2}\, ds$ is the distribution function of a normally distributed random variable with mean 0 and variance 1.

Proof.

$$\begin{aligned} G(\mathbf{x}^{(2)}) - G(\mathbf{x}^{(1)}) &= \sum_{k=1}^{n} (x_{2k} - x_{1k}) Y_k = \sum_{k=1}^{n} (x_{2k} - x_{1k})(x_{1k} + Z_k) \\ &= \gamma - L + \sum_{k=1}^{n} (x_{2k} - x_{1k}) Z_k. \end{aligned}$$

Thus $G(\mathbf{x}^{(2)}) - G(\mathbf{x}^{(1)})$ is normally distributed with mean $\gamma - L$ and variance $N \sum_{k=1}^{n} (x_{2k} - x_{1k})^2 = 2N(L - \gamma)$.

Hence

$$\begin{aligned} &P\{G(\mathbf{x}^{(2)}) - G(\mathbf{x}^{(1)}) \geq 0\} \\ &\quad = P\left\{\frac{G(\mathbf{x}^{(2)}) - G(\mathbf{x}^{(1)}) - (\gamma - L)}{[2N(L-\gamma)]^{1/2}} \geq \frac{-(\gamma - L)}{[2N(L-\gamma)]^{1/2}}\right\} \\ &\quad = F^*\left(\frac{\gamma - L}{[2N(L-\gamma)]^{1/2}}\right) \end{aligned}$$

and the result follows. Note that $|\gamma| \leq L$ by the Schwarz inequality for sums.

Now suppose that we have a code (u, n) for the time-discrete Gaussian channel, with the code words $\mathbf{x}^{(i)} = (x_{i1}, \ldots, x_{in})$ mutually orthogonal; that is, $\gamma(\mathbf{x}^{(i)}, \mathbf{x}^{(j)}) = 0$ for $i \neq j$. Assume that $\sum_{k=1}^{n} x_{ik}^2 = L$ for all $i = 1, 2, \ldots, u$ and that correlation detection is used. We may estimate the probability of error as follows. Suppose that $\mathbf{x}^{(1)}$ is transmitted (the argument will be the same for any code word). By definition of the correlation detector, an error in decoding implies that least one $G(\mathbf{x}^{(i)})$, $i > 1$, is $\geq G(x^{(1)})$. Thus, using Lemma 8.5.3, we have

$$\begin{aligned} p(e \mid \mathbf{x}^{(1)}) &\leq P[G(\mathbf{x}^{(i)}) \geq G(\mathbf{x}^{(1)}) \text{ for at least one } i \geq 2] \\ &\leq \sum_{i=2}^{u} P[G(\mathbf{x}^{(i)}) \geq G(\mathbf{x}^{(1)})] = (u-1)\, F^*(-\sqrt{L/2N}). \end{aligned}$$

Since $F^*(-t) = 1 - F^*(t)$,

$$p(e \mid \mathbf{x}^{(1)}) \le (u-1) \int_{\sqrt{L/2N}}^{\infty} (2\pi)^{-1/2} e^{-x^2/2}\, dx.$$

But $\int_a^\infty (2\pi)^{-1/2} e^{-x^2/2}\, dx \le (2\pi)^{-1/2} a^{-1} e^{-a^2/2}$ if $a > 0$. [This is a standard property of the normal distribution; see, for example, Feller (1950), Chapter 7.]

Thus the maximum probability of error cannot exceed

$$(u-1)\left(\frac{\pi L}{N}\right)^{-1/2} e^{-L/4N}. \tag{8.5.1}$$

We may use this result to construct explicit codes for the time-continuous Gaussian channel.

Theorem 8.5.4. Given a time-continuous Gaussian channel (with an associated fixed constant K) and a positive real number T, let $u = [e^{RT}]$, where $R < K/4$. Construct a code (u, u) for the time-discrete Gaussian channel with noise variance 1 such that the code words are mutually orthogonal and the Euclidean distance between each word and the origin is $(KT)^{1/2}$. (For example, we could take the ith component of the ith code word to be KT and all other components to be 0.) Assume that correlation detection is used. Now convert the code to a code for the time-continuous Gaussian channel, as described in the proof of Theorem 8.5.1. If β_T is the resulting maximum probability of error, then $\beta_T \to 0$ as $T \to \infty$.

Proof. We use (8.5.1) with $N = 1$, $L = KT$, and $u = [e^{RT}]$ to obtain

$$\beta_T \le (\pi KT)^{-1/2}\, e^{RT}\, e^{-KT/4},$$

and the result follows.

8.6. Band-limited channels

It is perhaps appropriate to close the book on a note of controversy. Many communication situations involve signals that are essentially "band-limited," that is, signals whose frequency spectra (Fourier transforms) are essentially confined to a finite interval. However, there is no agreement on a proper model for a channel whose inputs and outputs are band-limited. We shall examine and compare several suggested models.

Let $n(t)$, t real, be a stationary Gaussian process with zero mean, continuous covariance $R(\tau)$, and spectral density $N(\omega)$. The process is said to be *band-limited* if there is a finite interval $[-2\pi W, 2\pi W]$ such that

$N(\omega)$ vanishes outside the interval. W is called the *bandwidth* of the process. We might propose as a model for a band-limited channel the time-continuous Gaussian channel associated with the given process $n(t)$. However, the inputs to such a channel correspond to functions defined on the interval $[-T/2, T/2]$. In effect the signals to be transmitted are "time-limited," that is, vanish outside a finite interval.

Now a function in $L_2(-\infty, \infty)$ (other than the zero function) cannot be both time-limited and band-limited. For if the Fourier transform F of the function f vanishes outside the interval $[-B, B]$ then f is given by

$$f(t) = (2\pi)^{-1} \int_{-B}^{B} F(\omega) e^{i\omega t}\, d\omega. \tag{8.6.1}$$

The integral in (8.6.1) may be regarded as defining a function of the *complex* variable t; we may verify by direct differentiation that the function is analytic everywhere in the complex plane. If such a function vanishes for $|t| > T/2$, t real, then it must vanish everywhere, and hence f must be the zero function. Thus the inputs to the time-continuous Gaussian channel cannot correspond to band-limited signals.

Shannon (1948) takes the following approach. He considers "flat" band-limited noise; that is, he takes the spectral density to be $N(\omega) = N/2$, $-2\pi W \le \omega \le 2\pi W$; $N(\omega) = 0$, $|\omega| > 2\pi W$. The corresponding covariance function is

$$R(\tau) = NW \frac{\sin 2\pi W\tau}{2\pi W\tau}.$$

Shannon forms the allowable inputs to the channel in the following way. He takes a signal which is limited to the same frequency band as the noise, that is, a function $s \in L_2(-\infty, \infty)$ whose Fourier transform vanishes outside the interval $[-2\pi W, 2\pi W]$. The function s is assumed to satisfy the constraint

$$(2\pi)^{-1} \int_{-\infty}^{\infty} \frac{|S(\omega)|^2}{N(\omega)}\, d\omega \le KT,$$

where the integrand is defined to be zero when $N(\omega) = S(\omega) = 0$. If we set $K = 2M/N$ for convenience, the constraint becomes

$$(2\pi)^{-1} \int_{-\infty}^{\infty} |S(\omega)|^2\, d\omega \le MT,$$

or, using the Parseval relation,

$$T^{-1} \int_{-\infty}^{\infty} s^2(t)\, dt \leq M. \tag{8.6.2}$$

The signal transmitted through the channel is the restriction of the function s to the interval $[-T/2, T/2]$. The output is obtained from the input by adding $n(t)$, $|t| \leq T/2$. However, implicit in Shannon's development is the assumption that the "tail" of s, that is, the function $s(t)$, $|t| > T/2$, is not eliminated, but remains to interfere with signals which may be transmitted in the future.

To arrive at an expression for a channel capacity, Shannon reasons as follows. He first establishes the *sampling theorem*, which states that if $s \in L_2(-\infty, \infty)$ and the Fourier transform of s vanishes outside $[-2\pi W, 2\pi W]$, then

$$s(t) = \sum_{k=-\infty}^{\infty} s\left(\frac{k}{2W}\right) \frac{\sin(2\pi W t - k\pi)}{2\pi W t - k\pi},$$

where the series converges in $L_2(-\infty, \infty)$. Thus the function s is determined by its values at a sequence of "sample points" spaced $1/2W$ apart. Now the Fourier transform of $(\sin 2\pi W t)/2\pi W t$ is $(2W)^{-1}u(\omega)$, where $u(\omega) = 1$, $|\omega| \leq 2\pi W$; $u(\omega) = 0$, $|\omega| > 2\pi W$. Thus the Fourier transform of

$$\frac{\sin 2\pi W(t - k/2W)}{2\pi W(t - k/2W)} \quad \text{is} \quad \frac{e^{-i\omega k/2W}}{2W}\, u(\omega).$$

The Parseval relation shows that the functions

$$g_k(t) = \frac{\sin(2\pi W t - k\pi)}{2\pi W t - k\pi}, \qquad k = 0, \pm 1, \pm 2, \ldots,$$

are mutually orthogonal, and

$$\int_{-\infty}^{\infty} g_k^2(t)\, dt = (2\pi)^{-1} \int_{-\infty}^{\infty} \left| \frac{e^{-i\omega k/2W}}{2W}\, u(\omega) \right|^2 d\omega = (2W)^{-1}.$$

It follows that

$$\int_{-\infty}^{\infty} s^2(t)\, dt = (2W)^{-1} \sum_{k=-\infty}^{\infty} s^2\left(\frac{k}{2W}\right). \tag{8.6.3}$$

Now suppose that in forming code words, we restrict our attention to functions whose *sample values* vanish outside the interval $[-T/2, T/2]$.

Specifically, assume that the $s(k/2W)$ are 0 for $k/2W \leq -T/2$ or $k/2W > T/2$, so that there are at most $2WT$ nonzero samples. If we let the $2WT$ numbers $s(k/2W)$, $-T/2 < k/2W \leq T/2$, be labeled $x_1, \ldots, x_n$, with $n = 2WT$, then the constraint (8.6.2) becomes, with the aid of (8.6.3),

$$n^{-1} \sum_{i=1}^{n} x_i^2 \leq M.$$

Since the transmitted signal is determined by $2WT$ sample values, it is possible to decode by examining the output at $2WT$ points spaced $1/2W$ apart. But $R(\tau) = 0$ when $\tau = k/2W$, $k = \pm 1, \pm 2, \ldots$; thus for any t the random variables $n(t + k/2W)$, $k = 0, \pm 1, \pm 2, \ldots$, are orthogonal, hence independent. Since the variance of $n(t)$ is $R(0) = NW$, the situation is completely analogous to the case of a time-discrete Gaussian channel with average power limitation M and noise variance NW. Thus if $R_1 < \frac{1}{2} \log(1 + M/NW)$, or equivalently if $R = 2WR_1 < W \log(1 + M/NW)$, we can find $[e^{nR_1}] = [e^{RT}]$ code words and a corresponding decision scheme with a maximum probability of error which approaches zero as n (and hence T) approaches infinity. Since each code word is derived from a band-limited signal whose nonzero sample values are concentrated in the interval $[-T/2, T/2]$ there will be no interference with the transmission of signals whose nonzero sample values are restricted to the intervals $[T/2, 3T/2]$, $[3T/2, 5T/2]$, etc. Shannon concludes that the channel capacity is $W \log(1 + M/NW)$. However, since Shannon assumes a specific decoding procedure and a specific method of choosing code words, it is possible to conclude from his results only that the capacity is $\geq W \log(1 + M/NW)$; the evaluation of the actual capacity is an open problem. [Observe that Shannon's expression approaches M/N as $W \to \infty$; this is the capacity of the time-continuous Gaussian channel with $K = 2M/N$. [Recall that we set $K = 2M/N$ in the discussion preceding (8.6.2).] This fact may be justified physically if we observe that in the case of infinite bandwidth, we may solve the interference problem by using time-limited code words. Hence in the extreme case of infinite bandwidth (the so-called "white noise" case), the channel we are studying should be equivalent to the time-continuous Gaussian channel as defined in Section 8.5.]

Recently, Wyner has found a model for a band-limited channel in which the expression $W \log(1 + M/NW)$ appears as a true capacity. We shall briefly outline his results. Consider the time-continuous Gaussian channel with flat band-limited noise and $K = 2M/N$. The definition of a code is modified by placing another constraint on the code words. In addition to the restriction that $\sum_{k=1}^{\infty} x_k^2 \leq (2M/N)T$, it is required that $\sum_{k>2WT} x_k^2 = T\,\eta(T)$, where $\eta(T) \to 0$ as $T \to \infty$. Wyner shows that the allowable code words correspond to band-limited signals, truncated to the interval $[-T/2, T/2]$, with the restriction that the fraction of the signal energy which is concentrated in the interval must approach 1 as $T \to \infty$. [The argument uses the properties of band-limited signals derived by Slepian and Pollak (1961) and Landau and Pollak (1961, 1962.)]

If we examine the steps in the derivation that $C \geq K/2$ for the time-continuous Gaussian channel (the direct half of Theorem 8.5.1) for the special case $\alpha = 2W$, we see that the capacity of the channel just described is at least $W \log (1 + M/NW)$. The proof of the converse to the coding theorem and a further discussion of models for band-limited channels will appear in a forthcoming paper by Wyner.

8.7. Notes and remarks

The time-discrete Gaussian channel was introduced by Shannon (1948), who observed that the expression $\frac{1}{2} \log (1 + M/N)$ is the maximum information conveyed about the input X to the channel by the output Y, when X is restricted to be an absolutely continuous random variable with variance $\leq M$. (See the end of the proof of Theorem 8.4.3.) A complete analysis of the channel, including precise exponential bounds on the probability of error, was given by Shannon (1959). The approach we have taken here is that of Thomasian (1961), who gave a proof, similar to that of Theorem 3.6.4, that the probability of error approaches zero exponentially with increasing code-word length if the transmission rate is maintained below channel capacity. The proof of Theorem 8.4.4 is due to Wolfowitz (1961).

As indicated in Section 8.6, Shannon (1948) considered time-continuous channels with additive band-limited Gaussian noise. The general time-continuous Gaussian channel was first analyzed by Bethoux (1962) [see also Fortet (1961)], who, using a different approach from that of Section 8.5, showed that the capacity was at least $K/2$. The approach in the text is that of the author (1963, 1964), who proved that the capacity is in fact $K/2$ and noted that the probability of error approaches zero exponentially as $T \to \infty$ for transmission rates below capacity. (This follows directly from the corresponding exponential bound for the time-discrete Gaussian channel.) A proof of the strong converse for the time-continuous Gaussian channel has been given by Yoshihara (1964).

Lemma 8.3.1 may be used to advantage in certain extremal problems involving the uncertainty function. For example, if we restrict ourselves to the class of random variables having densities that vanish outside the finite interval $[a, b]$, and we take $q(x)$ to be the uniform density over $[a, b]$, we find that $H(X) \leq \log (b - a)$, with equality if and only if X is uniformly distributed between a and b. As another example, if we restrict ourselves to the class of nonnegative absolutely continuous random variables with finite mean $\mu = 1/\lambda$, and we take $q(x) = \lambda e^{-\lambda x}$ $(x \geq 0)$, we have the result that $H(X) \leq 1 - \log \lambda$, with equality if and only if X has the exponential density $\lambda e^{-\lambda x}$.

Some results on the explicit construction of codes for the time-discrete Gaussian channel are given by Slepian (1965).

An adaptation of an argument in Fano (1961), pp. 200–206, shows that the orthogonal coding technique of Theorem 8.5.4 actually realizes the channel capacity, i.e., the theorem holds under the hypothesis that $R < C = K/2$.

Appendix

We collect here enough material from Hilbert space theory to allow a proof of the Karhunen-Loève expansion theorem for random processes, and to obtain some specialized results that are needed for the analysis of the time-continuous Gaussian channel in Chapter 8. As a byproduct, we shall derive some results from the classical theory of integral equations. The material in the first three sections is standard, but does not all appear in a single place. The major references we have used are Liusternik and Sobolev (1961), Taylor (1958), and Riesz and Sz.-Nagy (1955). We assume the elementary facts about Hilbert space which can be found, for example, in Liusternik and Sobolev (1961), pp. 73–80.

1. Compact and symmetric operators on $L_2[a, b]$

We consider the real Hilbert space $L_2 = L_2[a, b]$ of all real-valued (Borel) measurable functions x on the finite or infinite interval $[a, b]$ such that $\int_a^b x^2(t)\, dt < \infty$. The *inner product* of two functions x and y is

$$(x, y) = \int_a^b x(t)y(t)\, dt$$

and the *norm* of a function x is

$$\|x\| = \left[\int_a^b x^2(t)\, dt\right]^{1/2}.$$

By the *Schwarz inequality*, $|(x, y)| \leq \|x\|\ \|y\|$ for all $x, y \in L_2$. Let A be an *operator* on L_2, that is, a function which assigns to each $x \in L_2$ another element $A(x) \in L_2$. A is said to be *linear* if $A(\alpha_1 x_1 + \alpha_2 x_2) = \alpha_1 A(x_1) + \alpha_2 A(x_2)$ for all $x_1, x_2 \in L_2$ and all real α_1, α_2. A is said to be *continuous* if $x_n \to x$ (that is, $\|x_n - x\| \to 0$) implies $A(x_n) \to A(x)$ for every sequence of points $x_n \in L_2$ converging to a point $x \in L_2$. The *norm* of the linear operator A is defined by $\|A\| = \sup_{\|x\|=1} \|Ax\|$; A is said to be *bounded* if $\|A\| < \infty$.

The results to be derived in this section will be valid in any Hilbert space, but since our main interest is in L_2, the results will be stated in this context.

Lemma 1.1. If A is a linear operator on L_2, then A is continuous if and only if A is bounded.

Proof. If A is unbounded, there is a sequence of points $x_n \in L_2$ such that $\|x_n\| = 1$ for all n, and $\|Ax_n\| \to \infty$. Let $y_n = x_n/\|Ax_n\|$. Then $y_n \to 0$, but since Ay_n has norm 1 for all n, $Ay_n \nrightarrow 0$. Thus A is not continuous.

Conversely, if A is discontinuous there is a sequence of points $x_n \to 0$ such that $Ax_n \nrightarrow 0$. Hence there is an $\varepsilon > 0$ and a subsequence of points $x_{n(k)}$ such that $\|Ax_{n(k)}\| \geq \varepsilon$ for all k. Let $y_k = x_{n(k)}/\|x_{n(k)}\|$. Then $\|y_k\| = 1$, but $\|Ay_k\| \geq \epsilon/\|x_{n(k)}\| \to \infty$. Consequently A is unbounded.

Note that if $x \neq 0$, then

$$\|Ax\| = \|x\| \left\| A\left(\frac{x}{\|x\|}\right) \right\|,$$

so that for all x we have $\|Ax\| \leq \|A\|\ \|x\|$; consequently

$$\|A\| = \sup_{\|x\| \leq 1} \|Ax\| = \sup_{x \neq 0} \left(\frac{\|Ax\|}{\|x\|}\right).$$

A linear operator on L_2 is said to be *compact* (also called *completely continuous*) if for every bounded sequence of points $x_n \in L_2$, (that is, every sequence such that for some $K < \infty$ we have $\|x_n\| \leq K$ for all n), the sequence of points Ax_n has a convergent subsequence. Note that *a compact linear operator is continuous.* For if A is discontinuous, then A is unbounded by Lemma 1.1, and hence there is a sequence of points $x_n \in L_2$ such that $\|x_n\| = 1$ and $\|Ax_n\| \to \infty$. Then $\{Ax_n\}$ could not possibly have a convergent subsequence, for such a subsequence would have to be bounded.

As an example, the *identity operator* $[A(x) = x]$ *on* L_2 *is not compact.* Let $x_n(t) = \pi^{-1/2} \sin nt$ on $[0, 2\pi]$; then $\|x_n\| = 1$ for all n, but

$$\begin{aligned}\|x_n - x_m\|^2 &= \|x_n\|^2 + \|x_m\|^2 - 2(x_n, x_m) \\ &= \|x_n\|^2 + \|x_m\|^2 = 2 \quad \text{if} \quad n \neq m.\end{aligned}$$

Thus the distance $\|x_n - x_m\|$ between x_n and x_m is $\sqrt{2}$ for $n \neq m$, and hence $\{x_n\}$ can have no convergent subsequence.

Now let A be a compact linear operator on L_2. For each real number λ, let $A_\lambda = A - \lambda I$, where I is the identity operator. We define two sets

$\rho(A)$ and $\sigma(A)$ as follows. We take $\rho(A)$ to be the set of all real λ such that the operator A_λ is both one to one and onto, that is, the set of all λ such that A_λ^{-1} exists. We take $\sigma(A)$ to be the set of all real λ not belonging to $\rho(A)$, that is, the complement of $\rho(A)$ with respect to the reals. [Note that we are not using the terminology "resolvent set" and "spectrum" for $\rho(A)$ and $\sigma(A)$ since we are dealing exclusively with real scalars.] We shall use the following two results without proof.

Theorem 1.2. If B is a continuous linear operator on L_2 and B^{-1} exists, then B^{-1} is continuous. [See Liusternik and Sobolev (1961), p. 88].

Theorem 1.3. If A is a compact linear operator on L_2 and λ is any nonzero real number, then A_λ is one to one if and only if it is onto.

Theorem 1.3 supports the contention that compact operators are a natural generalization of matrix operators on a finite-dimensional space. For a proof, see Taylor (1958), p. 279, Theorem 5.5E.

Lemma 1.4. Let A be a compact linear operator on L_2, and λ a nonzero real number. Then $\lambda \in \sigma(A)$ if and only if there is a sequence of points $x_n \in L_2$ such that $\|x_n\| = 1$ for all n and $A_\lambda x_n \to 0$.

Proof. Assume a sequence with $\|x_n\| \equiv 1$ and $A_\lambda x_n \to 0$. If $\lambda \notin \sigma(A)$, then A_λ^{-1} exists. Since A_λ^{-1} is continuous by Theorem 1.2, we have $x_n = A_\lambda^{-1} A_\lambda x_n \to 0$, contradicting $\|x_n\| = 1$. Conversely, if $\lambda \in \sigma(A)$ then A_λ^{-1} does not exist. By Theorem 1.3, A_λ is not one to one, hence there is a nonzero $x \in L_2$ such that $A_\lambda x = 0$. If we take $x_n = x/\|x\|$ for all n, then $\|x_n\| \equiv 1$ and $A_\lambda x_n \equiv 0$. The result follows.

A real number λ is called an *eigenvalue* of the linear operator A if $Ax = \lambda x$ for some nonzero x. Any $x \in L_2$ such that $Ax = \lambda x$ is called an *eigenvector* or *eigenfunction* corresponding to λ.

Theorem 1.5. If A is a compact linear operator on L_2 and $\lambda \in \sigma(A)$, $\lambda \neq 0$, then λ is an eigenvalue.

Proof. The result is immediate from Theorem 1.3. However, an alternate argument based on Lemma 1.4 is of interest. By Lemma 1.4, there is a sequence of points with $\|x_n\| \equiv 1$ and $A_\lambda x_n \to 0$. Let $y_n = A_\lambda x_n$; since $\lambda \neq 0$ we have $x_n = \lambda^{-1}(Ax_n - y_n)$. Since A is compact, $\{Ax_n\}$ has a convergent subsequence; since $y_n \to 0$, $\{x_n\}$ has a convergent subsequence. Suppose that $x_{n(k)} = \lambda^{-1}(Ax_{n(k)} - y_{n(k)}) \to x$. Now $y_{n(k)} \to 0$, and hence $\lambda^{-1}Ax_{n(k)} \to x$, or $Ax_{n(k)} \to \lambda x$. But $Ax_{n(k)} \to Ax$ by continuity of A; thus $Ax = \lambda x$. Since $x_{n(k)} \to x$, we have $\|x\| = 1$; in particular, $x \neq 0$. The theorem follows.

A linear operator A on L_2 is said to be *symmetric* (or *self-adjoint*) if $(x_1, Ax_2) = (Ax_1, x_2)$ for all $x_1, x_2 \in L_2$. The *quadratic form* of a symmetric linear operator A is a function from L_2 to the reals defined by

$$Q(x) = (Ax, x) = (x, Ax).$$

The *norm* of the quadratic form Q is defined by $\|Q\| = \sup_{\|x\|=1} |Q(x)|$.

Theorem 1.6. If Q is the quadratic form of the symmetric linear operator A on L_2, then $\|Q\| = \|A\|$.

Proof. We first note that

$$|Q(x)| = |(Ax, x)| = \|x\|^2 \left| \left(\frac{Ax}{\|x\|}, \frac{x}{\|x\|} \right) \right| \leq \|Q\| \; \|x\|^2.$$

Now by the Schwarz inequality,

$$|(Ax, x)| \leq \|Ax\| \; \|x\| \leq \|A\| \; \|x\|^2;$$

if $\|x\| = 1$ we obtain $|Q(x)| \leq \|A\|$; consequently $\|Q\| \leq \|A\|$. To prove $\|Q\| \geq \|A\|$ we must introduce a *bilinear form*

$$B(x_1, x_2) = (Ax_1, x_2) = (x_1, Ax_2).$$

We claim that

$$B(x_1, x_2) = \tfrac{1}{4}[Q(x_1 + x_2) - Q(x_1 - x_2)].$$

To see this, write

$$\begin{aligned} Q(x_1 + x_2) - Q(x_1 - x_2) &= (x_1 + x_2, Ax_1 + Ax_2) - (x_1 - x_2, Ax_1 - Ax_2) \\ &= 4(x_1, Ax_2). \end{aligned}$$

Thus

$$\begin{aligned} |B(x_1, x_2)| &\leq \tfrac{1}{4} \, |Q(x_1 + x_2)| + \tfrac{1}{4} \, |Q(x_1 - x_2)| \\ &\leq \tfrac{1}{4} \, \|Q\| \, [\|x_1 + x_2\|^2 + \|x_1 - x_2\|^2] = \tfrac{1}{2} \, \|Q\| \, (\|x_1\|^2 + \|x_2\|^2). \end{aligned}$$

(The last step may be verified by direct expansion; it is an instance of the so-called *parallelogram law* in Hilbert space, that is, the statement that the sum of the squares of the diagonals of a parallelogram equals twice the sum of the squares of the sides.) Therefore if $\|x_1\| = \|x_2\| = 1$ then $|B(x_1, x_2)| \leq \|Q\|$. Now

$$\begin{aligned} |(x_1, Ax_2)| &= \left| \left(\frac{x_1}{\|x_1\|}, \frac{Ax_2}{\|x_2\|} \right) \right| \; \|x_1\| \; \|x_2\| \\ &\leq \|Q\| \; \|x_1\| \; \|x_2\|. \end{aligned}$$

Let $x_1 = Ax_2$; then

$$\|Ax_2\|^2 \leq \|Q\| \; \|Ax_2\| \; \|x_2\| \quad \text{or} \quad \|Ax_2\| \leq \|Q\| \; \|x_2\|.$$

If $\|x_2\| = 1$, then $\|Ax_2\| \leq \|Q\|$. Thus $\|A\| \leq \|Q\|$.

Theorem 1.7. Let A be a compact symmetric linear operator on L_2. Then A has at least one eigenvalue.

Proof. Let $\lambda_0 = \sup_{\|x\|=1} Q(x)$, $\mu_0 = \inf_{\|x\|=1} Q(x)$, where Q is the quadratic form of A. By replacing A by $-A$ if necessary, we may assume $\lambda_0 \geq 0$. [Note that $\|Q\| = \max(\lambda_0, |\mu_0|)$.] We claim that if $\lambda_0 \geq |\mu_0|$, then λ_0 is an eigenvalue, and if $|\mu_0| \geq \lambda_0$ then μ_0 is an eigenvalue. First suppose that $\lambda_0 \geq |\mu_0|$. If $\lambda_0 = 0$, then $A = 0$; hence every vector in L_2 is an eigenvector for the eigenvalue $\lambda = 0$. Thus assume $\lambda_0 \neq 0$. By definition of λ_0, there is a sequence of points $x_n \in L_2$ such that $\|x_n\| = 1$ for all n and $Q(x_n) \to \lambda_0$. Since $\|A\| = \|Q\|$ by Theorem 1.6, and $\|Q\| = \lambda_0$ by the assumption $\lambda_0 \geq |\mu_0|$, we have

$$\|Ax_n - \lambda_0 x_n\|^2 = \|Ax_n\|^2 - 2\lambda_0 Q(x_n) + \lambda_0^2 \leq \lambda_0^2 - 2\lambda_0 Q(x_n) + \lambda_0^2 \to 0.$$

By Lemma 1.4 and Theorem 1.5, λ_0 is an eigenvalue. If $|\mu_0| \geq \lambda_0$, a similar argument shows that μ_0 is an eigenvalue.

Theorem 1.8. Let $\lambda \neq 0$ be a fixed eigenvalue of the compact linear operator A on L_2. Let E_λ be the subspace of eigenvectors corresponding to λ, that is, $E_\lambda = \{x : Ax = \lambda x\}$. Then E_λ is finite dimensional.

Proof. We shall show that every bounded sequence in E_λ has a convergent subsequence. If $\{x_n\}$ is such a sequence, then $Ax_n = \lambda x_n$, or $x_n = \lambda^{-1} A x_n$. Since A is compact, $\{Ax_n\}$ has a convergent subsequence, and therefore so does $\{x_n\}$.

Now if E_λ were infinite dimensional, we could construct (for example, by the Gram-Schmidt process) an infinite sequence $e_1, e_2, \ldots,$ of mutually orthonormal vectors [that is, $(e_n, e_m) = 0$ for $n \neq m$ and $\|e_n\| = 1$ for all n] in E_λ. The sequence $\{e_n\}$ has no convergent subsequence since

$$\begin{aligned} \|e_n - e_m\|^2 &= \|e_n\|^2 - 2(e_n, e_m) + \|e_m\|^2 \\ &= \|e_n\|^2 + \|e_m\|^2 = 2 \quad \text{for} \quad n \neq m; \end{aligned}$$

thus the distance between e_n and e_m is $\sqrt{2}$ for $n \neq m$, so that there can be no convergent subsequence. It follows that E_λ must be finite dimensional.

We remark that E_λ is a closed set since if $Ax_n = \lambda x_n$ and $x_n \to x$ then by continuity of A we have $Ax = \lambda x$.

Lemma 1.9. Let A be a symmetric linear operator on L_2. Then eigenvectors corresponding to distinct eigenvalues of A are orthogonal, that is, if $Ax = \lambda x$ and $Ay = \mu y$ where $\lambda \neq \mu$, then $(x, y) = 0$.

Proof. Since A is symmetric, $(Ax, y) = (x, Ay)$. Hence $(\lambda x, y) = (x, \mu y)$, or $(\lambda - \mu)(x, y) = 0$. The result follows.

Theorem 1.10. Let A be a compact symmetric linear operator on L_2. Then the set of eigenvalues of A is either finite or countably infinite. In addition, the only possible limit point of eigenvalues is 0.

Proof. Suppose that c is a nonzero limit point of eigenvalues. Let $\{\lambda_n\}$ be a sequence of eigenvalues converging to c, and let x_n be an eigenvector corresponding to λ_n. We may assume that $\lambda_n \neq \lambda_m$ for $n \neq m$, and that $\|x_n\| \equiv 1$. Now

$$\|Ax_n - Ax_m\|^2 = \|\lambda_n x_n - \lambda_m x_m\|^2$$
$$= (\text{by Lemma 1.9})\ \lambda_n^2 + \lambda_m^2 \to 2c^2 > 0 \quad \text{as} \quad n, m \to \infty.$$

Therefore (as in the argument of Theorem 1.8) $\{Ax_n\}$ can have no convergent subsequence, contradicting the compactness of A. Thus 0 is the only possible limit point.

Now if S is the set of eigenvalues, then

$$S = (S \cap \{0\}) \cup \left(\bigcup_{n=1}^{\infty} ([n^{-1}, n] \cap S)\right) \cup \left(\bigcup_{n=1}^{\infty} ([-n, -n^{-1}] \cap S)\right).$$

If S is uncountable, then one of the intervals $[n^{-1}, n]$ or $[-n, -n^{-1}]$ must have uncountably many points of S. But a bounded infinite set of real numbers has a limit point, and hence there is a nonzero limit point of eigenvalues, a contradiction.

Lemma 1.11. If λ is an eigenvalue of the linear operator A, then $|\lambda| \leq \|A\|$. (Note that if A is compact and symmetric, then by the argument of Theorem 1.7, either $\|A\|$ or $-\|A\|$ is an eigenvalue.)

Proof. If $Ax = \lambda x$ where $x \neq 0$, then $|\lambda| = \|Ax\|/\|x\| \leq \|A\|$.

Lemma 1.12. The inner product is a continuous linear functional, that is, if $\{x_n\}$ is a sequence of points in L_2 and $x_n \to x$, then for any $y \in L_2$ we have $(x_n, y) \to (x, y)$.

Proof. By the Schwarz inequality, $|(x, y) - (x_n, y)| = |(x - x_n, y)| \leq \|x - x_n\|\ \|y\| \to 0$.

The converse of Lemma 1.12 is also true. If f is a continuous linear functional on L_2, that is, a continuous linear mapping from L_2 to the reals, then f is given by an inner product, that is, there is an element $y \in L_2$ such that $f(x) = (x, y)$ for all $x \in L_2$. [Liusternik and Sobolev (1961), p. 110.]

The space *spanned* by the subset S of L_2 is defined as the smallest closed subspace of L_2 containing S.

Theorem 1.13. If A is a compact symmetric linear operator on L_2, then the eigenvectors of A span the entire space. In other words, if E is the smallest closed subspace which contains all the eigenvectors of A (corresponding to all the eigenvalues) then $E = L_2$.

Proof. Let $E^\perp$ be the orthogonal complement of E, that is,

$$E^\perp = \{x \in L_2\colon\ (x, y) = 0 \text{ for all } y \in E\}.$$

If $E \neq L_2$, then $E^\perp$ contains a nonzero element. [Liusternik and Sobolev (1961), pp. 75–76.] We note that $E^\perp$ is a closed subspace of L_2. For if $x_n \in E^\perp$ and $x_n \to x$ then for any $y \in E$ we have, by Lemma 1.12, $(x, y) = \lim_{n\to\infty} (x_n, y) = 0$. We now show that $E^\perp$ is invariant under A, that is, $x \in E^\perp$ implies $Ax \in E^\perp$. First of all if $x \in E^\perp$ and y is an eigenvector corresponding to the eigenvalue λ, then $(Ax, y) = (x, Ay) = (x, \lambda y) = \lambda(x, y) = 0$. Thus Ax is orthogonal to all eigenvectors. In general, if $y \in E$ there is a sequence $\{y_n\}$ of finite linear combinations of eigenvectors converging to y. By Lemma 1.12, $(Ax, y) = \lim_{n\to\infty} (Ax, y_n) = 0$; thus $Ax \in E^\perp$.

Since $E^\perp$ is a closed subspace of L_2 (hence is itself a Hilbert space), and $E^\perp$ is invariant under A, we may consider A as a linear operator on $E^\perp$. The restriction of A to $E^\perp$ is still compact and symmetric, and hence by Theorem 1.7 has at least one eigenvalue. Therefore there is at least one nonzero eigenvector $x \in E^\perp$. But we also have $x \in E$ (by definition of E). Since an element of $E^\perp$ is orthogonal to all elements of E, $(x, x) = 0$, which implies $x = 0$, a contradiction. It follows that $E = L_2$.

Theorem 1.14. (Expansion Theorem). Let A be a compact and symmetric linear operator on L_2. Let $\{e_n, n = 1, 2, \ldots\}$ be an orthonormal basis for the space spanned by the eigenvectors corresponding to the nonzero eigenvalues of A. [For example, we might construct such a basis by forming, for each nonzero eigenvalue λ, an orthonormal basis for the finite-dimensional space E_λ (Theorem 1.8), and then taking the union of all such bases. The resulting set of eigenvectors is countable by Theorem 1.10, and orthonormal by Lemma 1.9.]

Then if x is any element of L_2, and h is the projection of x on the space $E_0 = \{y\colon Ay = 0\}$ [see Liusternik and Sobolev (1961), pp. 75–76], then x can be represented as

$$x = h + \sum_{n=1}^{\infty} (x, e_n)e_n,$$

where the series converges in L_2.

If e_n is an eigenvector of the eigenvalue λ_n, then the continuity of A implies that

$$Ax = \sum_{n=1}^{\infty} (x, e_n)\lambda_n e_n.$$

Proof. We first show that $\sum_{n=1}^{\infty} (x, e_n)e_n$ converges in L_2. Using the orthonormality of the e_n we have, for any positive integer r,

$$0 \leq \left\| x - \sum_{k=1}^{r} (x, e_k)e_k \right\|^2 = \|x\|^2 - \sum_{k=1}^{r} |(x, e_k)|^2;$$

hence

$$\sum_{k=1}^{\infty} |(x, e_k)|^2 \leq \|x\|^2 < \infty \qquad \text{(Bessel's inequality)}.$$

Now if $m < n$,

$$\left\| \sum_{i=m}^{n} (x, e_i)e_i \right\|^2 = \sum_{i=m}^{n} |(x, e_i)|^2 \to 0 \quad \text{as} \quad m, n \to \infty.$$

Thus $\sum_{n=1}^{\infty} (x, e_n)e_n$ converges. Consequently we may write

$$x = \left[h + \sum_{n=1}^{\infty} (x, e_n)e_n \right] + \left[x - h - \sum_{n=1}^{\infty} (x, e_n)e_n \right].$$

If $y = x - h - \sum_{n=1}^{\infty} (x, e_n)e_n$, then y is orthogonal to all e_n. For $(h, e_n) = 0$ by Lemma 1.9, and $(\sum_{k=1}^{\infty} (x, e_k)e_k, e_n) = (x, e_n)$ by Lemma 1.12; thus $(y, e_n) = (x, e_n) - (x, e_n) = 0$. Furthermore, y is orthogonal to all elements of E_0. For if $h' \in E_0$ then $(y, h') = (x, h') - (h, h') = (x - h, h')$. Since h is the projection of x on E_0, $x - h$ is orthogonal to E_0, hence $(x - h, h') = 0$. Therefore y is orthogonal to all eigenvectors, so by Theorem 1.13, y is orthogonal to all elements of L_2. In particular, $(y, y) = 0$ and thus $y = 0$. The result follows.

2. *Integral operators*

We shall consider integral operators on $L_2[a, b]$, a, b *finite*, defined by

$$y(t) = \int_a^b R(t, \tau)x(\tau)\, d\tau, \qquad a \leq t \leq b, \tag{2.1}$$

where R is a real-valued function which is continuous in both variables.

Note that the formula (2.1) does in fact define a linear operator on L_2. For let $x \in L_2$, and $M = \max_{t,\tau \in [a,b]} |R(t, \tau)|$. Then by the Schwarz inequality,

$$|y(t)|^2 \leq \int_a^b R^2(t, \tau)\, d\tau \int_a^b x^2(\tau)\, d\tau \leq M^2(b - a)\, \|x\|^2.$$

Hence

$$\|y\|^2 = \int_a^b y^2(t)\, dt \leq M^2(b - a)^2\, \|x\|^2 < \infty.$$

In fact if $x \in L_2$ then y is continuous for all $t \in [a, b]$. For we have $|R(t, \tau)x(\tau)| \leq M\,|x(\tau)|$, which is integrable over $[a, b]$ by the Schwarz inequality:

$$\int_a^b |x(\tau)|\,d\tau = \int_a^b 1 \cdot |x(\tau)|\,d\tau \leq (b-a)^{1/2}\,\|x\| < \infty.$$

Thus we may apply the Lebesgue dominated convergence theorem [for example, Loève (1955), pp. 125–126] to obtain

$$\lim_{t\to t_0} y(t) = \lim_{t\to t_0} \int_a^b R(t, \tau)x(\tau)\,d\tau = \int_a^b \left[\lim_{t\to t_0} R(t, \tau)\right] x(\tau)\,d\tau$$
$$= \int_a^b R(t_0, \tau)x(\tau)\,d\tau = y(t_0).$$

In particular if e is an eigenfunction of (2.1) corresponding to the eigenvalue λ, then $\lambda e(t) = \int_a^b R(t, \tau)e(\tau)\,d\tau$. Hence:

The eigenfunctions of (2.1) corresponding to nonzero eigenvalues are continuous. (2.2)

We remark that any function continuous on $[a, b]$ has a maximum on $[a, b]$ and hence belongs to $L_2[a, b]$.

Theorem 2.1. The integral operator (2.1) is compact.

Proof. We use the *Arzela-Ascoli theorem* [for example, Liusternik and Sobolev (1961), p. 39], which states that if $\{y_k, k = 1, 2, \ldots\}$ is a sequence of continuous functions on $[a, b]$, and the y_k are *uniformly bounded* (that is, $|y_k(s)| \leq L < \infty$ for all k, s) and *equicontinuous* (given $\varepsilon > 0$, there is a $\delta > 0$, where δ depends only on ε, not on k, such that if $s_1, s_2 \in [a, b]$ and $|s_1 - s_2| < \delta$, then $|y_k(s_1) - y_k(s_2)| < \varepsilon$ for all k), then $\{y_k\}$ has a uniformly convergent subsequence.

Now let $\{x_n\}$ be a bounded sequence in L_2; say $\|x_n\| < N$ for all n. We must show that $\{Ax_n\}$ has an L_2-convergent subsequence, where A is the integral operator (2.1).

Let

$$y_n(t) = (Ax_n)(t) = \int_a^b R(t, \tau)x_n(\tau)\,d\tau.$$

Then [see the discussion after (2.1)] $|y_n(t)| \leq MN(b-a)^{1/2} < \infty$. Hence the functions y_n are uniformly bounded. To prove equicontinuity, we write

$$|y_k(t_1) - y_k(t_2)|^2 = \left| \int_a^b [R(t_1, \tau) - R(t_2, \tau)]x_k(\tau)\, d\tau \right|^2;$$

by the Schwarz inequality,

$$|y_k(t_1) - y_k(t_2)|^2 \leq \left(\int_a^b |R(t_1, \tau) - R(t_2, \tau)|^2\, d\tau \right) \|x_k\|^2. \tag{2.3}$$

Now R is uniformly continuous on $[a, b] \times [a, b]$; therefore given $\varepsilon > 0$ there is a $\delta > 0$ such that if $|t_1 - t_2| < \delta$ and $|\tau_1 - \tau_2| < \delta$, then $|R(t_1, \tau_1) - R(t_2, \tau_2)| < \varepsilon N^{-1}(b - a)^{-1/2}$. In particular, if $|t_1 - t_2| < \delta$, then $|R(t_1, \tau) - R(t_2, \tau)| < \varepsilon N^{-1}(b - a)^{-1/2}$ for all τ. Thus by (2.3), $|t_1 - t_2| < \delta$ implies $|y_k(t_1) - y_k(t_2)|^2 \leq \varepsilon^2 N^{-2}(b - a)^{-1}(b - a)\, \|x_k\|^2 < \varepsilon^2$ for all k.

Thus the functions y_n are equicontinuous. By the Arzela-Ascoli theorem, there is a subsequence $y_{n(i)}$ converging uniformly to a continuous function y. But

$$\|y_{n(i)} - y\|^2 = \int_a^b |y_{n(i)}(t) - y(t)|^2\, dt \to 0 \quad \text{as} \quad i \to \infty$$

since the uniform convergence allows the limit to be taken inside the integral sign; thus $y_{n(i)} \to y$ in L_2 and the result follows.

Theorem 2.2. Let R be a real-valued function on $[a, b] \times [a, b]$ which is *symmetric* [$R(t, \tau) = R(\tau, t)$ for all t, τ] as well as continuous, so that the associated operator A defined by (2.1) is symmetric as well as compact. Then all eigenvalues of A are nonnegative if and only if

$$\int_a^b \int_a^b R(t, \tau)x(t)x(\tau)\, dt\, d\tau \geq 0 \quad \text{for all continuous functions } x \text{ on } [a, b]. \tag{2.4}$$

Proof. Suppose that the double integral is nonnegative for all continuous x. Let x be an eigenfunction corresponding to the nonzero eigenvalue λ; x is continuous by (2.2). The double integral (2.4) becomes $\lambda \int_a^b x^2(t)\, dt = \lambda\, \|x\|^2$. We may take $x \neq 0$; it follows that $\lambda > 0$.

Conversely, suppose that all eigenvalues of A are nonnegative. If x is continuous on $[a, b]$, then $x \in L_2[a, b]$; and hence by Theorem 1.14, $x = h + \sum_{n=1}^{\infty} c_n e_n$ where the e_n are an orthonormal basis for the space spanned by the eigenvectors corresponding to the nonzero eigenvalues, $c_n = (x, e_n)$, and $Ah = 0$. Now the double integral (2.4) may be written as (x, Ax); by Lemma 1.12 we have

$$(x, Ax) = (x, Ah) + \sum_{n=1}^{\infty} c_n(x, Ae_n)$$
$$= (x, Ah) + \sum_{n=1}^{\infty} c_n\lambda_n(x, e_n) = \sum_{n=1}^{\infty} \lambda_n c_n{}^2 \geq 0.$$

The theorem is proved.

Note that in Theorem 2.2, the phrase "for all continuous x on $[a, b]$" in (2.4) may be replaced by "for all $x \in L_2[a, b]$"; essentially the same proof may be used.

Theorem 2.3. Let A be the integral operator (2.1) corresponding to a function R which is *nonnegative definite* as well as continuous and symmetric. (R is said to be nonnegative definite if

$$\sum_{i,j=1}^{n} x(t_i)R(t_i, t_j)x(t_j) \geq 0$$

for all possible choices of $t_1, \ldots, t_n \in [a, b]$, and all possible real-valued functions x on $[a. b]$.) Then all eigenvalues of A are nonnegative.

Proof. If x is continuous, then

$$\int_a^b\int_a^b R(t, \tau)x(t)x(\tau)\, dt\, d\tau$$

is an ordinary Riemann integral; the approximating Riemann sums are of the form $\sum_{i,j=1}^{n} x(t_i)R(t_i, t_j)x(t_j)$. Since R is nonnegative definite, the approximating sums are all nonnegative, and hence the integral is nonnegative. The result follows from Theorem 2.2.

Theorem 2.4. (Mercer's Theorem). Let A be the integral operator (2.1) corresponding to a function R that is continuous, symmetric, and nonnegative definite. Let $\{e_n, n = 1, 2, \ldots\}$ be an orthonormal basis for the space spanned by the eigenvectors corresponding to the nonzero (hence positive) eigenvalues of A. If the basis is taken so that e_n is an eigenvector corresponding to the eigenvalue λ_n, then

$$R(t, \tau) = \sum_{n=1}^{\infty} \lambda_n e_n(t)e_n(\tau), \qquad t, \tau \in [a, b] \tag{2.5}$$

where the series converges absolutely, converges to $R(t, \tau)$ uniformly in both variables, and also converges to $R(t, \tau)$ in $L_2([a, b] \times [a, b])$, that is,

$$\int_a^b \int_a^b |R(t, \tau) - \sum_{k=1}^{n} \lambda_k e_k(t) e_k(\tau)|^2 \, dt \, d\tau \to 0 \quad \text{as} \quad n \to \infty.$$

Proof. We break the proof into several parts.

a. $R(t, \tau) = \sum_{n=1}^{\infty} \lambda_n e_n(t) e_n(\tau)$, where the series converges in L_2 in each variable separately.

Fix t and apply Theorem 1.14 to $R(t, \tau)$. We obtain

$$R(t, \tau) = h(t, \tau) + \sum_{n=1}^{\infty} c_n(t) e_n(\tau), \tag{2.6}$$

where

$$\int_a^b R(t, \tau) h(t, \tau) \, d\tau = 0, \qquad c_n(t) = \int_a^b R(t, \tau) e_n(\tau) \, d\tau = \lambda_n e_n(t),$$

and the series converges in $L_2[a, b]$. It follows from (2.6) and Lemma 1.12 that

$$\int_a^b R(t, \tau) h(t, \tau) \, d\tau = \int_a^b h^2(t, \tau) \, d\tau + \sum_{n=1}^{\infty} c_n(t) \int_a^b h(t, \tau) e_n(\tau) \, d\tau.$$

By Lemma 1.9,

$$\int_a^b h(t, \tau) e_n(\tau) \, d\tau = 0, \qquad n = 1, 2, \ldots .$$

Hence $0 = \int_a^b h^2(t, \tau) \, d\tau$. Therefore for each t, $h(t, \cdot) = 0$ in $L_2[a, b]$ and the result (a) is proved.

b. Let $R_n(t, \tau) = R(t, \tau) - \sum_{i=1}^{n} \lambda_i e_i(t) e_i(\tau)$; then $R_n(t, t) \geq 0$ for all t. By (a),

$$R_n(t, \tau) = \sum_{i=n+1}^{\infty} \lambda_i e_i(t) e_i(\tau),$$

with L_2 convergence in each variable separately.

If $x \in L_2[a, b]$ then, using Lemma 1.12, we obtain

$$\int_a^b \int_a^b R_n(t, \tau) x(t) x(\tau) \, dt \, d\tau = \sum_{i=n+1}^{\infty} \lambda_i \, |(x, e_i)|^2 \geq 0. \tag{2.7}$$

Now if $R_n(t_0, t_0) < 0$ then by continuity, $R_n(t, \tau) \leq -\varepsilon < 0$ for some positive ε and for t, τ in some neighborhood U of (t_0, t_0) described by

$t_0 - \alpha < t < t_0 + \alpha$, $t_0 - \alpha < \tau < t_0 + \alpha$. Now define $x(t) = 1$ for $t_0 - \alpha < t < t_0 + \alpha$; $x(t) = 0$ elsewhere. Then $x \in L_2[a, b]$ and

$$\int_a^b \int_a^b R_n(t, \tau)x(t)x(\tau)\, dt\, d\tau = \iint_U R_n(t, \tau)\, dt\, d\tau \leq -4\alpha^2\varepsilon < 0,$$

contradicting (2.7).

c. The series $\sum_{n=1}^{\infty} \lambda_n e_n(t)e_n(\tau)$ converges absolutely; the convergence is uniform in τ for each fixed t, and uniform in t for each fixed τ.

By (b),

$$R_n(t, t) = R(t, t) - \sum_{i=1}^{n} \lambda_i e_i^2(t) \geq 0,$$

hence

$$\sum_{i=1}^{\infty} \lambda_i e_i^2(t) \leq R(t, t) \leq M < \infty.$$

Now by the Schwarz inequality for sums,

$$\left|\sum_{i=m}^{n} \lambda_i e_i(t)e_i(\tau)\right|^2 = \left|\sum_{i=m}^{n} \sqrt{\lambda_i}e_i(t)\sqrt{\lambda_i}e_i(\tau)\right|^2 \leq \sum_{i=m}^{n} \lambda_i e_i^2(t) \sum_{i=m}^{n} \lambda_i e_i^2(\tau)$$

$$\leq \sum_{i=m}^{n} \lambda_i e_i^2(t) \sum_{i=1}^{\infty} \lambda_i e_i^2(\tau) \leq M \sum_{i=m}^{n} \lambda_i e_i^2(t) \to 0 \quad \text{as} \quad m, n \to \infty,$$

uniformly in τ for each fixed t. The assertion (c) follows.

d. The series $\sum_{n=1}^{\infty} \lambda_n e_n(t)e_n(\tau)$ converges pointwise to $R(t, \tau)$.

Fix t. By (a), $\sum_{n=1}^{\infty} \lambda_n e_n(t)e_n(\tau) = R(t, \tau)$ with L_2 convergence. By (c), $\sum_{n=1}^{\infty} \lambda_n e_n(t)e_n(\tau) =$ some function $S(t, \tau)$ with uniform, hence L_2, convergence. It follows that for each t, $R(t, \tau) = S(t, \tau)$ for almost every τ. But $R(t, \cdot)$ is continuous and $S(t, \cdot)$, the uniform limit of continuous functions, is also continuous. Thus $R(t, \tau) = S(t, \tau)$ for *every* τ, that is, $R(t, \tau) \equiv S(t, \tau)$.

e. (Dini's Theorem). Given a sequence of real-valued functions $\{g_n, n = 1, 2, \ldots\}$ continuous on $[a, b]$ and a real-valued function g continuous on $[a, b]$ such that $g_n(x) \leq g_{n+1}(x)$ for all x and all n, and $\lim_{n\to\infty} g_n(x) = g(x)$ for all $x \in [a, b]$. Then $g_n \to g$ uniformly on $[a, b]$. The same conclusion holds if the sequence is monotone decreasing instead of increasing.

Given $\varepsilon > 0$, let $U_n = \{x \in [a, b]\colon\ |g_n(x) - g(x)| < \varepsilon\}$. Since $g_n(x) \to g(x)$ for all x, $\bigcup_{n=1}^{\infty} U_n = [a, b]$. Thus we have a family of open sets that cover $[a, b]$; by the Heine-Borel theorem, a finite subfamily covers $[a, b]$; that is, $\bigcup_{n=1}^{N} U_n = [a, b]$ for some N. By the monotonicity of the sequence $\{g_n\}$, $U_n \subset U_{n+1}$ for all n, and hence $\bigcup_{n=1}^{N} U_n = U_N$.

Consequently $U_N = [a, b]$, so that $|g_N(x) - g(x)| < \varepsilon$ for all x. By monotonicity, $|g_n(x) - g(x)| < \varepsilon$ for all x if $n \geq N$; the result follows.

Proof of Mercer's Theorem. By (d), $R(t, t) = \sum_{n=1}^{\infty} \lambda_n e_n^2(t)$ with pointwise convergence. Let $g_n(t) = \sum_{i=1}^{n} \lambda_i e_i^2(t)$. Then g_n is continuous, $g_n(t) \leq g_{n+1}(t)$ for all t, and $g_n(t) \to R(t, t)$ for all t. By (e), $g_n(t) \to R(t, t)$ uniformly on $[a, b]$. Now as in (c),

$$\left| \sum_{i=m}^{n} \lambda_i e_i(t) e_i(\tau) \right|^2 \leq M \sum_{i=m}^{n} \lambda_i e_i^2(t) \to 0 \quad \text{as} \quad m, n \to \infty,$$

uniformly in t *and* τ. The proof is complete.

3. *The Karhunen-Loeve theorem*

Let $X(t)$, $t \in [a, b]$ (a, b finite), be a continuous-parameter second-order random process with zero mean and covariance function $R(t, \tau)$. In other words, for each $t \in [a, b]$, $X(t)$ is a (real-valued) random variable; $E[X(t)] = 0$ for all $t \in [a, b]$, and $E[X(t)X(\tau)] = R(t, \tau)$ for $t, \tau \in [a, b]$. Note that R is nonnegative definite, since

$$\sum_{i,j=1}^{n} c_i R(t_i, t_j) c_j = E\left[\left(\sum_{i=1}^{n} c_i X(t_i)\right)^2\right] \geq 0.$$

If g is a real-valued function on $[a, b]$, we define $\int_a^b g(t)X(t)\,dt$ as follows. Let $\Delta: a = t_0 < t_1 < \cdots < t_n = b$ be a partition of $[a, b]$, with $|\Delta| = \max_{1 \leq i \leq n} |t_i - t_{i-1}|$. Define

$$I(\Delta) = \sum_{k=1}^{n} g(t_k) X(t_k)(t_k - t_{k-1}).$$

If the family of random variables $\{I(\Delta)\}$ converges in mean square to a random variable I as $|\Delta| \to 0$ [that is, $E[|I(\Delta) - I|^2] \to 0$ as $|\Delta| \to 0$] then we say that $g(t)X(t)$ is *integrable* over $[a, b]$. I is called the *integral* of $g(t)X(t)$ over $[a, b]$ and we write $I = \int_a^b g(t)X(t)\,dt$.

If Ω (with an appropriate sigma-algebra of subsets and an appropriate probability measure) is the probability space on which the $X(t)$ are defined, then mean square convergence is convergence in the Hilbert space $L_2(\Omega)$ of random variables on Ω with finite second moments.

Theorem 3.1. If g is continuous on $[a, b]$ and the covariance function R is continuous on $[a, b] \times [a, b]$, then $g(t)X(t)$ is integrable over $[a, b]$.

Proof. Let

$$\Delta : a = t_0 < t_1 < \cdots < t_n = b, \qquad \Delta' : a = \tau_0 < \tau_1 < \cdots < \tau_m = b$$

be two partitions of $[a, b]$. Then

$$I(\Delta)I(\Delta') = \sum_{k=1}^{n}\sum_{j=1}^{m} g(t_k)g(\tau_j)X(t_k)X(\tau_j)(t_k - t_{k-1})(\tau_j - \tau_{j-1}).$$

Thus

$$E[I(\Delta)I(\Delta')] = \sum_{k=1}^{n}\sum_{j=1}^{m} g(t_k)g(\tau_j)R(t_k, \tau_j)(t_k - t_{k-1})(\tau_j - \tau_{j-1}). \tag{3.1}$$

The right side of (3.1) is an approximating sum to a Riemann integral. Thus

$$\lim_{|\Delta|, |\Delta'| \to 0} E[I(\Delta)I(\Delta')] = \int_a^b\int_a^b g(t)g(\tau)R(t, \tau)\, dt\, d\tau. \tag{3.2}$$

By (3.2),

$$E[(I(\Delta) - I(\Delta'))^2] = E[I^2(\Delta)] - 2E[I(\Delta)I(\Delta')] + E[I^2(\Delta')] \to 0 \quad \text{as} \quad |\Delta|, |\Delta'| \to 0.$$

Thus $\{I(\Delta)\}$ is a generalized Cauchy sequence in $L_2(\Omega)$, and hence $I(\Delta)$ converges in mean square to a random variable I as $|\Delta| \to 0$. The theorem is proved.

Lemma 3.2. If g and h are continuous on $[a, b]$, and R is continuous on $[a, b] \times [a, b]$, then

$$E\left[\int_a^b g(t)X(t)\, dt \int_a^b h(\tau)X(\tau)\, d\tau\right] = \int_a^b\int_a^b g(t)h(\tau)R(t, \tau)\, dt\, d\tau$$

Furthermore, $E\left[\int_a^b g(t)X(t)\, dt\right] = E\left[\int_a^b h(\tau)X(\tau)\, d\tau\right] = 0.$

Proof. As in Theorem 3.1, let Δ and Δ' be two partitions of $[a, b]$. Let

$$I(\Delta) = \sum_{k=1}^{n} g(t_k)X(t_k)(t_k - t_{k-1}), \qquad J(\Delta') = \sum_{j=1}^{m} h(\tau_j)X(\tau_j)(\tau_j - \tau_{j-1}),$$

$$I = \int_a^b g(t)X(t)\, dt, \qquad J = \int_a^b h(\tau)X(\tau)\, d\tau.$$

By Theorem 3.1, $I(\Delta) \rightarrow I$, $J(\Delta') \rightarrow J$ in mean square as $|\Delta|, |\Delta'| \rightarrow 0$. It follows that $E[I(\Delta)J(\Delta')] \rightarrow E[IJ]$. For we have

$$\begin{aligned} |E[I(\Delta)J(\Delta') - IJ]| &\leq E\,[|I(\Delta)(J(\Delta') - J)|] + E[|(I(\Delta) - I)J|] \\ &\leq (E[I^2(\Delta)]E[(J(\Delta') - J)^2])^{1/2} \\ &\qquad + (E[J^2]E[(I(\Delta) - I)^2])^{1/2} \\ &\qquad\qquad \text{(by the Schwarz inequality)} \\ &\rightarrow \sqrt{E[I^2]} \cdot 0 + \sqrt{E[J^2]} \cdot 0 = 0. \end{aligned}$$

But as in (3.2),

$$E[I(\Delta)J(\Delta')] \rightarrow \int_a^b \int_a^b g(t)h(\tau)R(t, \tau)\, dt\, d\tau.$$

Similarly, $I(\Delta) \rightarrow I$ in mean square implies $E[I(\Delta)] \rightarrow E(I)$. Since $E[I(\Delta)] = 0$ for all Δ, the lemma follows.

Lemma 3.3. If h is continuous on $[a, b]$ and R is continuous on $[a, b] \times [a, b]$, then

$$E\left[X(t) \int_a^b h(\tau)X(\tau)\, d\tau\right] = \int_a^b h(\tau)R(t, \tau)\, d\tau, \qquad t \in [a, b].$$

Proof. Let $J(\Delta') = \sum_{j=1}^{m} h(\tau_j)X(\tau_j)(\tau_j - \tau_{j-1})$, and let

$$J = \int_a^b h(\tau)X(\tau)\, d\tau.$$

Then $J(\Delta') \rightarrow J$ in mean square as $|\Delta'| \rightarrow 0$. Just as in the argument of Lemma 3.2, $E[X(t)J(\Delta')] \rightarrow E[X(t)J]$. But [compare (3.2)]

$$E[X(t)J(\Delta')] \rightarrow \int_a^b h(\tau)R(t, \tau)\, d\tau,$$

and the lemma follows.

Theorem 3.4. (Karhunen-Loève Expansion). Let $X(t)$, $t \in [a, b]$, (a, b finite), be a continuous-parameter second-order random process with zero mean and continuous covariance function $R(t, \tau)$. Then we may write

$$X(t) = \sum_{k=1}^{\infty} Z_k e_k(t), \qquad a \leq t \leq b, \tag{3.3}$$

where the e_k are eigenfunctions of the integral operator (2.1) corresponding to R, and form an orthonormal basis for the space spanned by the eigenfunctions corresponding to the *nonzero* eigenvalues. The Z_k are given by $\int_a^b X(t)e_k(t)\,dt$, and are orthogonal random variables ($E[Z_kZ_j] = 0$ for $k \neq j$) with zero mean and variance λ_k, where λ_k is the eigenvalue corresponding to e_k.

The series $\sum_{k=1}^{\infty} Z_k e_k(t)$ converges in mean square to $X(t)$, *uniformly in* t, that is,

$$E\left[\left(X(t) - \sum_{k=1}^{n} Z_k e_k(t)\right)^2\right] \to 0 \quad \text{as} \quad n \to \infty,$$

uniformly for $t \in [a, b]$.

Proof. By Theorem 3.1, $\int_a^b X(t)e_k(t)\,dt$ exists and defines a random variable Z_k. By Lemma 3.2, $E[Z_k] = 0$ and

$$E(Z_kZ_j) = \int_a^b e_k(t)\left[\int_a^b R(t, \tau)e_j(\tau)\,d\tau\right] dt$$

$$= \lambda_j \int_a^b e_k(t)e_j(t)\,dt = \begin{matrix} 0 & \text{for} & k \neq j \\ \lambda_k & \text{for} & k = j \end{matrix}.$$

Now let $S_n(t) = \sum_{k=1}^{n} Z_k e_k(t)$. Then

$$E[(S_n(t) - X(t))^2] = E[S_n^2(t)] - 2E[S_n(t)X(t)] + E[X^2(t)]$$

$$= \sum_{k=1}^{n} \lambda_k e_k^2(t) - 2\sum_{k=1}^{n} E[X(t)Z_k]e_k(t) + R(t, t)$$

By Lemma 3.3,

$$E[X(t)Z_k] = \int_a^b R(t, \tau)e_k(\tau)\,d\tau = \lambda_k e_k(t).$$

Thus

$$E[(S_n(t) - X(t))^2] = R(t, t) - \sum_{k=1}^{n} \lambda_k e_k^2(t) \to 0 \quad \text{as} \quad n \to \infty,$$

uniformly in t, by Theorem 2.4. The result follows.

The Karhunen-Loève expansion assumes a special form when the process $X(t)$ is Gaussian.

Theorem 3.5. If $X(t)$ is a Gaussian process, then the Z_k of the Karhunen-Loève expansion of $X(t)$ are independent Gaussian random variables.

Proof. The approximating sum to $Z_k = \int_a^b X(t)e_k(t)\,dt$ is

$$I_k(\Delta) = \sum_{i=1}^{n} X(t_i)e_k(t_i)(t_i - t_{i-1}).$$

Since $X(t)$ is a Gaussian process, the $I_k(\Delta)$, $k = 1, 2, \ldots$, form a Gaussian sequence, that is, $I_{k_1}(\Delta), \ldots, I_{k_r}(\Delta)$ are jointly Gaussian for all finite sets of indices $\{k_1, \ldots, k_r\}$. Now the joint characteristic function of $I_{k_1}(\Delta), \ldots, I_{k_r}(\Delta)$ is

$$M(t_1, \ldots, t_n) = E\left[\exp\left(i\sum_{j=1}^{r} t_j I_{k_j}(\Delta)\right)\right]$$

$$= \exp\left(-\frac{1}{2}\sum_{j,m=1}^{r} t_j\sigma_{jm}(\Delta)t_m\right)$$

where

$$i = \sqrt{-1} \quad \text{and} \quad \sigma_{jm}(\Delta) = E[I_{k_j}(\Delta)I_{k_m}(\Delta)].$$

Since $I_k(\Delta)$ converges in mean square to Z_k as $|\Delta| \to 0$ ($k = 1, 2, \ldots$), it follows as in the argument of Lemma 3.2 that $E[I_{k_j}(\Delta)I_{k_m}(\Delta)] \to E[Z_{k_j}Z_{k_m}]$. Thus the joint characteristic function of $I_{k_1}(\Delta), \ldots, I_{k_r}(\Delta)$ approaches $\exp(-\frac{1}{2}\sum_{j,m=1}^{r} t_j\sigma_{jm}t_m)$ where $\sigma_{jm} = E[Z_{k_j}Z_{k_m}]$. Thus $Z_{k_1}, \ldots, Z_{k_r}$ are jointly Gaussian for each finite set of indices $\{k_1, \ldots, k_r\}$; that is, the Z_k, $k = 1, 2, \ldots$, form a Gaussian sequence. Since the Z_k are orthogonal random variables, they are independent.

Thus in the Gaussian case, the Karhunen-Loève expansion is a series of independent random variables. It follows (Loève, 1955, p. 251) that *for each fixed t*, the series (3.3) converges with probability 1, that is, for each t, the series converges pointwise to $X(t)$ for almost all points ω in the probability space Ω.

As an example of the Karhunen-Loève expansion, we take $R(t, \tau) = \min(t, \tau)$, with the expansion interval $= [0, 1]$. [If in addition we take $X(t)$ to be Gaussian, we obtain the *Brownian motion* process.] To find the eigenvalues of the integral operator associated with the covariance function, we must solve the integral equation

$$\int_0^1 \min(t, \tau)e(\tau)\,d\tau = \lambda e(t), \qquad 0 \le t \le 1,$$

or equivalently

$$\int_0^t \tau e(\tau)\,d\tau + t\int_t^1 e(\tau)\,d\tau = \lambda e(t), \qquad 0 \le t \le 1. \tag{3.4}$$

If we differentiate (3.4) with respect to t, we obtain

$$\int_t^1 e(\tau)\, d\tau = \lambda \frac{de(t)}{dt}. \tag{3.5}$$

Differentiating again with respect to t, we have

$$-e(t) = \lambda \frac{d^2e(t)}{dt^2}. \tag{3.6}$$

Thus any solution of the integral equation, that is, any eigenfunction, must be of the form

$$e(t) = A \sin \frac{t}{\sqrt{\lambda}} + B \cos \frac{t}{\sqrt{\lambda}}. \tag{3.7}$$

[Note that if $\lambda = 0$ then $e(t) \equiv 0$ by (3.6); thus 0 is not an eigenvalue.] Now if we let $t = 0$ in (3.4) we obtain $e(0) = 0$, and therefore $B = 0$ in (3.7). If we let $t = 1$ in (3.5) we obtain $de(t)/dt = 0$ when $t = 1$, and therefore

$$\cos \frac{1}{\sqrt{\lambda}} = 0, \quad \text{or} \quad \frac{1}{\sqrt{\lambda}} = \frac{(2n-1)\pi}{2}, \; n = 1, 2, \ldots.$$

Thus the eigenvalues are

$$\lambda_n = \frac{4}{(2n-1)^2\pi^2} \tag{3.8}$$

and the orthonormalized eigenfunctions are

$$e_n(t) = \sqrt{2} \sin (n - \tfrac{1}{2})\pi t. \tag{3.9}$$

[It may be verified by direct substitution that the functions (3.9) satisfy the integral equation (3.4).]

Finally, if we define $Z_k^* = Z_k/\sqrt{\lambda_k}$, where the Z_k are as given in Theorem 3.4, then the Karhunen-Loève expansion becomes

$$X(t) = \sqrt{2} \sum_{n=1}^{\infty} Z_n^* \frac{\sin (n - \frac{1}{2})\pi t}{(n - \frac{1}{2})\pi} \tag{3.10}$$

where the Z_n^* are a sequence of orthogonal random variables with mean 0 and variance 1.

If $X(t)$ is Gaussian, we may show that for almost all ω, the series (3.10) converges *uniformly* for $0 \le t \le 1$, and therefore may be used to define a process whose sample functions are continuous. For we have

$$\sum_{n=1}^{\infty} \frac{E[(Z_n^*)^2]}{(n - \frac{1}{2})^2\pi^2} = \sum_{n=1}^{\infty} (n - \tfrac{1}{2})^{-2}\pi^{-2} < \infty.$$

It follows that the series

$$\sum_{n=1}^{\infty} \frac{|Z_n^*|}{(n - \frac{1}{2})\pi}$$

converges in mean square, and consequently, being a series of independent random variables, converges with probability 1. The result now follows from the Weierstrass criterion for uniform convergence.

For a further discussion of the theory of second-order random processes, see Loève (1955), Chapter 10.

4. *Further results concerning integral operators determined by a covariance function*

In this section we shall need some standard facts about Fourier transforms on $L_2(-\infty, \infty)$. [See, for example, Goldberg (1961), Titchmarsh (1937).] Consider the Hilbert space of all complex-valued Borel measurable functions f defined on the entire real line, such that $\int_{-\infty}^{\infty} |f(t)|^2\, dt < \infty$; we denote this space by $L_2^*(-\infty, \infty)$. [The norm of f is

$$\|f\| = \left[\int_{-\infty}^{\infty} |f(t)|^2\, dt\right]^{1/2}$$

and the inner product of two functions f and g is

$$\int_{-\infty}^{\infty} f(t)\bar{g}(t)\, dt,$$

where the bar denotes complex conjugate.]

If $f \in L_2^*(-\infty, \infty)$, then the function F_N defined by

$$F_N(\omega) = \int_{-N}^{N} f(t)e^{-i\omega t}\, dt$$

converges in $L_2^*(-\infty, \infty)$ as $N \to \infty$ to a function F, that is, $\|F_N - F\| \to 0$. Conversely, the function f_N defined by

$$f_N(t) = (2\pi)^{-1} \int_{-N}^{N} F(\omega)e^{i\omega t}\, d\omega$$

converges in $L_2^*(-\infty, \infty)$ to f. F is called the *Fourier transform* of f, and f the *inverse Fourier transform* of F. The pair (f, F) is called a *Fourier* *transform pair*. A function in $L_2^*(-\infty, \infty)$ is uniquely determined by its Fourier transform.

If $f, g \in L_2^*(-\infty, \infty)$ with Fourier transforms F, G then the *Parseval relation* states that

$$\int_{-\infty}^{\infty} f(t)\bar{g}(t)\, dt = (2\pi)^{-1} \int_{-\infty}^{\infty} F(\omega)\bar{G}(\omega)\, d\omega;$$

consequently, $\|F\| = (2\pi)^{1/2} \|f\|$.

If $f, g \in L_2^*(-\infty, \infty)$, the *convolution* of f and g is the function h defined by

$$h(t) = \int_{-\infty}^{\infty} f(s)g(t-s)\, ds.$$

[By the Schwarz inequality, h exists but is not necessarily in $L_2^*(-\infty, \infty)$.] If F and G are the Fourier transforms of f and g respectively, and H is the product FG, then the *convolution theorem* is a name for a class of results which state that under certain conditions, (h, H) is a Fourier transform pair. In particular, if $f \in L_1^*(-\infty, \infty)$, that is, $\int_{-\infty}^{\infty} |f(t)|\, dt < \infty$, then $h \in L_2^*(-\infty, \infty)$ and the Fourier transform of h is H. Also, if $H \in L_2^*(-\infty, \infty)$, then so is h, and (h, H) is a Fourier transform pair.

Let $n(t)$, $-\infty < t < \infty$, be a second-order stationary random process with continuous covariance function $R(\tau) = E[n(t)n(t+\tau)]$. Assume that the process has a spectral density $N(\omega)$, that is, assume that $R(\tau)$ can be expressed as

$$(2\pi)^{-1} \int_{-\infty}^{\infty} N(\omega)e^{i\omega t}\, d\omega,$$

where N is a nonnegative function integrable over $(-\infty, \infty)$. Let T be a positive real number, and let A_T be the integral operator on $L_2[-T/2, T/2]$ corresponding to the covariance function; that is, the operator defined by

$$y(t) = \int_{-T/2}^{T/2} R(t-\tau)x(\tau)\, d\tau; \qquad -\frac{T}{2} \leq t \leq \frac{T}{2}. \tag{4.1}$$

We assume that 0 *is not an eigenvalue of* A_T. (Later we shall investigate conditions under which this result holds.) It follows (Theorems 1.14, 2.3) that the eigenfunctions corresponding to the positive eigenvalues of

A_T span the Hilbert space $L_2[-T/2, T/2]$. Let $\{e_n(t; T), -T/2 \leq t \leq T/2\}$ be an orthonormal basis of $L_2[-T/2, T/2]$, where $e_n(t; T)$ is taken as an eigenfunction corresponding to the positive eigenvalue $\lambda_n(T)$.

Let $G(\omega) = [N(\omega)]^{1/2}$ and let g be the inverse Fourier transform of G. (Since G is an even function, g is real valued.) Define

$$\varphi_n(t; T) = [\lambda_n(T)]^{-1/2} \int_{-T/2}^{T/2} g(t - s)e_n(s; T)\, ds, \qquad -\infty < t < \infty.$$

We shall establish several properties of the functions φ_n.

Lemma 4.1. For each T, the functions $\varphi_n(t; T)$ are orthonormal over $(-\infty, \infty)$.

Proof. For simplicity we write $\varphi_n(t)$ for $\varphi_n(t; T)$, $e_n(t)$ for $e_n(t; T)$, and λ_n for $\lambda_n(T)$. We have

$$\int_{-\infty}^{\infty} \varphi_n(t)\varphi_m(t)\, dt = \lambda_n^{-1/2}\, \lambda_m^{-1/2}$$

$$\times \int_{-\infty}^{\infty} \left[\int_{-T/2}^{T/2} g(t - s)e_n(s)\, ds \int_{-T/2}^{T/2} g(t - s')e_m(s')\, ds' \right] dt. \tag{4.2}$$

Since the Fourier transform of the function $h(t) = g(t - s)$ is $e^{-i\omega s}G(\omega)$, the Parseval relation implies that

$$\int_{-\infty}^{\infty} g(t - s)g(t - s')\, dt = (2\pi)^{-1} \int_{-\infty}^{\infty} |G(\omega)|^2\, e^{i\omega(s'-s)}\, d\omega = R(s' - s).$$

An application of the Schwarz inequality shows that

$$\int_{-\infty}^{\infty} |g(t - s)g(t - s')|\, dt \leq \|g\|^2.$$

It follows that Fubini's theorem may be applied to the integral (4.2), and we obtain

$$\int_{-\infty}^{\infty} \varphi_n(t)\varphi_m(t)\, dt = \lambda_n^{-1/2}\lambda_m^{-1/2} \int_{-T/2}^{T/2}\int_{-T/2}^{T/2} R(s' - s)e_n(s)e_m(s')\, ds\, ds'.$$

Since e_n is an eigenfunction corresponding to λ_n, the integral has the value 0 if $n \neq m$, and 1 if $n = m$. The lemma follows.

If L is a subspace of $L_2(-\infty, \infty)$ and $f \in L_2(-\infty, \infty)$ we will use the notation $f \perp L$ to indicate that f is orthogonal to L, that is,

$$\int_{-\infty}^{\infty} f(t)h(t)\, dt = 0 \quad \text{for} \quad \text{all } h \in L.$$

Lemma 4.2. For each T, let L_T be the subspace of $L_2(-\infty, \infty)$ spanned by the functions $\{\varphi_n(t; T)\}$. If $f \in L_2(-\infty, \infty)$ and $T' < T$, then

a. $f \perp L_T$ implies $f \perp L_{T'}$.

b. $L_{T'} \subset L_T$.

Proof. To say that $f \perp L_T$ is to say that

$$\int_{-\infty}^{\infty} f(t)\varphi_n(t; T)\, dt = 0, \qquad n = 1, 2, \ldots,$$

or

$$[\lambda_n(T)]^{-1/2} \int_{-\infty}^{\infty} f(t)\left[\int_{-T/2}^{T/2} g(t - s)e_n(s; T)\, ds\right] dt = 0, \qquad n = 1, 2, \ldots. \tag{4.3}$$

We may reverse the order of integration (invoking Fubini's theorem as in Lemma 4.1) to obtain

$$[\lambda_n(T)]^{-1/2} \int_{-T/2}^{T/2} e_n(s; T)\left[\int_{-\infty}^{\infty} g(t - s)f(t)\, dt\right] ds = 0, \qquad n = 1, 2, \ldots. \tag{4.4}$$

Thus the function $h(s) = \int_{-\infty}^{\infty} g(t - s)f(t)\, dt$, $-T/2 \le s \le T/2$, is orthogonal to all the $e_n(s; T)$. Since the e_n span $L_2[-T/2, T/2]$, h is orthogonal to itself so that $h = 0$ almost everywhere, that is, $\int_{-\infty}^{\infty} g(t - s)f(t)\, dt = 0$ for almost every $s \in [-T/2, T/2]$, hence for almost every $s \in [-T'/2, T'/2]$. It follows that (4.4), hence (4.3), holds with T replaced by T', and consequently $f \perp L_{T'}$, proving part (a) of the lemma.

We now prove (b). Given any $f \in L_2(-\infty, \infty)$, by the projection theorem we may write $f = f_1 + f_2$, where $f_1 \in L_T$ and $f_2 \perp L_T$. By (a), $f_2 \perp L_{T'}$. Using the projection theorem again, we have $f_1 = f_3 + f_4$, where $f_3 \in L_{T'}$, and $f_4 \perp L_{T'}$. If P_T is the linear operator on $L_2(-\infty, \infty)$ which assigns to a function its projection on L_T, we have $P_T f = f_1$ and $P_{T'}P_T f = P_{T'}f_1 = f_3$. But $f = f_3 + (f_2 + f_4)$, where $f_3 \in L_{T'}$ and $f_2 + f_4 \perp L_{T'}$, and thus $f_3 = P_{T'}f$. It follows that

$$P_{T'}P_T = P_{T'}. \tag{4.5}$$

Now assume $f \in L_{T'}$. By the Pythagorean relation, $\|f\|^2 = \|f_1\|^2 + \|f_2\|^2$. If f_2 is not zero, then $\|P_T f\| = \|f_1\| < \|f\|$; since projection cannot increase the norm, we have $\|P_{T'} P_T f\| \le \|P_T f\| < \|f\|$. But by (4.5), $\|P_{T'} P_T f\| = \|f\|$, a contradiction. Thus $f_2 = 0$ and therefore $f = f_1 \in L_T$, completing the proof.

Note that the above argument shows that the statements "$f \perp L_\beta$ implies $f \perp L_\alpha$" and "$L_\alpha \subset L_\beta$" are equivalent for any closed subspaces L_α, L_β of $L_2(-\infty, \infty)$.

Lemma 4.3. Let H be the subspace of $L_2(-\infty, \infty)$ consisting of those functions whose Fourier transforms vanish whenever $G(\omega)$ (equivalently, $N(\omega)$) vanishes. With L_T as in Lemma 4.2, let L_∞ be the subspace spanned by $\bigcup_T L_T$. Then $L_\infty = H$.

Proof. Fix T, and define

$$e_n^*(s) = e_n(s; T), \qquad |s| \le \frac{T}{2}$$

$$= 0, \qquad |s| > \frac{T}{2}.$$

Then $e_n^* \in L_2(-\infty, \infty)$; let E_n^* be the Fourier transform of e_n^*. The functions φ_n may then be written as

$$\varphi_n(t; T) = [\lambda_n(T)]^{-1/2} \int_{-\infty}^{\infty} g(t - s) e_n^*(s)\, ds.$$

Since the Fourier transform of a convolution is the product of the Fourier transforms, the Fourier transform of φ_n is $[\lambda_n(T)]^{-1/2} G(\omega) E_n^*(\omega)$. It follows that $\varphi_n \in H$ for all $n = 1, 2, \ldots,$ and all T, hence $L_T \subset H$ for all T, or $L_\infty \subset H$.

To show that $H \subset L_\infty$, we shall show that if $f \in L_2(-\infty, \infty)$, then $f \perp L_\infty$ implies $f \perp H$; the result then follows as in Lemma 4.2b. If $f \perp L_\infty$ then $f \perp L_T$ for every T; thus by the argument of Lemma 4.2a, $\int_{-\infty}^{\infty} g(s - t) f(t)\, dt = 0$ for almost every $s \in [-T/2, T/2]$. [Note that since $G(\omega)$ is real for all ω, g is an even function, so that $g(s - t) = g(t - s)$.] Since this holds for arbitrary T,

$$\int_{-\infty}^{\infty} g(s - t) f(t)\, dt = 0 \quad \text{for} \quad \text{almost all real numbers } s.$$

Again using the convolution theorem for Fourier transforms, we have

$$G(\omega)F(\omega) = 0 \quad \text{for} \quad \text{almost all } \omega, \tag{4.6}$$

where F is the Fourier transform of f.

Now let $f_1 \in H$, and let F_1 be the Fourier transform of f_1. By definition of H, $G(\omega) = 0$ implies $F_1(\omega) = 0$. But by (4.6), $G(\omega) \neq 0$ implies $F(\omega) = 0$. Thus $F(\omega)F_1(\omega) = 0$ for almost all ω, and consequently F and F_1 are orthogonal. By the Parseval relation, f and f_1 are orthogonal. Since f_1 is an arbitrary element of H, $f \perp H$ and the lemma is proved.

Lemma 4.4. Given a function $s \in H$ with Fourier transform S, define $F(\omega) = S(\omega)/G(\omega)$ if $G(\omega) \neq 0$; $F(\omega) = 0$ if $G(\omega) = 0$ [$= S(\omega)$]. Assume that $F \in L_2{}^*(-\infty, \infty)$, that is,

$$(2\pi)^{-1} \int_{-\infty}^{\infty} \frac{|S(\omega)|^2}{N(\omega)}\, d\omega < \infty.$$

Let f be the inverse Fourier transform of F (note $f \in H$). Define

$$s_n(T) = \int_{-T/2}^{T/2} s(t)e_n(t; T)\, dt, \quad f_n(T) = \int_{-\infty}^{\infty} f(t)\varphi_n(t; T)\, dt, \quad n = 1, 2, \ldots.$$

Then $[\lambda_n(T)]^{-1/2} s_n(T) = f_n(T)$, $n = 1, 2, \ldots.$

Proof. By the convolution theorem for Fourier transforms,

$$s(t) = \int_{-\infty}^{\infty} g(t - s)f(s)\, ds.$$

Thus

$$s_n(T) = \int_{-T/2}^{T/2} \left[\int_{-\infty}^{\infty} g(t - s)f(s)\, ds \right] e_n(t; T)\, dt.$$

Reversing the order of integration and using the fact that $g(t - s) = g(s - t)$, we obtain

$$s_n(T) = \int_{-\infty}^{\infty} f(s) \left[\int_{-T/2}^{T/2} g(s - t)e_n(t; T)\, dt \right] ds$$

$$= [\lambda_n(T)]^{1/2} \int_{-\infty}^{\infty} f(s)\varphi_n(s; T)\, ds$$

and the result follows.

Theorem 4.5. Let s be a function $\epsilon\ H$ with Fourier transform S. If $s_n(T)$ is as defined in Lemma 4.4, then

$$\lim_{T\to\infty}\sum_{n=1}^{\infty}\frac{|s_n(T)|^2}{\lambda_n(T)} = (2\pi)^{-1}\int_{-\infty}^{\infty}\frac{|S(\omega)|^2}{N(\omega)}\,d\omega \quad \text{(assuming the integral is finite)},$$

where the convergence as $T \to \infty$ is monotone from below. [The integrand $|S(\omega)|^2/N(\omega)$ is defined to be zero if $N(\omega) = S(\omega) = 0$.]

Proof. Let F, f, and $f_n(T)$ be as in Lemma 4.4. Define

$$h(t) = \sum_{n=1}^{\infty} f_n(T)\varphi_n(t; T).$$

[Note that since the φ_n are orthonormal and

$$\sum_{n=1}^{\infty} |f_n(T)|^2 \le \int_{-\infty}^{\infty} f^2(t)\,dt < \infty$$

(*Bessel's inequality*; see the argument of Theorem 1.14), the series $\sum_{n=1}^{\infty} f_n(T)\varphi_n(t; T)$ converges in $L_2(-\infty, \infty)$.] We claim that *h is the projection of f on L_T*. To see this, use the projection theorem to write $f = f_1 + f_2$ where $f_1 \in L_T$ and $f_2 \perp L_T$. Since f_2 is orthogonal to all the functions $\varphi_n(t; T)$, it follows that

$$f_n(T) = \int_{-\infty}^{\infty} f_1(t)\varphi_n(t; T)\,dt.$$

By definition of L_T, the $\varphi_n(t;\ T)$ are an orthonormal basis for L_T; hence

$$f_1(t) = \sum_{n=1}^{\infty} f_{1n}(T)\varphi_n(t; T)$$

where $f_{1n}(T) = \int_{-\infty}^{\infty} f_1(t)\varphi_n(t;\ T)\,dt = f_n(T)$ and the series converges in $L_2(-\infty, \infty)$. (Again this can be justified by an argument similar to that of Theorem 1.14.) Thus f_1, the projection of f on L_T, is in fact h.

Now write $P_T f$ for the projection of f on L_T. If $T' < T$, then $L_{T'} \subset L_T$ by Lemma 4.2b; the projection theorem then implies that $\|f - P_T f\| \le \|f - P_{T'} f\|$. Therefore $\lim_{T\to\infty} \|f - P_T f\|$ exists. If this limit is $\varepsilon > 0$, then given any T and any $f_1 \in L_T$ we would have $\|f - f_1\| \ge \|f - P_T f\| \ge \varepsilon$. This contradicts Lemma 4.3, which states that any function in H can be approximated arbitrarily closely (in the L_2 sense) by functions in $\bigcup_T L_T$.

Therefore $\lim_{T\to\infty} \|f - P_T f\| = 0$, and consequently $\lim_{T\to\infty} \|P_T f\|^2 = \|f\|^2$. But

$$\|P_T f\|^2 = \sum_{n=1}^{\infty} |f_n(T)|^2 = \sum_{n=1}^{\infty} \frac{|s_n(T)|^2}{\lambda_n(T)} \qquad \text{by Lemma 4.4.}$$

Furthermore if $T' < T$, then by (4.5), $\|P_{T'} f\| = \|P_{T'} P_T f\| \le \|P_T f\|$. The theorem follows.

We conclude with some conditions under which the zero eigenvalue is excluded.

Theorem 4.6. Let A_T be the integral operator defined by (4.1). If (a) the set on which $N(\omega) = 0$ has measure zero, or if (b) the covariance function R belongs to $L_2(-\infty,\infty)$, and the set on which $N(\omega) > 0$ is of positive measure and is contained in a finite interval, then 0 is not an eigenvalue of A_T. In particular, (a) implies that if $N(\omega)$ is a rational function of ω, then 0 is not an eigenvalue.

Proof. If $G(\omega) = [N(\omega)]^{1/2}$ and g is the inverse Fourier transform of G, then by the convolution theorem,

$$R(t) = \int_{-\infty}^{\infty} g(s)g(t-s)\,ds.$$

Hence

$$R(t-\tau) = \int_{-\infty}^{\infty} g(s)g(t-\tau-s)\,ds = \int_{-\infty}^{\infty} g(s'-\tau)g(s'-t)\,ds'.$$

If 0 is an eigenvalue then there is a function $f \in L_2[-T/2, T/2]$, $f \neq 0$, such that

$$\int_{-T/2}^{T/2} R(t-\tau)f(\tau)\,d\tau = 0, \qquad t \in \left[-\frac{T}{2}, \frac{T}{2}\right];$$

consequently

$$\int_{-T/2}^{T/2}\int_{-T/2}^{T/2} R(t-\tau)f(t)f(\tau)\,dt\,d\tau = 0.$$

But this integral is

$$\int_{-T/2}^{T/2}\int_{-T/2}^{T/2}\left[\int_{-\infty}^{\infty} g(s-t)g(s-\tau)\,ds\right] f(t)f(\tau)\,dt\,d\tau,$$

or, after the order of integration is interchanged,

$$\int_{-\infty}^{\infty}\left[\int_{-T/2}^{T/2} g(s-t)f(t)\,dt\right]^2 ds.$$

It follows that $\int_{-T/2}^{T/2} g(s-t)f(t)\,dt = 0$ for almost every s. If we define $f^*(t) = f(t)$, $-T/2 \leq t \leq T/2$, $f^*(t) = 0$ elsewhere, and if F^* is the Fourier transform of f^*, then the convolution theorem implies that $[N(\omega)]^{1/2}F^*(\omega) = 0$ for almost every ω. If N is positive almost everywhere, F^* must be zero almost everywhere, so that F^*, hence f, is 0 almost everywhere, a contradiction. This proves (a). To prove (b), again suppose that $\int_{-T/2}^{T/2} R(t-s)f(s)\,ds = 0, t \in [-T/2, T/2]$, for some function $f, f \not\equiv 0$. Define

$$f_1(t) = \int_{-T/2}^{T/2} R(t-s)f(s)\,ds, \qquad -\infty < t < \infty;$$

then $f_1 \in L_2(-\infty, \infty)$ and $f_1(t) = 0$, $-T/2 \leq t \leq T/2$. If F_1 and F^* are the Fourier transforms of f_1 and f^* respectively, where f^* is as defined in part (a), then the convolution theorem yields $F_1(\omega) = N(\omega)F^*(\omega)$, and hence F_1 vanishes outside a finite interval $[-B, B]$. But in this case f_1 is given explicitly by

$$f_1(t) = (2\pi)^{-1}\int_{-B}^{B} F_1(\omega)e^{i\omega t}\,d\omega. \tag{4.7}$$

We may use the formula (4.7) to define a function $\hat{f}_1$ which assigns to each *complex* number t the complex number $(2\pi)^{-1}\int_{-B}^{B} F_1(\omega)e^{i\omega t}\,d\omega$. Since the limits of integration are finite, it may be verified by direct differentiation that $\hat{f}_1$ is analytic everywhere. Since $\hat{f}_1$ agrees with f_1 on the real line and hence vanishes on the real interval $[-T/2, T/2]$, $\hat{f}_1$ is zero everywhere and hence so is f_1. Consequently $F_1(\omega) = 0$ for almost all ω, and hence $N(\omega) > 0$ implies $F^*(\omega) = 0$.

But since f^* vanishes outside a finite interval, the same argument as above shows that F^* is the restriction to the real line of a function which is analytic everywhere.

Finally, let D be the set on which $N(\omega) > 0$; by hypothesis, D has positive measure and is therefore uncountable, and consequently has a finite limit point. Since a function that is analytic everywhere and vanishes on such a set necessarily vanishes everywhere, it follows that $F^* = 0$, so that $f = 0$, a contradiction.

A result similar to Theorem 4.6 was proved by Root and Pitcher (1955); the other results of this section are due to Kelly, Reed, and Root (1960).

Values of $-\log_2 p$ and $-p \log_2 p$

$-\log_2 p$

p	0.00	0.01	0.02	0.03	0.04	0.05	0.06	0.07	0.08	0.09
0.0		6.643856	5.643856	5.058894	4.643856	4.321928	4.058894	3.836501	3.643856	3.473931
0.1	3.321928	3.184425	3.058894	2.943416	2.836501	2.736966	2.643856	2.556393	2.473931	2.395929
0.2	2.321928	2.251539	2.184425	2.120294	2.058894	2.000000	1.943416	1.888969	1.836501	1.785875
0.3	1.736966	1.689660	1.643856	1.599462	1.556393	1.514573	1.473931	1.434403	1.395929	1.358454
0.4	1.321928	1.286304	1.251539	1.217591	1.184425	1.152003	1.120294	1.089267	1.058894	1.029146
0.5	1.000000	0.971431	0.943416	0.915936	0.888969	0.862496	0.836501	0.810966	0.785875	0.761213
0.6	0.736966	0.713119	0.689660	0.666576	0.643856	0.621488	0.599462	0.577767	0.556393	0.535332
0.7	0.514573	0.494109	0.473931	0.454032	0.434403	0.415037	0.395929	0.377070	0.358454	0.340075
0.8	0.321928	0.304006	0.286304	0.268817	0.251539	0.234465	0.217591	0.200913	0.184425	0.168123
0.9	0.152003	0.136062	0.120294	0.104697	0.089267	0.074001	0.058894	0.043943	0.029146	0.014500

$-p \log_2 p$

p	0.00	0.01	0.02	0.03	0.04	0.05	0.06	0.07	0.08	0.09
0.0	0	0.066439	0.112877	0.151767	0.185754	0.216096	0.243534	0.268555	0.291508	0.312654
0.1	0.332193	0.350287	0.367067	0.382644	0.397110	0.410545	0.423017	0.434587	0.445308	0.455226
0.2	0.464386	0.472823	0.480573	0.487668	0.494134	0.500000	0.505288	0.510022	0.514220	0.517904
0.3	0.521090	0.523795	0.526034	0.527822	0.529174	0.530101	0.530615	0.530729	0.530453	0.529797
0.4	0.528771	0.527385	0.525646	0.523564	0.521147	0.518401	0.515335	0.511956	0.508269	0.504282
0.5	0.500000	0.495430	0.490577	0.485446	0.480043	0.474373	0.468441	0.462251	0.455808	0.449116
0.6	0.442179	0.435002	0.427589	0.419943	0.412068	0.403967	0.395645	0.387104	0.378347	0.369379
0.7	0.360201	0.350817	0.341230	0.331443	0.321458	0.311278	0.300906	0.290344	0.279594	0.268660
0.8	0.257542	0.246245	0.234769	0.223118	0.211293	0.199295	0.187129	0.174794	0.162294	0.149629
0.9	0.136803	0.123816	0.110671	0.097369	0.083911	0.070301	0.056538	0.042625	0.028563	0.014355

Solutions to Problems

1.1 $p(A) + p(B) = 1$

$$p(y_1) = 0.5p(A) + 0.3[1 - p(A)]$$
$$p(y_2) = 0.3p(A) + 0.5[1 - p(A)]$$
$$p(y_3) = 0.2p(A) + 0.2[1 - p(A)] = 0.2.$$
$$H(Y) = -\sum_{i=1}^{3} p(y_i) \log p(y_i) = -[0.3 + 0.2p(A)] \log [0.3 + 0.2p(A)]$$
$$-[0.5 - 0.2p(A)] \log [0.5 - 0.2p(A)] - 0.2 \log 0.2.$$
$$H(Y \mid X) = p(A)H(Y \mid X = A) + p(B)H(Y \mid X = B)$$
$$= H(0.5, 0.3, 0.2) = -0.5 \log 0.5 - 0.3 \log 0.3 - 0.2 \log 0.2.$$
$$I(X \mid Y) = H(X) - H(X \mid Y).$$

To maximize $I(X \mid Y)$, differentiate with respect to $p(A)$ and set the result equal to zero. The maximum occurs when $p(A) = 0.5$; hence

$$I_{\max} = -0.8 \log 0.4 + 0.5 \log 0.5 + 0.3 \log 0.3 = 0.036.$$

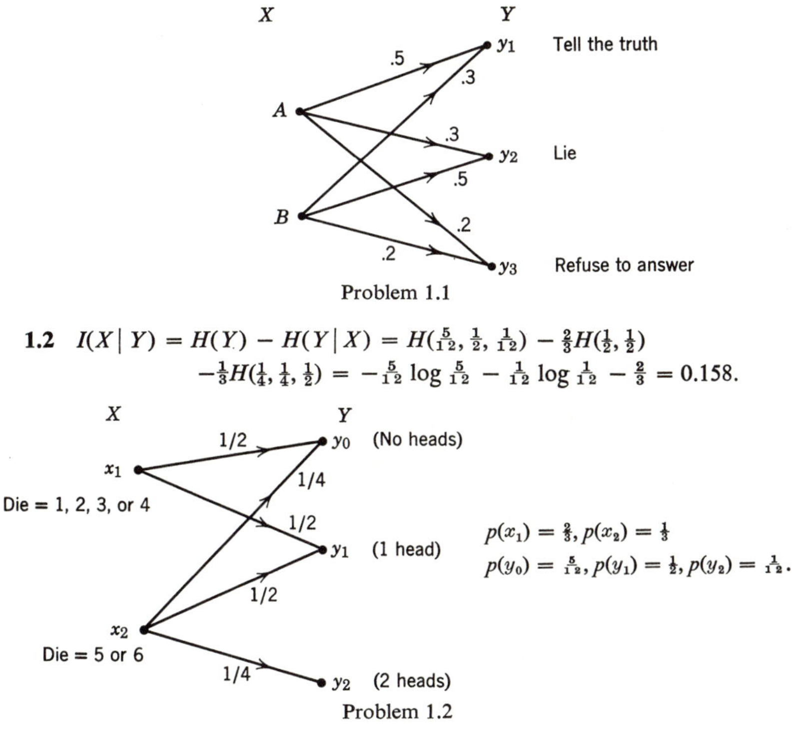

Problem 1.1

1.2 $I(X \mid Y) = H(Y) - H(Y \mid X) = H(\frac{5}{12}, \frac{1}{2}, \frac{1}{12}) - \frac{2}{3}H(\frac{1}{2}, \frac{1}{2})$

$$-\tfrac{1}{3}H(\tfrac{1}{4}, \tfrac{1}{4}, \tfrac{1}{2}) = -\tfrac{5}{12} \log \tfrac{5}{12} - \tfrac{1}{12} \log \tfrac{1}{12} - \tfrac{2}{3} = 0.158.$$

$p(x_1) = \frac{2}{3}, p(x_2) = \frac{1}{3}$

$p(y_0) = \frac{5}{12}, p(y_1) = \frac{1}{2}, p(y_2) = \frac{1}{12}.$

Problem 1.2

1.3 $p(x_1) = 0.75, \quad p(x_2) = 0.25 \qquad p(z_1 \mid x_1) = 0.46, \quad p(z_2 \mid x_1) = 0.54$
$p(y_1) = 0.2, \quad p(y_2) = 0.8 \qquad p(z_1 \mid x_2) = 0.7, \quad p(z_2 \mid x_2) = 0.3$
$p(z_1) = 0.52, \quad p(z_2) = 0.48$

a. $$I(X \mid Y) = H(Y) - H(Y \mid X) = H(0.2, 0.8) - 0.75H(0.1, 0.9) - 0.25H(0.5, 0.5) = 0.12$$

b. $$I(X \mid Z) = H(Z) - H(Z \mid X) = H(0.52, 0.48) - 0.75H(0.46, 0.54) - 0.25H(0.7, 0.3) = 0.03$$

c. First digit conveys $H(X) = H(0.75, 0.25) = 0.811$

Second digit conveys $H(Y \mid X) = 0.75H(0.1, 0.9) + 0.25H(0.5, 0.5) = 0.602$

Third digit conveys $H(Z \mid X, Y) = H(Z \mid Y)$ [since $p(z_k \mid x_i, y_j) = p(z_k \mid y_j)$]
$$= 0.2H(1) + 0.8H(0.4, 0.6)$$
$$= 0.8H(0.4, 0.6) = 0.777.$$

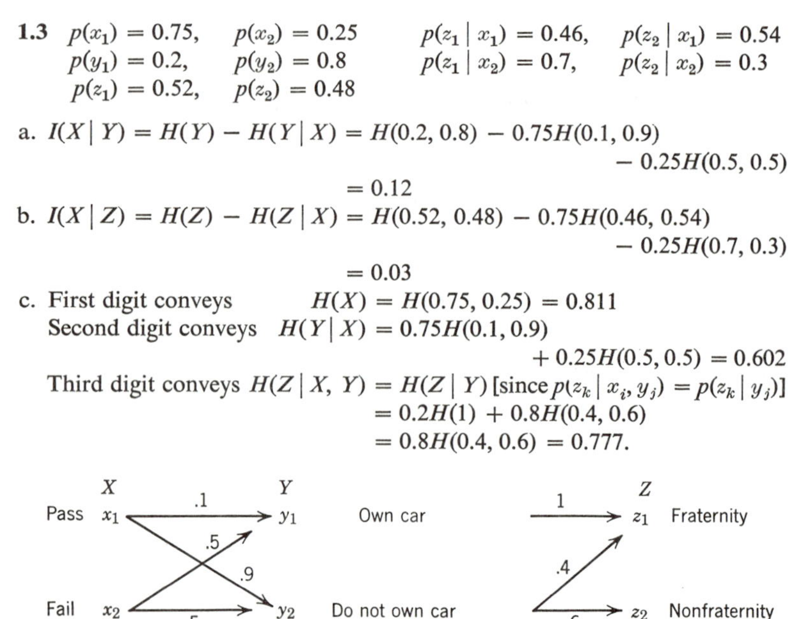

Problem 1.3

1.4 a. $H(Y \mid X) + H(Z \mid X)$
$$= -\sum_{i,j,k} p(x_i, y_j, z_k)[\log p(y_j \mid x_i) + \log p(z_k \mid x_i)]$$
$$= -\sum_i p(x_i) \sum_{j,k} p(y_j, z_k \mid x_i) \log p(y_j \mid x_i) p(z_k \mid x_i)$$

By Lemma 1.4.1, for each i,
$$-\sum_{j,k} p(y_j, z_k \mid x_i) \log p(y_j \mid x_i) p(z_k \mid x_i) \geq -\sum_{j,k} p(y_j, z_k \mid x_i) \log p(y_j, z_k \mid x_i)$$

with equality if and only if $p(y_j, z_k \mid x_i) = p(y_j \mid x_i)p(z_k \mid x_i)$ for all j, k. Thus $H(Y \mid X) + H(Z \mid X) \geq H(Y, Z \mid X)$ with equality if and only if
$$p(y_j, z_k \mid x_i) = p(y_j \mid x_i)p(z_k \mid x_i) \quad \text{for all } i, j, k.$$

b. $$H(Y, Z \mid X) = -\sum_{i,j,k} p(x_i, y_j, z_k) \log p(y_j, z_k \mid x_i)$$
$$= -\sum_{i,j,k} p(x_i, y_j, z_k) \log [p(y_j \mid x_i) p(z_k \mid x_i, y_j)]$$
$$= H(Y \mid X) + H(Z \mid X, Y)$$

c. $H(Z \mid X, Y) = H(Y, Z \mid X) - H(Y \mid X)$ [by (b)]
$\leq H(Y \mid X) + H(Z \mid X) - H(Y \mid X)$ [by (a)]

so $H(Z \mid X, Y) \leq H(Z \mid X)$ with equality if and only if
$$p(y_j, z_k \mid x_i) = p(y_j \mid x_i)p(z_k \mid x_i) \quad \text{for all } i,j,k.$$

1.5 By Lemma 1.4.1,

$$-\sum_{i=1}^{n} a_i \log a_i \le -\sum_{i=1}^{n} a_i \log\left(\frac{a_i x_i}{\sum_{j=1}^{n} a_j x_j}\right)$$

or

$$-\sum_{i=1}^{n} a_i \log a_i \le -\sum_{i=1}^{n} a_i \log a_i - \sum_{i=1}^{n} a_i \log x_i + \left(\sum_{i=1}^{n} a_i\right)\log\left(\sum_{j=1}^{n} a_j x_j\right).$$

Thus

$$\sum_{i=1}^{n} a_i \log x_i \le \log\left(\sum_{j=1}^{n} a_j x_j\right) \text{or } \log(x_1^{a_1}\cdots x_n^{a_n}) \le \log\left(\sum_{j=1}^{n} a_j x_j\right)$$

and the desired inequality follows. By Lemma 1.4.1, equality holds if and only if

$$a_i = \frac{a_i x_i}{\sum_{j=1}^{n} a_j x_j} \quad \text{for all } i,$$

or

$$x_i = \sum_{j=1}^{n} a_j x_j = \text{constant} \quad \text{for all } i.$$

1.6 $H(p_1', \ldots, p_M') = -(p_1 - \Delta p)\log(p_1 - \Delta p) - (p_2 + \Delta p)\log(p_2 + \Delta p)$

$$-\sum_{i=3}^{M} p_i \log p_i$$

$$= -p_1 \log(p_1 - \Delta p) - p_2 \log(p_2 + \Delta p) - \sum_{i=3}^{M} p_i \log p_i$$

$$+ \Delta p \log\frac{p_1 - \Delta p}{p_2 + \Delta p}$$

$$\ge -p_1 \log(p_1 - \Delta p) - p_2 \log(p_2 + \Delta p) - \sum_{i=3}^{M} p_i \log p_i$$

since the last term is ≥ 0

$$> -\sum_{i=1}^{M} p_i \log p_i = H(p_1, \ldots, p_M) \qquad \text{by Lemma 1.4.1}$$

1.7 $p_i' = \sum_{j=1}^{M} a_{ij} p_j$; hence

$$H(p_i', \ldots, p_M') = -\sum_{i=1}^{M} p_i' \log p_i' = -\sum_{i=1}^{M}\sum_{j=1}^{M} a_{ij} p_j \log p_i'$$

$$= -\sum_{j=1}^{M} p_j \left[\sum_{i=1}^{M} a_{ij} \log p_i'\right].$$

Now by problem 1.5,

$$(p_1')^{a_{1j}} \cdots (p_M')^{a_{Mj}} \le \sum_{i=1}^{M} a_{ij} p_i'. \tag{1}$$

(Note $\sum_{i=1}^{M} a_{ij} = 1$.) Hence, taking logarithms,

$$\sum_{i=1}^{M} a_{ij} \log p_i' \le \log\left(\sum_{i=1}^{M} a_{ij} p_i'\right).$$

Consequently

$$H(p_1', \ldots, p_M') \geq -\sum_{j=1}^{M} p_j \log \left(\sum_{i=1}^{M} a_{ij} p_i' \right).$$

But $\sum_{j=1}^{M} (\sum_{i=1}^{M} a_{ij} p_i') = \sum_{i=1}^{M} p_i' (\sum_{j=1}^{M} a_{ij}) = 1$ so Lemma 1.4.1 yields

$$-\sum_{j=1}^{M} p_j \log \left(\sum_{i=1}^{M} a_{ij} p_i' \right) \geq -\sum_{j=1}^{M} p_j \log p_j \tag{2}$$

so that $H(p_1', \ldots, p_M') \geq H(p_1, \ldots, p_M)$. Clearly if $(p_1', \ldots, p_M')$ is a rearrangement of $(p_1, \ldots, p_M)$, then $H(p_1', \ldots, p_M') = H(p_1, \ldots, p_M)$. On the other hand, if the uncertainties are equal, then equality must hold in both (1) and (2), the equality in (1) holding for all j. Equality in (1) implies (see Problem 1.5) that for each j, all p_i' corresponding to positive a_{ij} are equal, that is,

$$p_i' = \sum_{i=1}^{M} a_{ij} p_i' = \text{a constant } c_j \tag{3}$$

whenever $a_{ij} > 0$. (Note if $p_i' = c_j$ then $\sum_{i=1}^{M} a_{ij} p_i' = c_j \sum_{i=1}^{M} a_{ij} = c_j = p_i'$.) But equality in (2) implies

$$p_j = \sum_{i=1}^{M} a_{ij} p_i' \quad \text{for} \quad \text{all } j. \tag{4}$$

Therefore, by (3) and (4), $a_{ij} > 0$ implies $p_i' = p_j$. Since for each i, there is at least one j such that $a_{ij} > 0$, each p_i' is equal to some p_j; hence $(p_1', \ldots, p_M')$ is a rearrangement of $(p_1, \ldots, p_M)$.

In Problem 1.6 we have

$$\begin{aligned} p_1' &= p_1 - \Delta p = \alpha p_1 + \beta p_2 \\ p_2' &= p_2 + \Delta p = \beta p_1 + \alpha p_2 \\ p_i' &= p_i, \qquad i = 3, \ldots, M \end{aligned}$$

where $\alpha = (p_1 - \Delta p - p_2)/(p_1 - p_2)$, $\beta = \Delta p/(p_1 - p_2)$. Since $\alpha + \beta = 1$, the matrix

$$A = \begin{bmatrix} \alpha & \beta & 0 & \cdots & & 0 \\ \beta & \alpha & 0 & \cdots & & 0 \\ 0 & 0 & 1 & \cdots & & 0 \\ 0 & 0 & 0 & \cdots & & 0 \\ \cdot & & & & & \\ \cdot & & & & & \\ \cdot & & & & & \\ 0 & 0 & 0 & \cdots & 1 & 0 \\ 0 & 0 & 0 & \cdots & 0 & 1 \end{bmatrix}$$

is doubly stochastic, and therefore Problem 1.6 is a special case of this result.

1.8 $H(X, Y) = H(X) + H(Y \mid X) = H(Y) + H(X \mid Y)$. But $H(Y \mid X) = 0$ since Y is determined by X; hence $H(Y) \leq H(X)$. $H(Y) = H(X)$ if and only if $H(X \mid Y) = 0$, that is, if and only if X is determined by Y, that is, for each y_j

there is an x_i such that $p(x_i \mid y_j) = 1$. Hence $H(Y) = H(X)$ if and only if g is one to one on $\{x_1, \ldots, x_M\}$, that is, $x_i \neq x_j$ implies $g(x_i) \neq g(x_j)$.

1.9 $P\{Z = z_k \mid X = x_i\} = P\{Y = z_k - x_i \mid X = x_i\}$. Given $X = x_i$, as z_k runs over all possible values of Z, $z_k - x_i$ runs over all possible values of Y; hence

$$H(Z \mid X = x_i) = H(Y \mid X = x_i) \quad \text{for} \quad \text{each } i.$$

Consequently $H(Z \mid X) = H(Y \mid X)$. The remaining conclusions of part (a) follow from Theorem 1.4.5. For part (b), take $Y = -X$; then $H(Z) = 0$ so in general

$$H(Z) < H(X), \qquad H(Z) < H(Y).$$

1.10 For $k = 2$, this is the ordinary grouping axiom 3. The general assertion follows by induction, since if the result has been established for $k - 1$, then by induction hypothesis,

$$\begin{aligned} H(p_1, \ldots, p_{r_1}; \cdots; p_{r_{k-1}+1}, \ldots, p_{r_k}) &= H(p_1 + \cdots + p_{r_1} \\ &\quad + p_{r_1+1} + \cdots + p_{r_2}; p_{r_2+1} \cdots + p_{r_3}; \cdots; p_{r_{k-1}+1} + \cdots + p_{r_k}) \\ &\quad + \left(\sum_{i=1}^{r_2} p_i\right) H\left(\frac{p_1}{\sum_{i=1}^{r_2} p_i}, \ldots, \frac{p_{r_2}}{\sum_{i=1}^{r_2} p_i}\right) + \sum_{i=3}^{k} (p_{r_{i-1}+1} + \cdots + p_{r_i}) \\ &\quad \times H\left(\frac{p_{r_{i-1}+1}}{\sum_{j=r_{i-1}+1}^{r_i} p_j}, \ldots, \frac{p_{r_i}}{\sum_{j=r_{i-1}+1}^{r_i} p_j}\right) \end{aligned}$$

Applying the ordinary grouping axiom, the second term on the right becomes

$$\begin{aligned} &\left(\sum_{i=1}^{r_2} p_i\right) \left[H\left(\frac{\sum_{i=1}^{r_1} p_i}{\sum_{i=1}^{r_2} p_i}, \frac{\sum_{i=r_1+1}^{r_2} p_i}{\sum_{i=1}^{r_2} p_i}\right) \right. \\ &\left. + \frac{\sum_{i=1}^{r_1} p_i}{\sum_{i=1}^{r_2} p_i} H\left(\frac{p_1}{\sum_{i=1}^{r_1} p_i}, \ldots, \frac{p_{r_1}}{\sum_{i=1}^{r_1} p_i}\right) + \frac{\sum_{i=r_1+1}^{r_2} p_i}{\sum_{i=1}^{r_2} p_i} H\left(\frac{p_{r_1+1}}{\sum_{i=r_1+1}^{r_2} p_i}, \ldots, \frac{p_{r_2}}{\sum_{i=r_1+1}^{r_2} p_i}\right)\right]. \end{aligned}$$

Thus

$$\begin{aligned} &H(p_1, \ldots, p_{r_1}; \cdots; p_{r_{k-1}+1}, \ldots, p_{r_k}) \\ &\quad = H(p_1 + \cdots + p_{r_2}; p_{r_2+1} + \cdots + p_{r_3}; \cdots; p_{r_{k-1}+1} + \cdots + p_{r_k}) \\ &\quad\quad + \left(\sum_{i=1}^{r_2} p_i\right) H\left(\frac{\sum_{i=1}^{r_1} p_i}{\sum_{i=1}^{r_2} p_i}, \frac{\sum_{i=r_1+1}^{r_2} p_i}{\sum_{i=1}^{r_2} p_i}\right) \\ &\quad\quad + \sum_{i=1}^{k} (p_{r_{i-1}+1} \cdots + p_{r_i}) H\left(\frac{p_{r_{i-1}+1}}{\sum_{j=r_{i-1}+1}^{r_i} p_j}, \ldots, \frac{p_{r_i}}{\sum_{j=r_{i-1}+1}^{r_i} p_j}\right). \end{aligned}$$

Again using the induction hypothesis, the sum of the first two terms on the right in this equation is

$$H(\underbrace{p_1 + \cdots + p_{r_1}, p_{r_1+1} + \cdots p_{r_2}}_{\substack{\text{group 1} \\ \text{(2 terms)}}}, \underbrace{p_{r_2+1} + \cdots + p_{r_3}}_{\substack{\text{group 2} \\ \text{(1 term)}}}, \ldots, \underbrace{p_{r_{k-1}+1} + \cdots p_{r_k}}_{\substack{\text{group } k-1 \\ \text{(1 term)}}})$$

and the result follows.

The above argument uses the ordinary grouping axiom but none of the other axioms for the uncertainty function. If we use all four axioms, the derivation is easier: Let $q_j = \sum_{i=r_{j-1}+1}^{r_j} p_i$. Then the right-hand side of the statement to be proved is

$$\begin{aligned}
&-\sum_{j=1}^{k} q_j \log q_j - \sum_{i=1}^{k} q_i \sum_{m=r_{i-1}+1}^{r_i} \frac{p_m}{q_i} \log \frac{p_m}{q_i} \\
&= -\sum_{j=1}^{k} q_j \log q_j - \sum_{i=1}^{k} \sum_{m=r_{i-1}+1}^{r_i} (p_m \log p_m - p_m \log q_i) \\
&= -\sum_{j=1}^{k} q_j \log q_j - \sum_{t=1}^{r_k} p_t \log p_t + \sum_{i=1}^{k} q_i \log q_i \\
&= H(p_1, \ldots, p_{r_k}).
\end{aligned}$$

1.11 Let $h'(p) = h(p) + C \log p$. Then $\sum_{i=1}^{M} p_i h'(p_i) = 0$. Taking all $p_i = 1/n$ we have $h'(1/n) = 0$, $n = 1, 2, \ldots$. If r/s is a rational number, then, taking

$$p_1 = \frac{r}{s} \qquad p_2 = p_3 = \cdots = p_{s-r+1} = \frac{1}{s}$$

we have

$$\frac{r}{s} h'\left(\frac{r}{s}\right) + \frac{s-r}{s} h'\left(\frac{1}{s}\right) = 0;$$

hence $h'(r/s) = 0$. The result now follows by continuity.

1.12 Let $f(n) = h(1/n)$. Then the requirements (a) and (b) imply $f(nm) = f(n) + f(m)$, and $n < m \Rightarrow f(n) < f(m)$, $n, m = 1, 2, \ldots$. The proof of Theorem 1.2.1 in the text shows that $f(n) = C \log_b n$ where $C > 0$, $b > 1$.

If p is a rational number r/s, then

$$h\left(\frac{1}{s}\right) = h\left(\frac{r}{s}\frac{1}{r}\right) = h\left(\frac{r}{s}\right) + h\left(\frac{1}{r}\right).$$

Thus

$$h\left(\frac{r}{s}\right) = f(s) - f(r) = C \log \frac{s}{r} = -C \log \frac{r}{s}.$$

Hence $h(p) = -C \log p$ for rational p. The general assertion follows by continuity.

2.1 a.

S_0	S_1	S_2	S_3	S_4	S_5	S_6
010	1	100	11	00	01	0
0001		1110		110	011	10
0110		01011			110	001
1100					0	110
00011						0011
00110						(0110) code word
11110						
101011						

The code is not uniquely decipherable.

$A_0 = 0001$ $A_1 = 1$ $A_2 = 01011$ $A_3 = 11$ $A_4 = 110$ $A_5 = 0$ $A_6 = 0110$
$W_0 = 00011$ $W_1 = 101011$ $W_2 = 010$ $W_3 = 11110$ $W_4 = 1100$ $W_5 = 00110$ $W_6 = 0110$

$$\text{Ambiguous sequence} = A_0W_1W_4W_6 = 0001 \quad 101011 \quad 1100 \quad 0110$$
$$= W_0W_2W_3W_5 = 00011 \quad 010 \quad 11110 \quad 00110$$

S_0	S_1	S_2	S_3	S_4	S_5	S_6	S_7	S_8	S_9	S_{10}
abc	*d*	*ba*	*ce*	*ac*	*cd*	*eac*	*ac*	*c*	*eac*	*ac*
abcd	*abd*			*ab*	*c*	*eab*	*ab*	*cd*	*eab*	*ab*
e							*d*	*ba*	*ce*	*d*
dba										
bace										
ceac										
ceab										
eabd										

$$S_7 = S_{10}; \text{ hence } S_i = S_{i+3}, \qquad i \geq 7$$

None of the sets S_i contains a code word, so the code is uniquely decipherable.

2.2 a. Let ω_i be the number of code words of length i ($\omega_1 = 1$, $\omega_2 = 2$, $\omega_n = 0$, $n \geq 3$). Then a message whose coded form has exactly k letters must begin with a code word of length 1 or a code word of length 2; hence

$$N(k) = \omega_1 N(k-1) + \omega_2 N(k-2), \qquad k \geq 3.$$

Now $N(2) = \omega_1 N(1) + \omega_2$; hence if we define $N(0) = 1$, the above equation is valid for $k \geq 2$. Thus we must solve the linear homogeneous difference equation $N(k+2) - \omega_1 N(k+1) - \omega_2 N(k) = 0$, $k = 0, 1, \ldots$, subject to $N(0) = 1$, $N(1) = 1$. Assuming a solution of the form $N(k) = \lambda^k$ we obtain

$$\lambda^{k+2} - \omega_1\lambda^{k+1} - \omega_2\lambda^k = 0 \quad \text{or} \quad \lambda^k(\lambda^2 - \lambda - 2) = 0.$$

The two nonzero roots are $\lambda_1 = 2$, $\lambda_2 = -1$. Hence

$$N(k) = A \cdot 2^k + B(-1)^k, \qquad k = 0, 1, \ldots.$$

Since $N(0) = A + B = 1$, $N(1) = 2A - B = 1$, we have

$$A = \tfrac{2}{3}, \; B = \tfrac{1}{3}.$$

So $N(k) = (\frac{2}{3})2^k + (\frac{1}{3})(-1)^k, \qquad k = 0, 1, \ldots.$

b. As in (a), $N(k) = \omega_1 N(k-1) + \omega_2 N(k-2) + \omega_3 N(k-3)$ ($\omega_1 = 1$, $\omega_2 = 1$, $\omega_3 = 2$, $\omega_n = 0$, $n \geq 4$) or

$$N(k+3) - N(k+2) - N(k+1) - 2N(k) = 0, \qquad k \geq 0.$$

The assumption $N(k) = \lambda^k$ yields the characteristic equation

$$\lambda^3 - \lambda^2 - \lambda - 2 = (\lambda - 2)(\lambda^2 + \lambda + 1) = 0$$

with roots

$$\lambda_1 = 2, \qquad \lambda_2 = -\tfrac{1}{2} + \tfrac{1}{2}i\sqrt{3} = e^{i\cdot 2\pi/3}, \qquad \lambda_3 = -\tfrac{1}{2} - \tfrac{1}{2}i\sqrt{3} = e^{-i2\pi/3}.$$

Thus $N(k) = A \cdot 2^k + B \cos(2\pi/3)k + C \sin(2\pi/3)k$, $k = 0, 1, \ldots.$ The boundary conditions are

$$N(0) = 1 = A + B$$
$$N(1) = 1 = 2A - \tfrac{1}{2}B + \tfrac{1}{2}\sqrt{3}\,C$$
$$N(2) = 2 = 4A - \tfrac{1}{2}B - \tfrac{1}{2}\sqrt{3}\,C.$$

These equations have the solution $A = \frac{4}{7}$, $B = \frac{3}{7}$, $C = \sqrt{3}/21$, so

$$N(k) = (\tfrac{4}{7})2^k + \tfrac{3}{7}\cos\frac{2\pi}{3}k + \frac{\sqrt{3}}{21}\sin\frac{2\pi}{3}k, \qquad k = 0, 1, \ldots.$$

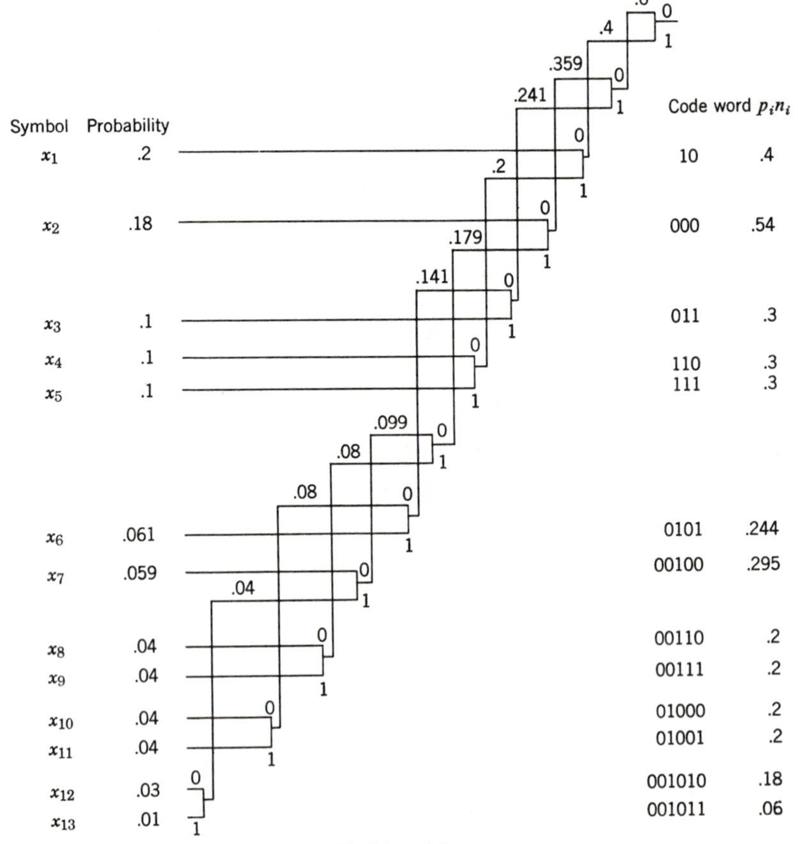

Problem 2.3

2.3 The process of constructing the code is indicated schematically in the accompanying figure. The average code-word length is $\bar{n} = \sum_{i=1}^{13} p_i n_i = 3.419$. The uncertainty may be computed to be $H(X) = 3.355$.

2.4 By Theorem 2.5.2 there is a binary code whose average word length is less than $H(X) + 1$; since the Huffman code minimizes the average word length for a given set of probabilities, the result is immediate.

2.5 Add fictitious symbols with zero probability to the given set until the number of symbols is congruent to 1, modulo $(D - 1)$. An optimal code for the enlarged set will then yield an optimal code for the original set. Now we shall prove that given any optimal code we may construct another code with exactly the same word lengths such that, among the words of maximal length, at least D words agree in all places but the last. Therefore without loss of generality we may assume that an optimal code has this property. In particular there are at least D words of maximal length, and consequently (note that Lemma 2.6.2a holds just as in the binary case) the D least probable symbols have code words of equal length. The proof for the binary case now goes through word for word, except that we combine symbols D at a time instead of 2 at a time.

To prove the required assertion, pick any word W of maximal length n_M. Let Q be the sequence consisting of all digits of W except the last. Say the code alphabet is $\{0, 1, \ldots, D - 1\}$ and $W = Q0$. If there are more than D words of maximal length we may, without changing the instantaneous property of the code, take any $(D - 1)$ words of length n_M other than W and replace them by $Q1, Q2, \ldots, Q(D - 1)$; thus it suffices to consider the case in which there are at most D words of maximal length; we may assume without loss of generality that these words are $Q0, Q1, \ldots, Qi$, $i < D$. We may assume that $i > 0$ since if $i = 0$, $Q0$ may be replaced by Q.

Now as in the proof of Theorem 2.3.1, the number of terminal points on the tree of order D and size n_M excluded by a code with word lengths $n_1, \ldots, n_M$ is $T = \sum_{i=1}^{M} D^{(n_M - n_i)}$. Since $D \equiv 1 \bmod (D - 1)$ and (after the fictitious symbols are added) $M \equiv 1 \bmod (D - 1)$, it follows that $T \equiv 1 \bmod (D - 1)$. If R is any sequence of length $n_M - 1$ other than Q, then if R is not a code word and no code word is a prefix of R we may replace one of the words Qj by R. The resulting code is still instantaneous and has an average word length at most that of the original code. (Possibly Qj corresponds to a symbol with probability zero so that the average word length is unchanged.) Thus without loss of generality we may assume that every such sequence R is either a code word or has a code word as a prefix. In either case, the terminal points corresponding to the sequences Rj, $j = 0, 1, \ldots, D - 1$ are all excluded. Therefore the only terminal points not excluded are those corresponding to $Q(i + 1), \ldots, Q(D - 1)$. Since the total number of terminal points is $D^{n_M} \equiv 1 \bmod (D - 1)$, the number of terminal points not excluded is congruent to zero modulo $(D - 1)$. Hence $D - 1 - i = 0$ or $D - 1$; the second case is impossible since $i > 0$; consequently $i = D - 1$ and the proof is complete.

For the given numerical example, we first group 2 symbols at a time (or add one fictitious symbol and group 3 at a time) and afterward we combine 3 symbols at a time. The code-word lengths are 1, 2, 2, 2, 2, 2, 3, 3.

3.1 Since

$$p(y_j) = \sum_{i=1}^{3} p(x_i)p(y_j \mid x_i) \quad \text{and} \quad p(x_i \mid y_j) = \frac{p(x_i)p(y_j \mid x_i)}{p(y_j)}$$

we compute

$$\begin{array}{lll} p(x_1 \mid y_1) = 2/3, & p(x_2 \mid y_1) = 1/9, & p(x_3 \mid y_1) = 2/9 \\ p(x_1 \mid y_2) = 1/2, & p(x_2 \mid y_2) = 3/8, & p(x_3 \mid y_2) = 1/8 \\ p(x_1 \mid y_3) = 2/7, & p(x_2 \mid y_3) = 2/7, & p(x_3 \mid y_3) = 3/7 \end{array}$$

Thus the ideal observer is $g(y_1) = x_1, g(y_2) = x_1, g(y_3) = x_3$. The probability of error is

$$\sum_{j=1}^{3} p(y_j)p(e \mid y_j) = \sum_{j=1}^{3} p(y_j)P\{X \neq g(y_j) \mid y_j\}$$

$$p(e) = \frac{3}{8}\left(\frac{1}{9} + \frac{2}{9}\right) + \frac{1}{3}\left(\frac{3}{8} + \frac{1}{8}\right) + \frac{7}{24}\left(\frac{2}{7} + \frac{2}{7}\right) = \frac{11}{24}.$$

3.2 If y_j is received, the probability of correct transmission is $p(e' \mid y_j) = \sum_{i=1}^{M} q_{ji} P\{X = x_i \mid y_j\}$. (Necessarily $q_{ji} \geq 0$ for all i, j and $\sum_{i=1}^{M} q_{ji} = 1, j = 1, \ldots, L$.)

If $p(x_{i_0} \mid y_j) = \max_{1 \leq i \leq M} p(x_i \mid y_j)$, then

$$p(e' \mid y_j) \leq \sum_{i=1}^{M} q_{ji}\, p(x_{i_0} \mid y_j) = p(x_{i_0} \mid y_j);$$

equality can be achieved by taking $q_{ji_0} = 1, q_{ji} = 0, i \neq i_0$.

Thus $p(e' \mid y_j)$ is maximized for each j, and hence the overall probability of correct transmission is maximized, by the above choice of q_{ji}; however, this choice is precisely the ideal observer.

3.3

$$I(X \mid Y) = H(X) - H(X \mid Y)$$
$$I(X \mid Z) = H(X) - H(X \mid Z)$$

Now

$$p(x \mid y, z) = \frac{p(x, y, z)}{p(y, z)} = \frac{p(x, y)p(z \mid x, y)}{p(y)p(z \mid y)}.$$

The essential consequence of the cascading process is that $p(z \mid x, y) = p(z \mid y)$ for all x, y, z; in fact, this condition should be used as the definition of a cascade combination. It follows that $p(x \mid y, z) = p(x \mid y)$, hence $H(X \mid Y, Z) = H(X \mid Y)$. Thus

$$I(X \mid Y) - I(X \mid Z) = H(X \mid Z) - H(X \mid Y, Z) \geq 0$$

by Problem 1.4.

Since $I(X \mid Y) \geq I(X \mid Z)$ for any input distribution, the capacity of K_1 $\geq$ the capacity of the cascade. Now write

$$I(Y \mid Z) = H(Z) - H(Z \mid Y)$$
$$I(X \mid Z) = H(Z) - H(Z \mid X).$$

Since $p(z \mid y) = p(z \mid x, y)$, $H(Z \mid Y) = H(Z \mid X, Y)$. Thus

$$I(Y \mid Z) - I(X \mid Z) = H(Z \mid X) - H(Z \mid X, Y) \geq 0,$$

so that the capacity of $K_2 \geq$ the capacity of the cascade.

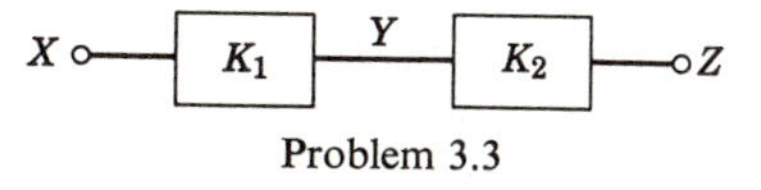

Problem 3.3

3.4 We may regard Y as the input and X the output of a discrete memoryless channel; the output X may be thought of as being applied as the input of a deterministic channel with output Z (see diagram). The conditions for cascading are satisfied, that is, $p(z \mid y, x) = p(z \mid x)$; the result then follows from the argument of Problem 3.3. Alternately,

$$\begin{aligned} H(Y \mid Z) &= E[-\log p(Y \mid Z)] = E[-\log p(Y \mid g(X))] \\ &= -\sum_{i,j} p(x_i, y_j) \log P\{Y = y_j \mid Z = g(x_i)\} \\ &= \sum_i p(x_i)[-\sum_j p(y_j \mid x_i) \log P\{Y = y_j \mid Z = g(x_i)\}]. \end{aligned}$$

By Lemma 1.4.1,

$$H(Y \mid Z) \geq \sum_i p(x_i)[-\sum_j p(y_j \mid x_i) \log p(y_j \mid x_i)] = H(Y \mid X).$$

Y — K_1 — X — K_2 — $Z = g(X)$

Problem 3.4. Channel matrix of $K_1 = [p(x \mid y)]$.

3.5 $p_n = (1 - \beta)p_{n-1} + \beta(1 - p_{n-1})$ or $p_n - (1 - 2\beta)p_{n-1} = \beta$. The solution to this difference equation is $p_n = A(1 - 2\beta)^n + B$ where the "particular solution" B is determined by $B - (1 - 2\beta)B = \beta$ or $B = \frac{1}{2}$, and A is determined by taking $n = 0$: $A + B = p_0$. Thus

$$p_n = \tfrac{1}{2} + (p_0 - \tfrac{1}{2})(1 - 2\beta)^n \rightarrow \tfrac{1}{2} \quad \text{as} \quad n \rightarrow \infty.$$

Now the channel matrix of the cascade combination of n channels is

$$\Pi_n = \begin{array}{c} \\ 0 \\ 1 \end{array} \begin{array}{c} \begin{array}{cc} 0 & 1 \end{array} \\ \begin{bmatrix} p_n^* & 1 - p_n^* \\ 1 - p_n^* & p_n^* \end{bmatrix} \end{array}$$

where $p_n^* = \frac{1}{2} + (p_0 - \frac{1}{2})(1 - 2\beta)^n$, evaluated when $p_0 = 1$.

The capacity of the n-stage combination is

$$C_n = 1 - H(p_n^*, 1 - p_n^*) \rightarrow 1 - H(\tfrac{1}{2}, \tfrac{1}{2}) = 0 \quad \text{as} \quad n \rightarrow \infty.$$

p_{n-1} 0 —($1-\beta$)→ 0 p_n; 0 —β→ 1; 1 —β→ 0; $1 - p_{n-1}$ 1 —($1-\beta$)→ 1 $1 - p_n$

Problem 3.5

3.6 Let Π_i = channel matrix of K_i, $i = 1, 2$. We have

$$p(z \mid x) = \sum_y p(y \mid x)p(z \mid y).$$

Thus the matrix of the cascade combination is $\Pi_1\Pi_2$; similarly if channels $K_1, K_2, \ldots, K_n$ are cascaded (in that order) the matrix of the cascade combination is the product $\Pi_1\Pi_2 \cdots \Pi_n$.

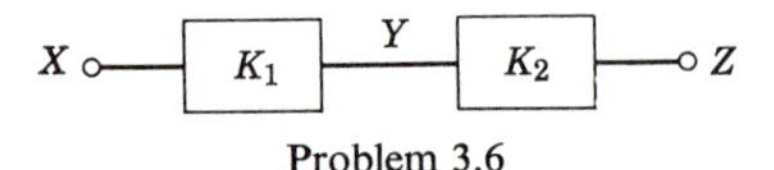

Problem 3.6

Remark Concerning Problems 3.5 and 3.6

If X_i is the output of the ith channel [= the input of the $(i + 1)$th channel] of a cascade combination, the conditions for cascading are expressed by requiring $p(x_k \mid x_0, \ldots, x_{k-1}) = p(x_k \mid x_{k-1})$ for all $x_0, \ldots, x_k$, $k = 1, 2, \ldots, n$.

3.7 Without loss of generality assume $\beta > \alpha$. We may use Theorem 3.3.3. We have

$$\Pi = \begin{bmatrix} \alpha & 1 - \alpha \\ \beta & 1 - \beta \end{bmatrix}, \qquad \Pi^{-1} = (\alpha - \beta)^{-1} \begin{bmatrix} 1 - \beta & \alpha - 1 \\ -\beta & \alpha \end{bmatrix}.$$

Let $H_\alpha = H(\alpha, 1 - \alpha) = H(Y \mid X = x_1)$, $H_\beta = H(\beta, 1 - \beta) = H(Y \mid X = x_2)$. We must show that the hypothesis of Theorem 3.3.3 is satisfied. We have [see (3.3.4)]

$$d_1 = -\frac{1 - \beta}{\beta - \alpha} \exp_2\left[\frac{(1 - \beta)H_\alpha + (\alpha - 1)H_\beta}{\beta - \alpha}\right] + \frac{\beta}{\beta - \alpha} \exp_2\left(\frac{-\beta H_\alpha + \alpha H_\beta}{\beta - \alpha}\right).$$

Thus $d_1 > 0$ if and only if

$$\log \frac{1 - \beta}{\beta} < \frac{-\beta H_\alpha + \alpha H_\beta - (1 - \beta)H_\alpha - (\alpha - 1)H_\beta}{\beta - \alpha}, \quad \text{or} \quad \log \frac{1 - \beta}{\beta} < \frac{H_\beta - H_\alpha}{\beta - \alpha}.$$

Upon expansion, this becomes $-\alpha \log \alpha - (1 - \alpha) \log (1 - \alpha) < -\alpha \log \beta - (1 - \alpha) \log (1 - \beta)$, which is true by Lemma 1.4.1. Consequently $d_1 > 0$. A similar argument shows that $d_2 > 0$.

Thus by (3.3.5),

$$C = \log\left[\exp_2\left(\frac{(1 - \beta)H_\alpha + (\alpha - 1)H_\beta}{\beta - \alpha}\right) + \exp_2\left(\frac{-\beta H_\alpha + \alpha H_\beta}{\beta - \alpha}\right)\right]$$

3.8 Let the input alphabets of K_1 and K_2 be $x_1, \ldots, x_r$ and $x_{r+1}, \ldots, x_M$ respectively; let the output alphabets of K_1 and K_2 be $y_1, \ldots, y_s$ and $y_{s+1}, \ldots,$

y_L. If $p(x)$ is any input distribution, and $p = \sum_{i=1}^{r} p(x_i)$, then

$$H(X) = -\sum_{i=1}^{M} p(x_i) \log p(x_i)$$
$$= -p \log p - (1 - p) \log (1 - p) + pH_1(X) + (1 - p)H_2(X)$$

by the grouping axiom 3 of Section 1.2

where

$$H_1(X) = -\sum_{i=1}^{r} \frac{p(x_i)}{p} \log \frac{p(x_i)}{p}$$

is the input uncertainty of K_1 under the distribution $\{p(x_i)/p, i = 1, \ldots, r\}$ and

$$H_2(X) = -\sum_{i=r+1}^{M} \frac{p(x_i)}{1 - p} \log \frac{p(x_i)}{1 - p}$$

is the input uncertainty of K_2 under the distribution $\{p(x_i)/(1 - p), i = r + 1, \ldots, M\}$. (Alternately

$$H(X) = -\sum_{i=1}^{r} p \frac{p(x_i)}{p} \log \left[\left(\frac{p(x_i)}{p}\right)p\right] - \sum_{i=r+1}^{M} \frac{(1 - p)p(x_i)}{1 - p} \log \left[\frac{p(x_i)}{1 - p}(1 - p)\right]$$

leading to the same expression as above.) Now

$$H(X \mid Y) = -\sum_{i=1}^{M} \sum_{j=1}^{L} p(x_i, y_j) \log p(x_i \mid y_j)$$
$$= -\sum_{i=1}^{r} \sum_{j=1}^{s} p \frac{p(x_i)}{p} p(y_j \mid x_i) \log p(x_i \mid y_j)$$
$$- \sum_{i=r+1}^{M} \sum_{j=s+1}^{L} (1 - p) \left[\frac{p(x_i)}{1 - p}\right] p(y_j \mid x_i) \log p(x_i \mid y_j).$$

Now if $1 \leq i \leq r$, $1 \leq j \leq s$, the conditional probability $p(x_i \mid y_j) = P\{X = x_i \mid Y = y_j\}$ is

$$p(x_i \mid y_j) = \frac{p(x_i)p(y_j \mid x_i)}{p(y_j)} = \frac{p[p(x_i)/p]\, p(y_j \mid x_i)}{p \sum_{i=1}^{M} \frac{p(x_i)}{p} p(y_j \mid x_i)}$$
$$= P\{X_1 = x_i \mid Y_1 = y_j\},$$

that is, the conditional probability that the input of K_1 is x_i given that the output of K_1 is y_j, under the distribution $\{p(x_i)/p, i = 1, 2, \ldots, r\}$.

Thus

$$H(X \mid Y) = pH_1(X \mid Y) + (1 - p)H_2(X \mid Y), \text{and } I(X \mid Y) = H(X) - H(X \mid Y)$$
$$= H(p, 1 - p) + pI_1(X \mid Y) + (1 - p)I_2(X \mid Y),$$

where the subscript i denotes that the indicated quantity is calculated for K_i under the appropriate input distribution. Now for a given p, we are completely free to choose the probabilities $p(x_i)/p$, $i = 1, \ldots, r$, and $p(x_i)/1 - p$, $i = r + 1, \ldots, M$. We can do no better than to choose the probabilities so that $I_1(X \mid Y) = C_1$, $I_2(X \mid Y) = C_2$.

Thus it remains to maximize $H(p, 1 - p) + pC_1 + (1 - p)C_2$. Differentiating, we obtain

$$-1 - \log p + 1 + \log(1 - p) + C_1 - C_2 = 0, \text{ or } p = \frac{2^{C_1}}{2^{C_1} + 2^{C_2}}.$$

Hence

$$\begin{aligned} C &= H\left(\frac{2^{C_1}}{2^{C_1} + 2^{C_2}}, \frac{2^{C_2}}{2^{C_1} + 2^{C_2}}\right) + \frac{C_1 \cdot 2^{C_1} + C_2 \cdot 2^{C_2}}{2^{C_1} + 2^{C_2}} \\ &= \frac{2^{C_1}}{2^{C_1} + 2^{C_2}} \log(2^{C_1} + 2^{C_2}) + \frac{2^{C_2}}{2^{C_1} + 2^{C_2}} \log(2^{C_1} + 2^{C_2}). \end{aligned}$$

Therefore $C = \log(2^{C_1} + 2^{C_2})$.

3.9 Let X_i be the input to K_i, and Y_i the corresponding output. By definition of the product channel,

$$H(Y_1, Y_2 \mid X_1, X_2) = H(Y_1 \mid X_1) + H(Y_2 \mid X_2).$$

Now $H(Y_1, Y_2) \leq H(Y_1) + H(Y_2)$ with equality if and only if Y_1 and Y_2 are independent. Therefore $I(X_1, X_2 \mid Y_1, Y_2) \leq I(X_1 \mid Y_1) + I(X_2 \mid Y_2) \leq C_1 + C_2$.

If we take X_1 and X_2 to be independent, with X_i having the distribution that achieves channel capacity for K_i, then Y_1 and Y_2 are independent (by the same argument as in the proof of the fundamental theorem), and equality is achieved in the above expression.

Thus $C = C_1 + C_2$.

3.10 Channel (a) is the sum of

$$\begin{bmatrix} 1 - \beta & \beta \\ \beta & 1 - \beta \end{bmatrix} \quad \text{and} \quad [1].$$

By Problem 3.8, $2^C = 2^{1 - H(\beta, 1 - \beta)} + 2^0$, or $C = \log(1 + 2^{1 - H(\beta, 1 - \beta)})$.

Channel (b) is symmetric; hence

$$C = \log 4 + p \log \tfrac{1}{2}p + (1 - p) \log \tfrac{1}{2}(1 - p),$$

or $C = 1 - H(p, 1 - p) =$ the capacity of a binary symmetric channel with error probability p.

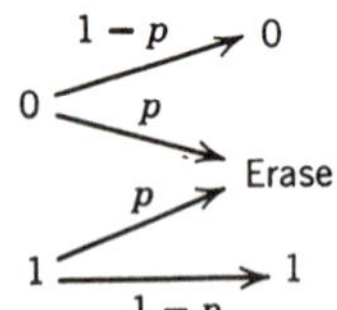

Problem 3.10

For channel (c), see the accompanying figure. If $P\{X = 0\} = \alpha$ then $H(X) = H(\alpha, 1 - \alpha)$.

$$H(X \mid Y) = p(\text{erase})H(X \mid Y = \text{erase}) = pH(\alpha, 1 - \alpha).$$

Thus $I(X \mid Y) = (1 - p)H(\alpha, 1 - \alpha)$, which is a maximum for $\alpha = \frac{1}{2}$. Therefore $C = 1 - p$.

3.11 a. Proceed exactly as in the proof of the fundamental lemma, with the set A replaced by $A^* = A \cap F^*$, where $F^* = \{(\mathbf{x}, \mathbf{y}) : \mathbf{x} \in F\}$. Observe that all code words selected in the algorithm must belong to F, since if $\mathbf{x} \notin F$, $A_{\mathbf{x}}^*(= \{\mathbf{y}:(\mathbf{x}, \mathbf{y}) \in A^*\})$ is empty. We obtain a code (s, n, λ) such that $\lambda \le s \cdot 2^{-a} + P\{(\mathbf{X}, \mathbf{Y}) \notin A^*\}$. Since

$$P\{(\mathbf{X}, \mathbf{Y}) \notin A^*\} \le P\{(\mathbf{X}, \mathbf{Y}) \notin A\} + P\{(\mathbf{X}, \mathbf{Y}) \notin F^*\} = P\{(\mathbf{X}, \mathbf{Y}) \notin A\} + P\{\mathbf{X} \notin F\},$$

the result follows.

b. We shall prove that the maximum value of C_0, that is, the least upper bound of all permissible rates of transmission, is

$$C_0^* = H[(1 - r)(1 - \beta) + r\beta, (1 - r)\beta + r(1 - \beta)] - H(\beta, 1 - \beta),$$

which is the information processed by the binary symmetric channel when the input distribution is $p(1) = r, p(0) = 1 - r$. Notice that as $r \to \frac{1}{2}$, $C_0 \to C = 1 - H(\beta, 1 - \beta)$.

We shall first prove a *coding theorem*, which states that given $R < C_0^*$, there exist codes $([2^{nR}], n, \lambda_n)$, with all words belonging to F_n, such that $\lim_{n\to\infty} \lambda_n = 0$. We use the results of part (a), with $F = F_n$.

Let $\mathbf{X} = (X_1, \ldots, X_n)$ where the X_i are independent and $P\{X_i = 1\} = p < r$, $P\{X_i = 0\} = 1 - p$. Then $P\{\mathbf{X} \notin F_n\} = P\{n^{-1} \sum_{i=1}^n X_i > r\} \to 0$ as $n \to \infty$ by the weak law of large numbers. Let us take $a = kn$. Then as in the proof of the fundamental theorem, $P\{(\mathbf{X}, \mathbf{Y}) \notin A\} = P\{n^{-1} \sum_{i=1}^n U_i \le k\}$, where the U_i are independent, identically distributed random variables with expected value

$$I(X_i \mid Y_i) = H(Y_i) - H(Y_i \mid X_i) = H[(1 - p)(1 - \beta) + p\beta,$$
$$(1 - p)\beta + p(1 - \beta)] - H(\beta, 1 - \beta) \to C_0^* \quad \text{as} \quad p \to r.$$

Given $R < C_0^*$, we take $s = [2^{nR}]$, $k = \frac{1}{2}(R + C_0^*)$, and p sufficiently close to r so that $I(X_i \mid Y_i) > k$. Then by choice of s and $a = kn$, $s \cdot 2^{-a} \to 0$ as $n \to \infty$. Also $P\{(\mathbf{X}, \mathbf{Y}) \notin A\} \to 0$ as $n \to \infty$ by the weak law of large numbers. The result follows.

We now prove a *weak converse*, which states that any code (s, n), all of whose words belong to F_n, must satisfy

$$\log s \le \frac{nC_0^* + \log 2}{1 - \overline{p(e)}},$$

where $\overline{p(e)}$ is the average probability of error of the code. As in Theorem 3.7.3, this implies that if we require that all code words belong to F_n, it is not possible to maintain a transmission rate above C_0^* with an arbitrarily small probability of error.

Given a code (s, n) with all words in F_n. We can proceed just as in the proof of Theorem 3.7.3 (with C_0^* replacing C) except for one difficulty. Although we have $I(\mathbf{X} \mid \mathbf{Y}) \le \sum_{i=1}^n I(X_i \mid Y_i)$ as in (3.7.4), it is not true that $I(X_i \mid Y_i) \le C_0^*$ for all i. For example, if $s = 2$, $n = 5$, $r = \frac{2}{5}$, and the words of the code are $\mathbf{w}_1 = 11000$ and $\mathbf{w}_2 = 10000$, then $P\{X_2 = 1\} = P\{X_2 = 0\} = \frac{1}{2}$, and therefore $I(X_2 \mid Y_2) = C > C_0^*$. However, we may prove that $\sum_{i=1}^n I(X_i \mid Y_i) \le nC_0^*$. Thus $I(\mathbf{X} \mid \mathbf{Y}) \le nC_0^*$, so that the proof in the text will go through. To see this, let X be a symbol chosen with uniform probability from among all sn digits of the code vocabulary $\mathbf{w}_1, \ldots, \mathbf{w}_s$. Since X may be selected by first choosing a digit j at random, and then selecting a code word at random and examining the jth digit of the word, we have

$$P\{X = \alpha\} = n^{-1} \sum_{j=1}^{n} P\{X_j = \alpha\}, \qquad \alpha = 0, 1.$$

By Theorem 3.3.1, $I(X \mid Y) \geq n^{-1} \sum_{j=1}^{n} I(X_j \mid Y_j)$. But since all words belong to F_n, $P\{X = 1\} \leq r$, and hence $I(X \mid Y) \leq C_0^*$. The result follows.

3.12 The information $I(X \mid Y)$ processed by a channel may be expressed as a function of the input probabilities $p(x_1), \ldots, p(x_M)$. [See (3.3.6).] The domain of I is the subset of Euclidean M-space on which all $p(x_i) \geq 0$ and

$$\sum_{i=1}^{M} p\,(x_i) = 1.$$

Thus the domain is a closed and bounded subset of Euclidean M-space. Since such a set is compact, and a continuous function defined on a compact set attains a maximum value on the set, the result follows.

3.13 The easiest example is that of a useless channel. In this case $Y_1, \ldots, Y_n$ are independent regardless of the input distribution. To see this, write

$$P\{Y_1 = \beta_1, \ldots, Y_n = \beta_n\} = \sum_{\alpha_1, \ldots, \alpha_n} P\{X_1 = \alpha_1, \ldots, X_n = \alpha_n\} P\{Y_1 = \beta_1, \ldots, Y_n = \beta_n \mid X_1 = \alpha_1, \ldots, X_n = \alpha_n\}.$$

Since the channel is memoryless the second factor in the summand is

$$\prod_{i=1}^{n} P\{Y_i = \beta_i \mid X_i = \alpha_i\}.$$

Since the channel is useless, $P\{Y_i = \beta_i \mid X_i = \alpha_i\} = P\{Y_i = \beta_i\}$. It follows that $P\{Y_1 = \beta_1, \ldots, Y_n = \beta_n\} = \prod_{i=1}^{n} P\{Y_i = \beta_i\}$, and therefore $Y_1, \ldots, Y_n$ are independent.

4.1 There are $2^m - 1$ nonzero binary sequences of length m. Hence for a given m, the maximum n is $2^m - 1$. Note that since $2^m = 1 + n$, the Hamming bound (Theorem 4.5.2) for a single-error correcting code is met. For $m = 4$ we have

$$A = \begin{bmatrix} 1 & 0 & 1 & 0 & 1 & 0 & 1 & 0 & 1 & 0 & 1 & 0 & 1 & 0 & 1 \\ 0 & 1 & 1 & 0 & 0 & 1 & 1 & 0 & 0 & 1 & 1 & 0 & 0 & 1 & 1 \\ 0 & 0 & 0 & 1 & 1 & 1 & 1 & 0 & 0 & 0 & 0 & 1 & 1 & 1 & 1 \\ 0 & 0 & 0 & 0 & 0 & 0 & 0 & 1 & 1 & 1 & 1 & 1 & 1 & 1 & 1 \end{bmatrix}, \qquad n = 16 - 1 = 15.$$

4.2 a. It is possible to correct e-tuple errors and detect $(e + 1)$-tuple errors if and only if every error pattern of weight $\leq e$ yields a distinct corrector and in addition no error pattern of weight $e + 1$ has the same corrector as an error pattern of weight $\leq e$. (For then an $(e + 1)$-tuple error would be confused with an error pattern of weight $\leq e$ and would therefore not be detected.) This is precisely the condition that every set of $2e + 1$ columns of the parity check matrix be linearly independent.

b. Since the original code corrects e-tuple errors, every set of $2e$ columns of A is linearly independent. Now suppose

$$\sum_{k=1}^{2e+1} \lambda_{i_k} \mathbf{c}(r_{i_k}) = \mathbf{0} \qquad \text{(modulo 2)},$$

where $\lambda_j = 0$ or 1 and $\mathbf{c}(r_j)$ is the jth column of A_0. Since $2e + 1$ is odd, and each column of A_0 has a "1" in its bottom element, there must be an even number of nonzero λ's, in particular there are at most $2e$ nonzero λ's. This implies a linear dependence among $2e$ or fewer columns of A, a contradiction.

c. Augment a parity check matrix of a single-error correcting code to obtain

$$A = \begin{bmatrix} 1 & 0 & 0 & 0 & 0 & 1 & 1 & 0 & 1 & 1 & 0 & 1 & 1 \\ 0 & 1 & 0 & 0 & 0 & 1 & 0 & 1 & 1 & 0 & 1 & 1 & 1 \\ 0 & 0 & 1 & 0 & 0 & 0 & 1 & 1 & 1 & 0 & 0 & 0 & 1 \\ 0 & 0 & 0 & 1 & 0 & 0 & 0 & 0 & 0 & 1 & 1 & 1 & 1 \\ 1 & 1 & 1 & 1 & 1 & 1 & 1 & 1 & 1 & 1 & 1 & 1 & 1 \end{bmatrix}.$$

d. The condition is $2^m > \sum_{i=0}^{2e} \binom{n-1}{i}$. In view of part (a), the proof proceeds exactly as in Theorem 4.5.3.

4.3 a.
$$\begin{matrix} 0 & 0 & 0 & 0 & 0 & 0 \\ 1 & 0 & 0 & 1 & 0 & 1 \\ 1 & 1 & 1 & 0 & 1 & 0 \\ 0 & 1 & 1 & 1 & 1 & 1 \end{matrix}$$

b. Diagonalize the parity check matrix A (see Section 4.5) to obtain a new parity check matrix

$$A' = \begin{bmatrix} 1 & 0 & 0 & 0 & 1 & a_{16} \\ 0 & 1 & 0 & 0 & 1 & a_{16} + a_{26} \\ 0 & 0 & 1 & 0 & 1 & a_{16} + a_{36} \\ 0 & 0 & 0 & 1 & 0 & a_{26} + a_{36} + a_{46} \end{bmatrix}.$$

If the last column of A' is

$$\begin{bmatrix} 1 \\ 1 \\ 0 \\ 1 \end{bmatrix} \quad \text{or} \quad \begin{bmatrix} 1 \\ 0 \\ 1 \\ 1 \end{bmatrix} \quad \text{or} \quad \begin{bmatrix} 0 \\ 1 \\ 1 \\ 1 \end{bmatrix},$$

then every set of 3 columns of A' is linearly independent and the code will correct single errors and detect double errors. We obtain

$$\begin{bmatrix}1\\1\\0\\1\end{bmatrix} \quad \text{by taking } a_{16} = 1, \quad a_{26} = 0, \quad a_{36} = 1, \quad a_{46} = 0;$$

$$\begin{bmatrix}1\\0\\1\\1\end{bmatrix} \quad \text{by taking } a_{16} = 1, \quad a_{26} = 1, \quad a_{36} = 0, \quad a_{46} = 0;$$

$$\begin{bmatrix}0\\1\\1\\1\end{bmatrix} \quad \text{by taking } a_{16} = 0, \quad a_{26} = 1, \quad a_{36} = 1, \quad a_{46} = 1.$$

Now in A', $\mathbf{c}(r_1) + \mathbf{c}(r_2) + \mathbf{c}(r_3) + \mathbf{c}(r_5) = \mathbf{0}$ regardless of the choice of the elements of column 6; hence not every set of 4 columns of A' (hence of A) is linearly independent so that not all double errors can be corrected.

4.4 Let $\mathbf{w}$ be a nonzero code word of minimum weight, and let h be the weight of $\mathbf{w}$. Since $\text{dist}(\mathbf{w}_0, \mathbf{w}) = h$, (where $\mathbf{w}_0$ is the zero code word), the minimum distance is $\leq h$. Suppose the minimum distance is $h_1 < h$. Then there are distinct code words $\mathbf{w}_i, \mathbf{w}_j$ such that $\text{dist}(\mathbf{w}_i, \mathbf{w}_j) = h_1$. But $\text{dist}(\mathbf{w}_i, \mathbf{w}_j) = \text{dist}(\mathbf{w}_i + \mathbf{w}_i, \mathbf{w}_i + \mathbf{w}_j) = \text{dist}(\mathbf{w}_0, \mathbf{w}_i + \mathbf{w}_j) = \text{weight of } (\mathbf{w}_i + \mathbf{w}_j)$. Thus the weight of $\mathbf{w}_i + \mathbf{w}_j$ is $h_1 < h$, a contradiction.

4.5 a. An error in decoding will be made if and only if an error of magnitude $\leq e$ is confused with the "no-error" condition. This will happen if and only if some set of e or fewer columns of the parity check matrix adds to zero. Hence e-tuple (and all smaller) errors can be detected if and only if every set of e columns of the parity check matrix is linearly independent.

b. To say that columns $k_1, k_2, \ldots, k_i$ of a matrix A add to zero is to say that $A\mathbf{w}^T = \mathbf{0}$ where $\mathbf{w}$ is a row vector with ones in positions $k_1, \ldots, k_i$ and zeros elsewhere. Thus the equivalent condition on the code words is that every nonzero code word have weight $> e$. (Alternately, the minimum distance between code words is $> e$.)

c. An error pattern $\mathbf{z}$ is detectable if and only if $\mathbf{z}$ does not yield the same corrector as the zero vector, that is, if and only if $A\mathbf{z}^T \neq \mathbf{0}$, that is, if and only if $\mathbf{z}$ is not a code word. Therefore the number of detectable errors is 2^n minus the number of code words, or $2^n - 2^k$.

4.6 a. If $\mathbf{w}$ is a code word, then $\mathbf{w} = \sum_{i=1}^{k} \lambda_i \mathbf{u}_i$ (modulo 2) where the $\mathbf{u}_i$ are the rows of G^*, and $\lambda_i = 0$ or 1. Thus

$$\mathbf{w} = [\lambda_1, \ldots, \lambda_k]G^*,$$

and therefore

$$[I_m \mid B^T]\mathbf{w}^T = [I_m \mid B^T]G^{*T}\begin{bmatrix}\lambda_1\\ \cdot\\ \cdot\\ \cdot\\ \lambda_k\end{bmatrix}$$

$$= [I_m \mid B^T]\left[\frac{B^T}{I_k}\right]\begin{bmatrix}\lambda_1\\ \cdot\\ \cdot\\ \cdot\\ \lambda_k\end{bmatrix} = (B^T + B^T)\begin{bmatrix}\lambda_1\\ \cdot\\ \cdot\\ \cdot\\ \lambda_k\end{bmatrix} = \mathbf{0}$$

Thus $[I_m \mid B^T]\mathbf{w}^T = \mathbf{0}$ for every code word $\mathbf{w}$. Since rank $[I_m \mid B^T] = m$, $[I_m \mid B^T]\mathbf{w}^T = \mathbf{0}$ has $2^k = 2^{n-m}$ solutions; thus the vectors $\mathbf{w}$ satisfying $[I_m \mid B^T]\mathbf{w}^T = \mathbf{0}$ are precisely the code words of the given code, so that $[I_m \mid B^T]$ is a parity check matrix.

b. (i) Diagonalize G to obtain

$$G^* = \begin{bmatrix}1 & 0 & 1 & 0 & 0 & 0\\ 1 & 1 & 0 & 1 & 0 & 0\\ 1 & 1 & 0 & 0 & 1 & 0\\ 0 & 1 & 0 & 0 & 0 & 1\end{bmatrix}. \qquad \text{Hence } A = \begin{bmatrix}1 & 0 & 1 & 1 & 1 & 0\\ 0 & 1 & 0 & 1 & 1 & 1\end{bmatrix}$$

(ii) $G^* = G = [1 \quad 1 \quad 1 \quad 1 \quad 1]$ so $A = \begin{bmatrix}1 & 0 & 0 & 0 & 1\\ 0 & 1 & 0 & 0 & 1\\ 0 & 0 & 1 & 0 & 1\\ 0 & 0 & 0 & 1 & 1\end{bmatrix}$.

4.7 Let $A = [I_m \mid A_1]$. Delete row 1 and column 1 of A (or equally well, row j and column j where $1 \le j \le m$). If columns i and j of the reduced matrix are the same, then columns i and j of the original matrix A agree in all digits except the first, where they must disagree. It follows that $\mathbf{c}(r_1) + \mathbf{c}(r_i) + \mathbf{c}(r_j) = \mathbf{0}$, contradicting the hypothesis (see Problem 4.2) that every set of 3 columns of A is linearly independent. Furthermore, no column of the reduced matrix can be zero. For if column i of the reduced matrix is zero, then column i of the original matrix would be zero or else would coincide with column 1; each possibility contradicts the hypothesis.

4.8 a. Take

$$A = \begin{bmatrix}1 & 0 & 0 & 1 & 1 & 0 & 1\\ 0 & 1 & 0 & 1 & 0 & 1 & 1\\ 0 & 0 & 1 & 0 & 1 & 1 & 1\end{bmatrix},$$

$n = 7$, $k = 4$, $m = 3$; $2^m = 1 + n$, $2^k = 16 =$ number of code words. The Hamming bound is met, hence the code is lossless.

b. The Hamming bound is

$$2^m = \sum_{i=0}^{e} \binom{n}{i} \quad \text{or } (e = 2), \quad 2^m = 1 + n + \frac{n(n-1)}{2}.$$

Since $m = 4$, $n^2 + n - 30 = 0$ or $n = 5$. Then $k = n - m = 1$

$$A = \begin{bmatrix} 1 & 0 & 0 & 0 & 1 \\ 0 & 1 & 0 & 0 & 1 \\ 0 & 0 & 1 & 0 & 1 \\ 0 & 0 & 0 & 1 & 1 \end{bmatrix}$$

(compare Problem 4.6b).

c. If $n = 2^m - 1$ the Hamming code meets the Hamming bound and is lossless. In general, let $2^{m-1} \le n < 2^m$. [Note that by construction of the code (Problem 4.1), n cannot be less than 2^{m-1}.] Given any binary column vector **c** of length m, if **c** is a column of the parity check matrix A then **c** is the corrector associated with a single error. If **c** is not a column of A then **c** is the modulo 2 sum of the column vector corresponding to 2^{m-1} and a column vector **c**′ corresponding to a binary number less than 2^{m-1}. For example if $m = 4$, $n = 10$ then

$$A = \begin{bmatrix} 1 & 0 & 1 & 0 & 1 & 0 & 1 & 0 & 1 & 0 \\ 0 & 1 & 1 & 0 & 0 & 1 & 1 & 0 & 0 & 1 \\ 0 & 0 & 0 & 1 & 1 & 1 & 1 & 0 & 0 & 0 \\ 0 & 0 & 0 & 0 & 0 & 0 & 0 & 1 & 1 & 1 \end{bmatrix}$$

The vector

$$\mathbf{c} = \begin{bmatrix} 1 \\ 1 \\ 0 \\ 1 \end{bmatrix}$$

is not a column of A, but

$$\mathbf{c} = \begin{bmatrix} 0 \\ 0 \\ 0 \\ 1 \end{bmatrix} + \begin{bmatrix} 1 \\ 1 \\ 0 \\ 0 \end{bmatrix} = \mathbf{c}(r_8) + \mathbf{c}(r_3).$$

Thus all correctors correspond either to single or to double errors, so the code is close-packed.

4.9 $B(\lambda, \beta) = -H(\lambda, 1 - \lambda) - \lambda \log \beta - (1 - \lambda) \log (1 - \beta)$

$$\frac{\partial B}{\partial \lambda} = -\log \frac{1 - \lambda}{\lambda} + \log \frac{1 - \beta}{\beta}$$

Now $1 - R = H(\lambda, 1 - \lambda)$ so $dR/d\lambda = \log [\lambda/(1 - \lambda)]$. Thus

$$\frac{\partial B}{\partial R} = \frac{\partial B/\partial \lambda}{dR/d\lambda} = 1 - \frac{\log (1 - \beta)/\beta}{\log (1 - \lambda)/\lambda}$$

Since

$$\frac{\lambda_c}{1 - \lambda_c} = \left(\frac{\beta}{1 - \beta}\right)^{1/2}, \quad \frac{\partial B}{\partial R} = 1 - 2 = -1 \quad \text{at} \quad \lambda = \lambda_c.$$

Therefore the tangent to $B(\lambda, \beta)$ at $\lambda = \lambda_c$ has slope -1, so that the tangent must be

$$1 - R + B(\lambda_c, \beta) - H(\lambda_c, 1 - \lambda_c) = H(\lambda, 1 - \lambda) + B(\lambda_c, \beta) - H(\lambda_c, 1 - \lambda_c),$$

which is the lower bound of (4.7.28) for $R < R_c$.

4.10 Given a code (s, n) with average probability of error less than λ, choose any number $k > 1$ such that $k\lambda < 1$. Then it is not possible that as many as s/k code words $\mathbf{w}$ have $p(e \mid \mathbf{w}) \geq k\lambda$, for if this were the case, we would have

$$\sum_{i=1}^{s} p(e \mid \mathbf{w}_i) \geq \frac{s}{k}(k\lambda) = s\lambda,$$

and consequently the average probability of error would be $\geq \lambda$, a contradiction. Thus we may construct a subcode having at least $s - s/k$ words and a maximum probability of error $< k\lambda$. If $s = [2^{nR}]$, $R > C$, then there is a $\delta > 0$ such that for sufficiently large n, $s(1 - k^{-1}) \geq 2^{n(C+\delta)}$. (Compare Lemma 3.5.3.) In other words, we are able to construct a sequence of codes that allows us to maintain a transmission rate above channel capacity with a maximum probability of error $< k\lambda < 1$. This contradicts the strong converse (Theorem 4.8.1).

4.11 Let B_i be the decoding set of $\mathbf{w}_i$ $(i = 1, 2, \ldots, s)$ in the original code. Since the first and second words are assigned the same probability, the average probability of error is unchanged if we replace B_2 by $B_1 \cup B_2$ and B_1 by the empty set $\emptyset$. Now consider the new code in which $\mathbf{w}_1'$ replaces $\mathbf{w}_1$. If we take the decoding set of $\mathbf{w}_1'$ to be $\emptyset$, the decoding set of $\mathbf{w}_2$ to be $B_1 \cup B_2$, and the decoding set of $\mathbf{w}_i$ to be $\mathbf{B}_i$, $i > 2$, then the average probability of error is the same as that of the original code. But the ideal-observer decision scheme has an average probability of error at least as small as the scheme just given. The result follows.

5.1 a.
$$T = \begin{bmatrix} 0 & 1 & 0 & 0 \\ 0 & 0 & 1 & 0 \\ 0 & 0 & 0 & 1 \\ 1 & 1 & 0 & 0 \end{bmatrix}, \qquad \text{period} = 15.$$

Choosing

$$\mathbf{x} = \begin{bmatrix} 0 \\ 0 \\ 0 \\ 1 \end{bmatrix}$$

we obtain

$$A = \begin{bmatrix} 0 & 0 & 0 & 1 & 0 & 0 & 1 & 1 & 0 & 1 & 0 & 1 & 1 & 1 & 1 \\ 0 & 0 & 1 & 0 & 0 & 1 & 1 & 0 & 1 & 0 & 1 & 1 & 1 & 1 & 0 \\ 0 & 1 & 0 & 0 & 1 & 1 & 0 & 1 & 0 & 1 & 1 & 1 & 1 & 0 & 0 \\ 1 & 0 & 0 & 1 & 1 & 0 & 1 & 0 & 1 & 1 & 1 & 1 & 0 & 0 & 0 \end{bmatrix}$$

b.
$$T = \begin{bmatrix} 0 & 1 & 0 & 0 \\ 0 & 0 & 1 & 0 \\ 0 & 0 & 0 & 1 \\ 1 & 1 & 1 & 1 \end{bmatrix}, \qquad \text{period} = 5.$$

With

$$\mathbf{x} = \begin{bmatrix} 0 \\ 0 \\ 0 \\ 1 \end{bmatrix} \quad \text{we have } A = \begin{bmatrix} 0 & 0 & 0 & 1 & 1 \\ 0 & 0 & 1 & 1 & 0 \\ 0 & 1 & 1 & 0 & 0 \\ 1 & 1 & 0 & 0 & 0 \end{bmatrix}.$$

Since there are 3 cycles of length 5, we may get different parity check matrices for different choices of $\mathbf{x}$. However, all such matrices define the same code (see Lemma 5.3.3).

c.
$$T = \begin{bmatrix} 0 & 1 & 0 & 0 & 0 \\ 0 & 0 & 1 & 0 & 0 \\ 0 & 0 & 0 & 1 & 0 \\ 0 & 0 & 0 & 0 & 1 \\ 1 & 0 & 0 & 0 & 1 \end{bmatrix}$$

There is one cycle of length 21, one cycle of length 7, one cycle of length 3, and the trivial cycle of length 1. Hence the period of T is 21. If

$$\mathbf{x} = \begin{bmatrix} 0 \\ 0 \\ 0 \\ 0 \\ 1 \end{bmatrix},$$

then

$$A = A_1 = \begin{bmatrix} 0&0&0&0&1&1&1&1&1&0&1&0&1&0&0&1&1&0&0&0&1 \\ 0&0&0&1&1&1&1&1&0&1&0&1&0&0&1&1&0&0&0&1&0 \\ 0&0&1&1&1&1&1&0&1&0&1&0&0&1&1&0&0&0&1&0&0 \\ 0&1&1&1&1&1&0&1&0&1&0&0&1&1&0&0&0&1&0&0&0 \\ 1&1&1&1&1&0&1&0&1&0&0&1&1&0&0&0&1&0&0&0&0 \end{bmatrix}.$$

If

$$\mathbf{x} = \begin{bmatrix} 0 \\ 1 \\ 1 \\ 1 \\ 0 \end{bmatrix}, \quad \text{then} \quad A = A_2 = \begin{bmatrix} 0&1&1&1&0&0&1 \\ 1&1&1&0&0&1&0 \\ 1&1&0&0&1&0&1 \\ 1&0&0&1&0&1&1 \\ 0&0&1&0&1&1&1 \end{bmatrix} \qquad (A_2 \text{ has rank } 3).$$

If

$$\mathbf{x} = \begin{bmatrix} 1 \\ 1 \\ 0 \\ 1 \\ 1 \end{bmatrix}, \quad \text{then} \quad A = A_3 = \begin{bmatrix} 1&1&0 \\ 1&0&1 \\ 0&1&1 \\ 1&1&0 \\ 1&0&1 \end{bmatrix} \qquad (A_3 \text{ has rank } 2).$$

5.2 By Lemma 5.3.3a, $f(x) = 1 + x^4 + x^5$ is a multiple of the generator polynomial $g(x)$. Now $x^5 + x^4 + 1 = (x^3 + x + 1)(x^2 + x + 1)$; since the rank of the parity check matrix A_2 of Problem 5.1c is 3, $g(x)$ is of degree 3, and thus $g(x) = x^3 + x + 1$. We may therefore generate the code by taking

$$T = \begin{bmatrix} 0&1&0 \\ 0&0&1 \\ 1&1&0 \end{bmatrix} \quad \text{and} \quad \mathbf{x} = \begin{bmatrix} 0 \\ 0 \\ 1 \end{bmatrix}.$$

The corresponding parity check matrix is

$$\begin{bmatrix} 0&0&1&0&1&1&1 \\ 0&1&0&1&1&1&0 \\ 1&0&1&1&1&0&0 \end{bmatrix}$$

which is of the form (5.3.1). It follows that $h(x) = 1 + x + x^2 + x^4$.

5.3 Let A be as in Table 5.4.1. If $\mathbf{w} = (r_0, r_1, \ldots r_{n-1})$ is a code word then

$$A_0\mathbf{w}^T = \begin{bmatrix} \mathbf{x} & T\mathbf{x} & \cdots & T^{n-1}\mathbf{x} \\ \mathbf{x} & T^3\mathbf{x} & \cdots & T^{3(n-1)}\mathbf{x} \\ & \cdot & & \\ & \cdot & & \\ & \cdot & & \\ \mathbf{x} & T^{2e-1}\mathbf{x} & \cdots & T^{(2e-1)(n-1)}\mathbf{x} \end{bmatrix} \begin{bmatrix} r_0 \\ r_1 \\ \cdot \\ \cdot \\ \cdot \\ r_{n-1} \end{bmatrix} = \begin{bmatrix} 0 \\ \cdot \\ \cdot \\ \cdot \\ 0 \end{bmatrix}$$

If A_0 is premultiplied by the matrix

$$\begin{bmatrix} T & & & & & \\ & T^3 & & & & \\ & & T^5 & & 0 & \\ & & & \cdot & & \\ & & & & \cdot & \\ & & & & & \cdot \\ & 0 & & & & T^{2e-1} \end{bmatrix} = B,$$

we obtain (noting that $T^{sn}\mathbf{x} = \mathbf{x}$, $s = 1, 2, \ldots, 2e - 1$)

$$BA_0\mathbf{w}^T = \begin{bmatrix} T\mathbf{x} & T^2\mathbf{x} & \cdots & T^{n-1}\mathbf{x} & \mathbf{x} \\ T^3\mathbf{x} & T^6\mathbf{x} & \cdots & T^{3(n-1)}\mathbf{x} & \mathbf{x} \\ \cdot & & & & \\ \cdot & & & & \\ \cdot & & & & \\ T^{2e-1}\mathbf{x} & T^{4e-2}\mathbf{x} & \cdots & T^{(2e-1)(n-1)}\mathbf{x} & \mathbf{x} \end{bmatrix} \begin{bmatrix} r_0 \\ r_1 \\ \cdot \\ \cdot \\ \cdot \\ r_{n-1} \end{bmatrix} = \begin{bmatrix} 0 \\ 0 \\ \cdot \\ \cdot \\ \cdot \\ 0 \end{bmatrix}.$$

Thus A_0 is obtained from BA_0 by a cyclic shift of columns. Examination of the equation $BA_0\mathbf{w}^T = \mathbf{0}$ shows that $(r_{n-1}r_0 \cdots r_{n-2})$ is a code word.

5.4 $e = 2$, $n = 31 = 2^q - 1$, so $q = 5$. From Table 5.2.1 we may take

$$T = \begin{bmatrix} 0 & 1 & 0 & 0 & 0 \\ 0 & 0 & 1 & 0 & 0 \\ 0 & 0 & 0 & 1 & 0 \\ 0 & 0 & 0 & 0 & 1 \\ 1 & 0 & 1 & 0 & 0 \end{bmatrix}$$

with characteristic polynomial $\lambda^5 + \lambda^2 + 1$. If

$$\mathbf{x} = \begin{bmatrix} 1 \\ 0 \\ 0 \\ 0 \\ 0 \end{bmatrix},$$

the corresponding parity check matrix is

$$A_0 = \left[\begin{array}{ccccccccccccccccccccccccccccccc}
1&0&0&0&0&1&0&0&1&0&1&1&0&0&1&1&1&1&1&0&0&0&1&1&0&1&1&1&0&1&0\\
0&0&0&0&1&0&0&1&0&1&1&0&0&1&1&1&1&1&0&0&0&1&1&0&1&1&1&0&1&0&1\\
0&0&0&1&0&0&1&0&1&1&0&0&1&1&1&1&1&0&0&0&1&1&0&1&1&1&0&1&0&1&0\\
0&0&1&0&0&1&0&1&1&0&0&1&1&1&1&1&0&0&0&1&1&0&1&1&1&0&1&0&1&0&0\\
0&1&0&0&1&0&1&1&0&0&1&1&1&1&1&0&0&0&1&1&0&1&1&1&0&1&0&1&0&0&0\\
1&0&0&0&0&1&1&0&0&1&0&0&1&1&1&1&1&0&1&1&1&0&0&0&1&0&1&0&1&1&0\\
0&0&0&1&0&1&0&1&1&0&1&0&0&0&0&1&1&0&0&1&0&0&1&1&1&1&1&0&1&1&1\\
0&1&1&1&1&1&0&1&1&1&0&0&0&1&0&1&0&1&1&0&1&0&0&0&0&1&1&0&0&1&0\\
0&0&0&0&1&1&0&0&1&0&0&1&1&1&1&1&0&1&1&1&0&0&0&1&0&1&0&1&1&0&1\\
0&0&1&0&1&0&1&1&0&1&0&0&0&0&1&1&0&0&1&0&0&1&1&1&1&1&0&1&1&1&0
\end{array}\right]$$

5.5 a. There are $7 - 3 = 4$ check digits, hence the code may be generated by a nonsingular 4 by 4 matrix T of the form (5.1.1), and a vector

$$\mathbf{x} = \begin{bmatrix} 0 \\ 0 \\ 0 \\ 1 \end{bmatrix}.$$

(See Corollary 5.3.4.1.) If T has characteristic polynomial $a_0 + a_1\lambda + \cdots + a_{m-1}\lambda^{m-1} + \lambda^m$, then $(a_0a_1 \cdots a_{m-1}10 \cdots 0)$ together with $k - 1$ cyclic shifts generates the code. By shifting the given word $\mathbf{w} = (1010011)$ we find that (1110100) is a code word and that (1110100), (0111010), (0011101) generate the code.

It follows that an appropriate choice of T is

$$\begin{bmatrix} 0 & 1 & 0 & 0 \\ 0 & 0 & 1 & 0 \\ 0 & 0 & 0 & 1 \\ 1 & 1 & 1 & 0 \end{bmatrix}$$

With

$$\mathbf{x} = \begin{bmatrix} 0 \\ 0 \\ 0 \\ 1 \end{bmatrix},$$

the parity check matrix is

$$A = [\mathbf{x} \quad T\mathbf{x} \quad \cdots \quad T^6\mathbf{x}] = \begin{bmatrix} 0 & 0 & 0 & 1 & 0 & 1 & 1 \\ 0 & 0 & 1 & 0 & 1 & 1 & 0 \\ 0 & 1 & 0 & 1 & 1 & 0 & 0 \\ 1 & 0 & 1 & 1 & 0 & 0 & 0 \end{bmatrix}.$$

b. By Corollary 5.3.4.1, $g(x)$ is the minimal polynomial of T, that is, $g(x) = 1 + x + x^2 + x^4$. Since the above parity check matrix is of the form (5.3.1), we obtain $h(x) = 1 + x + x^3$.

5.6 a. The corrector associated with a given error pattern is $T^i\mathbf{x}$ for some i. (By the remarks after Theorem 5.2.6, every matric polynomial in T is a power of T.) If the error burst is translated by t digits, the associated corrector is $T^tT^i\mathbf{x} = T^{t+i}\mathbf{x}$.

b. The corrector associated with a burst of length $\leq q$ beginning at digit 1 is $p(T)\mathbf{x}$ where $p(T)$ is a matric polynomial of degree $\leq q - 1$. If $p_1(T)\mathbf{x} = p_2(T)\mathbf{x}$, then if $p_1(T) \neq p_2(T), p_1(T) - p_2(T)$ is nonsingular by Lemma 5.2.3 and Theorem 5.2.6, hence $\mathbf{x} = \mathbf{0}$, a contradiction. Thus $p_1(T) = p_2(T)$ so that the corrector determines the error burst uniquely.

The above argument holds under the weaker assumption that T has an irreducible minimal polynomial.

5.7 Clearly, single errors are correctible.

Assume a double adjacent error. The associated corrector is

$$\begin{bmatrix} (I + T)T^i\mathbf{x} \\ 0 \end{bmatrix}.$$

This could not coincide with the corrector of a single error, since the last digit of the latter would be 1. If $(I + T)T^i\mathbf{x} = (I + T)T^j\mathbf{x}$, $0 \leq i, j \leq n - 1 = 2^q - 2$ then by Lemma 5.2.3, $I + T$ is nonsingular, hence $T^i\mathbf{x} = T^j\mathbf{x}$. Another application of Lemma 5.2.3 yields $T^i = T^j$, hence $i = j$. Thus distinct double adjacent errors have distinct correctors.

5.8 We must show that the corrector determines uniquely (a) the length of the burst, (b) the burst-error pattern, and (c) the starting position of the burst. Let

$$\mathbf{c} = \begin{bmatrix} \mathbf{c}_2 \\ \hline \mathbf{c}_1 \end{bmatrix} \begin{matrix} \updownarrow q_2 \\ \updownarrow q_1 \end{matrix}$$

be the corrector corresponding to an error burst of width $q_1' \le q_1$. Since $q_2 = 2q_1 - 1$, $\mathbf{c}_2$ will contain a block of zeros of width $2q_1 - 1 - q_1' \ge q_1 - 1$ (considering the first component of $\mathbf{c}_2$ as following the last component); that is,

$$\mathbf{c}_2 = \begin{bmatrix} 0 \\ \cdot \\ \cdot \\ \cdot \\ 0 \\ 1 \\ b_1 \\ \cdot \\ \cdot \\ \cdot \\ b_{q_1'-2} \\ 1 \\ 0 \\ \cdot \\ \cdot \\ \cdot \\ 0 \end{bmatrix} \updownarrow q_2$$

Hence,

a. The length of the burst is q_2 minus the number of zeros in the block.

b. The burst pattern is $1b_1b_2 \cdots b_{q_1'-2}1$. No other allowable burst pattern could produce the same $\mathbf{c}_2$, for such a pattern would have to begin with one of the b_i and end at or after the "1" before b_1. The length of the burst would then be $\ge q_1 + 1$.

c. The starting position of the burst is the position of the first nonzero entry after the block of zeros $+ tq_2$ where t is an integer and $0 \le t < n/q_2$.

We claim that t is determined by $\mathbf{c}_1$ and therefore the corrector specifies the error pattern uniquely. For (see Problem 5.6) the correctors $\mathbf{c}_1 = \mathbf{c}_{1j}$ produced by a fixed burst-error pattern of width $\le q_1$ beginning at digit j ($j = 1, 2, \ldots, 2^{q_1} - 1$) are distinct. (If $\mathbf{c}_{1j}$ and $\mathbf{c}_{1k}$ are two such correctors, with $j < k$, then $\mathbf{c}_{1j} = T^a\mathbf{x}$ and $\mathbf{c}_{1k} = T^{a+k-j}\mathbf{x}$; hence $\mathbf{c}_{1j} = \mathbf{c}_{1k}$ implies $T^{k-j}\mathbf{x} = \mathbf{x}$, or $T^{k-j} = I$, with $0 < k - j \le 2^{q_1} - 2$, a contradiction.)

Thus $\mathbf{c}_2$ tells us that the starting position of the burst is $i + tq_2$ and $\mathbf{c}_1$ tells us that the starting position is $j + u(2^{q_1} - 1)$, where i and j are known and $t < n/q_2$, $u < n/(2^{q_1} - 1)$. If there were more than one possible value for t (and hence u) we would have

$$i + tq_2 = j + u(2^{q_1} - 1)$$
$$i + t'q_2 = j + u'(2^{q_1} - 1)$$

Hence $(t - t')q_2 = (u - u')(2^{q_1} - 1) = r$. Assuming that $t > t'$ and hence $u > u'$, r is a multiple of both q_2 and $2^{q_1} - 1$ and $r < n$, contradicting the fact that n is the least common multiple. The result follows.

5.9 We illustrate the calculation for T^5. The characteristic polynomial of T^5 is of degree 4; hence the minimal polynomial is of degree ≤ 4. Let $S = T^5$. Since $T^4 + T + I = 0$,

$$S = TT^4 = T(T + I) = T^2 + T.$$

The "coordinates" of S with respect to I, T, T^2, T^3 are therefore (0110). Now

$$\begin{aligned} S^2 &= T^{10} = (T^2 + T)^2 &&= T^4 + T^2 = T^2 + T + I = (1110) \\ S^3 &= T^{15} = I &&= (1000) \\ S^4 &= T^{20} = T^5 = T^2 + T = (0110) \end{aligned}$$

We shall try to find b_0, b_1, b_2, b_3 such that

$$b_0 I + b_1 S + b_2 S^2 + b_3 S^3 + b_4 S^4 = 0;$$

that is,

$$b_0(1000) + b_1(0110) + b_2(1110) + b_3(1000) + b_4(0110) = (0000).$$

This yields

$$\begin{aligned} b_0 + b_2 + b_3 &= 0 \\ b_1 + b_2 + b_4 &= 0. \end{aligned}$$

We may take $b_3 = b_4 = 0$, $b_0 = b_1 = b_2 = 1$. Thus $S^2 + S + I = 0$. Clearly, no polynomial of degree 1 can be satisfied by S. Hence the minimal polynomial of S is $\lambda^2 + \lambda + 1$.

A similar calculation shows that the minimal polynomial of T^3 is $\lambda^4 + \lambda^3 + \lambda^2 + \lambda + 1$.

5.10 (Solution due to A. Wyner and B. Elspas.) Let $f(\lambda)$ be the minimal polynomial of T. Write

$$\lambda^k - 1 = q_k(\lambda) f(\lambda) + r_k(\lambda), \qquad k = 1, 2, \ldots, 2^q - 1$$

where the degree of $r_k(\lambda)$ is less than the degree of $f(\lambda) = q$. If $r_k(\lambda) = 0$ for some k we may replace λ by T to obtain

$$T^k - I = q_k(T) f(T) = 0$$

and we are finished.

If $r_k(\lambda)$ is the constant polynomial $1(= -1)$ for some k, then

$$T^k - I = q_k(T) f(T) - I = -I, \quad \text{or} \quad T^k = 0,$$

which is impossible since T is nonsingular. Thus assume that none of the $r_k(\lambda)$ is 0 or 1. There are 2^q distinct polynomials of degree $< q$, leaving $2^q - 2$ when 0 and 1 are removed. Since there are $2^q - 1$ of the $r_k(\lambda)$, we must have $r_{k_1}(\lambda) = r_{k_2}(\lambda)$ for some $k_1 > k_2$. It follows that $T^{k_1} = T^{k_2}$, or $T^{k_1 - k_2} = I$, with $1 \leq k_1 - k_2 \leq 2^q - 1$.

6.1 a. To find the order, construct the following table:

Z_{t-1}	X_t: A	B
s_1	s_1	s_2
s_2	s_4	s_3
s_3	s_4	s_3
s_4	s_1	s_2

Z_{t-2}	$X_{t-1}X_t$: AA	AB	BA	BB
s_1	s_1	s_2	s_4	s_3
s_2	s_1	s_2	s_4	s_3
s_3	s_1	s_2	s_4	s_3
s_4	s_1	s_2	s_4	s_3

Thus X_{t-1} and X_t determine Z_t; hence the source is of order 2.

b. $P\{X_3 = A \mid Z_0 = s_1\} = P\{Z_3 = s_1 \mid Z_0 = s_1\} + P\{Z_3 = s_4 \mid Z_0 = s_1\}$
$= p_{11}^{(3)} + p_{14}^{(3)} = .512 + .208 = .72.$

c. $P\{X_{t+1} = A, X_{t+2} = B, X_{t+3} = B, X_{t+4} = A, X_{t+5} = A\}$
$=$ (since $X_{t+1} = A, X_{t+2} = B \Leftrightarrow Z_{t+2} = s_2$)
$P\{Z_{t+2} = s_2\}P\{X_{t+3} = B, X_{t+4} = X_{t+5} = A \mid Z_{t+2} = s_2\}$
$= P\{Z_{t+2} = s_2\}P\{Z_{t+3} = s_3, Z_{t+4} = s_4, Z_{t+5} = s_1 \mid Z_{t+2} = s_2\}$
$= w_2 p_{23} p_{34} p_{41} = \frac{1}{3}(.1)(.8)(.7) = .0187$

6.2

Z_{t-1}	X_t: A	B	C
s_1	s_1	s_2	
s_2		s_2	s_3
s_3	s_1		s_3

The source is of order 1.

$$\Pi = \begin{bmatrix} \frac{3}{4} & \frac{1}{4} & 0 \\ 0 & \frac{2}{3} & \frac{1}{3} \\ \frac{1}{4} & 0 & \frac{3}{4} \end{bmatrix}$$

$W\Pi = W$ becomes

$$\begin{aligned} \tfrac{3}{4}w_1 \qquad\quad + \tfrac{1}{4}w_3 &= w_1 \\ \tfrac{1}{4}w_1 + \tfrac{2}{3}w_2 \qquad\quad &= w_2 \\ \tfrac{1}{3}w_2 + \tfrac{3}{4}w_3 &= w_3 \end{aligned}$$

and

$$w_1 + w_2 + w_3 = 1$$

Thus $w_1 = w_3 = \frac{4}{11}$, $w_2 = \frac{3}{11}$.

$$H\{\mathbf{X}\} = \tfrac{4}{11}H(3/4, 1/4) + \tfrac{3}{11}H(2/3, 1/3) + \tfrac{4}{11}H(3/4, 1/4).$$

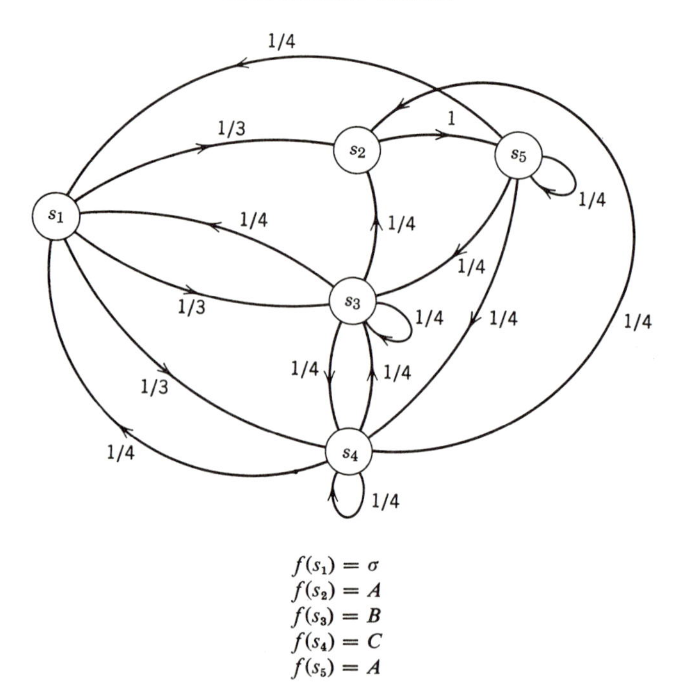

$$f(s_1) = \sigma$$
$$f(s_2) = A$$
$$f(s_3) = B$$
$$f(s_4) = C$$
$$f(s_5) = A$$

Problem 6.3.*a*,*b*.

6.3 Let σ = space. A possible state diagram is shown in the accompanying figure. If

$$\begin{aligned} X_t &= \sigma, \quad \text{then} & Z_t &= s_1 \\ X_t &= B, & Z_t &= s_3 \\ X_t &= C, & Z_t &= s_4 \\ X_{t-1} \neq A, X_t &= A & Z_t &= s_2 \\ X_{t-1} = X_t &= A & Z_t &= s_5 \end{aligned}$$

Thus X_{t-1} and X_t determine Z_t and the source is of order 2. The steady state probabilities are $w_1 = w_2 = \frac{1}{6}$, $w_3 = w_4 = w_5 = \frac{2}{9}$. The source uncertainty is

$$\begin{aligned} H\{\mathbf{X}\} &= \tfrac{1}{6}H(1/3, 1/3, 1/3) + \tfrac{1}{6}H(1) + \tfrac{2}{3}H(1/4, 1/4, 1/4, 1/4) \\ &= \tfrac{1}{6}\log 3 + \tfrac{2}{3}\log 4 = \tfrac{1}{6}\log 3 + \tfrac{4}{3} = 1.5975. \end{aligned}$$

A state diagram for the first-order approximation is shown in the accompanying figure. Note

$$\begin{aligned} p_{AA} = P\{X_t = A \mid X_{t-1} = A\} &= \frac{P\{X_{t-1} = A, X_t = A\}}{P\{X_{t-1} = A\}} \\ &= \frac{w_5}{w_2 + w_5} = \frac{4/18}{7/18} = \frac{4}{7}. \end{aligned}$$

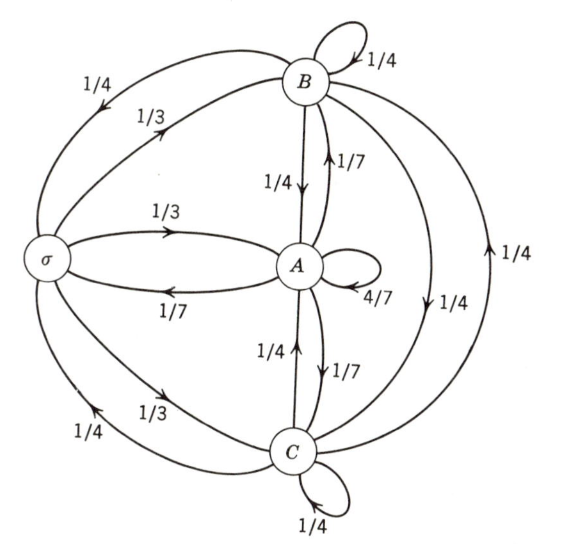

Problem 6.3.*c*.

The steady state probabilities are

$$w_\sigma = w_1 = 1/6 = P\{X_t = \sigma\}$$
$$w_B = w_3 = 2/9 = P\{X_t = B\}$$
$$w_C = w_4 = 2/9 = P\{X_t = C\}$$
$$w_A = w_2 + w_5 = 7/18 = P\{X_t = A\}.$$

The uncertainty of the first-order approximation is

$$H_1\{\mathbf{X}\} = H(X_t \mid X_{t-1}) = \tfrac{1}{6}H(1/3, 1/3, 1/3) + \tfrac{4}{9}H(1/4, 1/4, 1/4, 1/4)$$
$$+ \tfrac{7}{18}H(4/7, 1/7, 1/7, 1/7) = \tfrac{1}{6}\log 3 + \tfrac{8}{9} + \tfrac{7}{18}H(4/7, 1/7, 1/7, 1/7) = 1.8004,$$

which is larger than $H\{\mathbf{X}\}$, as expected.

6.4

$$P\{X_1 = \alpha_1, \ldots, X_n = \alpha_n\} = \sum_{\alpha_0 \in \Gamma} P\{X_0 = \alpha_0, X_1 = \alpha_1, \ldots, X_n = \alpha_n\}$$
$$= \sum_{\alpha_0 \in \Gamma} P\{X_0 = \alpha_0, \ldots, X_{n-1} = \alpha_{n-1}\} P\{X_n = \alpha_n \mid X_0 = \alpha_0, \ldots, X_{n-1} = \alpha_{n-1}\}$$
$$= \sum_{\alpha_0 \in \Gamma} P\{X_0 = \alpha_0, \ldots, X_{n-1} = \alpha_{n-1}\} p_{(\alpha_0 \ldots \alpha_{n-1})(\alpha_1 \ldots \alpha_n)}$$

Now in $\mathcal{S}_n$, $p_{(\alpha_0\alpha_1' \cdots \alpha_{n-1}')(\alpha_1\alpha_2 \cdots \alpha_n)} = 0$ unless $\alpha_i' = \alpha_i$, $i = 1, 2, \ldots, n-1$, so we may write the above expression as

$$P\{X_1 = \alpha_1, \ldots, X_n = \alpha_n\}$$
$$= \sum_{\alpha_0, \alpha_1', \cdots, \alpha'_{n-1} \in \Gamma} P\{X_0 = \alpha_0, X_1 = \alpha_1', \ldots, X_{n-1} = \alpha'_{n-1}\} p_{(\alpha_0\alpha_1' \ldots \alpha_{n-1}')(\alpha_1 \ldots \alpha_n)}$$

This is of the form

$$w_{(\alpha_1 \cdots \alpha_n)} = \sum_s w_s p_{s(\alpha_1 \cdots \alpha_n)}.$$

Hence the $w_{(\alpha_1 \cdots \alpha_n)} = P\{X_1 = \alpha_1, \ldots, X_n = \alpha_n\}$ form a stationary distribution.

6.5

$$P\{X_{i_1} = X_{i_2} = \cdots = X_{i_n} = 1\}$$
$$= P\{X_{i_1} = X_{i_2} = \cdots = X_{i_n} = -1\} = 1/2 \quad \text{for} \quad \text{all } i_i, \ldots, i_n$$

so the sequence is stationary.

Let α be the sequence consisting of the single digit 1. Then

$$n^{-1}N_\alpha^{\ n}(X_1, \ldots, X_n) = n/n = 1 \quad \text{with} \quad \text{probability } 1/2$$
$$= 0 \quad \text{with} \quad \text{probability } 1/2$$

Let $0 < \varepsilon < 1/2$; then $P\{|n^{-1}N_\alpha^{\ n} - P\{X_1 = \alpha\}| \geq \varepsilon\} = P\{|n^{-1}N_\alpha^{\ n} - 1/2| \geq \varepsilon\} = 1$ since $|n^{-1}N_\alpha^{\ n} - \frac{1}{2}| = \frac{1}{2}$.

Thus $n^{-1}N_\alpha^{\ n}$ cannot converge in probability to $P\{X_1 = \alpha\}$, and the sequence is not ergodic.

6.6

$$P\{X_{t-1} = A\} = P\{X_{t-1} = A, X_t = A\} + P\{X_{t-1} = A, X_t = B\}$$

and

$$P\{X_{t-1} = A\} = P\{X_t = A\} = P\{X_{t-1} = A, X_t = A\} + P\{X_{t-1} = B, X_t = A\}$$

The result follows.

6.7 The original source is of order 1, so that $H\{\mathbf{X}\} = H(X_n \mid X_{n-1})$. Now

$$H(Y_n \mid Y_0, Y_1, \ldots, Y_{n-1}) = \sum_{\alpha_0, \alpha_1, \ldots, \alpha_{n-1}} P\{Y_0 = \alpha_0, \ldots, Y_{n-1} = \alpha_{n-1}\}$$
$$\times H(Y_n \mid Y_0 = \alpha_0, \ldots, Y_{n-1} = \alpha_{n-1}).$$

Break the summation into 3 parts:

1. $\alpha_{n-2} \neq e, \alpha_{n-1} \neq e$:

$$H(Y_n \mid Y_0 = \alpha_0, \ldots, Y_{n-1} = \alpha_{n-1}) = 0, \quad \text{and}$$
$$P\{Y_{n-2} = \alpha_{n-2}, Y_{n-1} = \alpha_{n-1}\} = P\{X_{n-2} = \alpha_{n-2}, X_{n-1} = \alpha_{n-1}\}.$$

2. $\alpha_{n-2} \neq e, \alpha_{n-1} = e$:

$$H(Y_n \mid Y_0 = \alpha_0, \ldots, Y_{n-1} = \alpha_{n-1})$$
$$= H(X_n \mid X_{n-2} = \alpha_{n-2}) \text{ and } P\{Y_{n-2} = \alpha_{n-2}, Y_{n-1} = e\} = \tfrac{1}{3}P\{X_{n-2} = \alpha_{n-2}\}.$$

3. $\alpha_{n-2} = e, \alpha_{n-1} \neq e$:

$$H(Y_n \mid Y_0 = \alpha_0, \ldots, Y_{n-1} = \alpha_{n-1}) = H(X_n \mid X_{n-1} = \alpha_{n-1}) \quad \text{and}$$
$$P\{Y_{n-2} = e, Y_{n-1} = \alpha_{n-1}\} = \tfrac{1}{3}P\{X_{n-1} = \alpha_{n-1}\}.$$

Thus

$$H(Y_n \mid Y_0, Y_1, \ldots, Y_{n-1}) = \tfrac{1}{3}[H(X_n \mid X_{n-2}) + H(X_n \mid X_{n-1})]$$
$$= \tfrac{1}{3}[H(X_n \mid X_{n-2}) + H\{\mathbf{X}\}] = \text{constant} \quad \text{for} \quad n \geq 2.$$

Thus $H\{\mathbf{Y}\} = \frac{1}{3}[H(X_n \mid X_{n-2}) + H\{\mathbf{X}\}]$. Now

$$H\{\mathbf{X}\} = \tfrac{7}{16}H(5/7, 2/7) + \tfrac{9}{16}H(4/9, 5/9) \quad \text{and}$$

$$H(X_n \mid X_{n-2}) = w_A H(X_n \mid X_{n-2} = A) + w_B H(X_n \mid X_{n-2} = B)$$
$$\tfrac{7}{16}H(\tfrac{4}{49} + \tfrac{25}{63}, \tfrac{10}{49} + \tfrac{20}{63}) + \tfrac{9}{16}H(\tfrac{10}{63} + \tfrac{20}{81}, \tfrac{25}{63} + \tfrac{16}{81}).$$

This determines $H\{\mathbf{Y}\}$.

6.8 a. As in the proof of (6.6.3), write

$$E(V_n) = \sum_k kP\{V_n = k\} = \sum_{k<\beta-\varepsilon} kP\{V_n = k\} + \sum_{\beta-\varepsilon\le k\le\beta+\delta} kP\{V_n = k\} + \sum_{k>\beta+\delta} kP\{V_n = k\}.$$

Thus

$$E(V_n) \ge 0 + (\beta - \varepsilon)P\{\beta - \varepsilon \le V_n \le \beta + \delta\} + (\beta + \delta)P\{V_n > \beta + \delta\}$$
$$= (\beta - \varepsilon)(1 - P\{V_n < \beta - \varepsilon\} - P\{V_n > \beta + \delta\}) + (\beta + \delta)P\{V_n > \beta + \delta\},$$

and the result follows.

b. Let $n \to \infty$ in (6.6.3) and part (a) of this problem. We obtain

$$-\varepsilon \le -\beta + \liminf_{n\to\infty} E(V_n) \le -\beta + \limsup_{n\to\infty} E(V_n) \le \delta.$$

Since ε and δ are arbitrary positive numbers,

$$\lim_{n\to\infty} E(V_n) = \beta.$$

By part (a) of the proof of Theorem 6.6.1, $\beta = H\{\mathbf{X}\}$.

6.9 Let $\{X_n, n = 1, 2, \dots\}$ be the given information source. Let $\boldsymbol{\alpha} = (\alpha_1, \dots, \alpha_m)$ and $\boldsymbol{\beta} = (\beta_1, \dots, \beta_m)$ be any two states of the mth-order approximation $\mathcal{S}_m$. Consider the set of all sequences $\boldsymbol{\gamma} = (\gamma_1, \dots, \gamma_r)$, $r \ge m + 1$, such that $(\gamma_1, \dots, \gamma_m) = \boldsymbol{\alpha}$ and $(\gamma_{r-m+1}, \dots, \gamma_r) = \boldsymbol{\beta}$. We claim that we can find an r and a $\boldsymbol{\gamma}$ such that $p(\boldsymbol{\gamma}) = P\{X_1 = \gamma_1, \dots, X_r = \gamma_r\} > 0$. For by ergodicity, $n^{-1}N_{\boldsymbol{\alpha}}^n(X_1, \dots, X_n) \to p(\boldsymbol{\alpha})$, $n^{-1}N_{\boldsymbol{\beta}}^n(X_1, \dots, X_n) \to p(\boldsymbol{\beta})$ in probability; by definition of $\mathcal{S}_m$, $p(\boldsymbol{\alpha}) > 0$, $p(\boldsymbol{\beta}) > 0$. Therefore if $0 < \delta < \min(p(\boldsymbol{\alpha}), p(\boldsymbol{\beta}))$, then $P\{n^{-1}N_{\boldsymbol{\alpha}}^n \ge \delta$ and $n^{-1}N_{\boldsymbol{\beta}}^n \ge \delta\} \to 1$ as $n \to \infty$. This implies that there is a sequence $\boldsymbol{\gamma}$ with positive probability such that either $\boldsymbol{\gamma}$ begins with $\boldsymbol{\alpha}$ and ends with $\boldsymbol{\beta}$, or vice versa. In the former case, the claim is established. In the latter case, we observe that by ergodicity, the relative frequency of $\boldsymbol{\gamma}$ converges in probability to $p(\boldsymbol{\gamma}) > 0$, and if $\boldsymbol{\gamma}$ occurs sufficiently many times among the symbols $X_1, \dots, X_n$, $\boldsymbol{\alpha}$ must precede $\boldsymbol{\beta}$ somewhere in the sequence.

Now let $\mathbf{s}_1 = (\gamma_1, \dots, \gamma_m)$, $\mathbf{s}_2 = (\gamma_2, \dots, \gamma_{m+1})$, $\dots$, $\mathbf{s}_{r-m+1} = (\gamma_{r-m+1}, \dots, \gamma_r)$, and let $p(\mathbf{s}_i, \mathbf{s}_{i+1}) = P\{X_i = \gamma_i, \dots, X_{i+m-1} = \gamma_{i+m-1}, X_{i+m} = \gamma_{i+m}\}$, $i = 1, 2 \dots, r - m$. Since $X_1 = \gamma_1, \dots, X_r = \gamma_r \Leftrightarrow (X_1, \dots, X_m) = \mathbf{s}_1, (X_2, \dots, X_{m+1}) = \mathbf{s}_2, \dots, (X_{r-m+1}, \dots, X_r) = \mathbf{s}_{r-m+1}$, it follows that $p(\mathbf{s}_i) > 0$ and $p(\mathbf{s}_i, \mathbf{s}_{i+1}) > 0$ for all i. But by definition of $\mathcal{S}_m$, the transition probability from $\mathbf{s}_i$ to $\mathbf{s}_{i+1}$ in $\mathcal{S}_m$ is

$$p_{\mathbf{s}_i\mathbf{s}_{i+1}} = \frac{p(\mathbf{s}_i, \mathbf{s}_{i+1})}{p(\mathbf{s}_i)} = P\{X_{i+m} = \gamma_{i+m} \mid X_i = \gamma_i, \dots, X_{i+m-1} = \gamma_{i+m-1}\}$$

Since $\mathbf{s}_1 = \boldsymbol{\alpha}$, $\mathbf{s}_{r-m+1} = \boldsymbol{\beta}$, and $p_{\mathbf{s}_i\mathbf{s}_{i+1}} > 0$ for all i, $\boldsymbol{\beta}$ is reachable from $\boldsymbol{\alpha}$ in $\mathcal{S}_m$. Thus every state in the mth-order approximation is reachable from every other state, and hence $\mathcal{S}_m$ is indecomposable with the essential class consisting of all the states.

6.10 If the set S of all states is not essential, then there are two distinct states, let us call them s_1 and s_2, such that s_2 cannot be reached from s_1. Let S_1 be the set of states reachable from s_1. Then the states of S_1 form a Markov chain (if $s \in S_1$ and s' can be reached from s, then s' can be reached from s_1 and hence $s' \in S_1$; thus the submatrix of the transition matrix corresponding to the states of S_1 is a Markov matrix). Since $s_2 \notin S_1$, S_1 is a proper subset of S. Apply the above argument to S_1, and continue in this fashion until an essential set is found. The process must terminate since the original chain has only a finite number of states; if we come down to a chain containing only one state, the set consisting of that state alone is essential.

An infinite chain need not have an essential set. For example, take the set of states to be the integers, with $p_{i,i+1} = 1$ for all integers i (that is, at each transition, move one unit to the right). If state j can be reached from state i, then state i cannot be reached from state j, so no set of states can be essential.

6.11 For the source $\mathcal{S}$ shown in the accompanying figure, the order of every approximation $\mathcal{S}_n$ is 1.

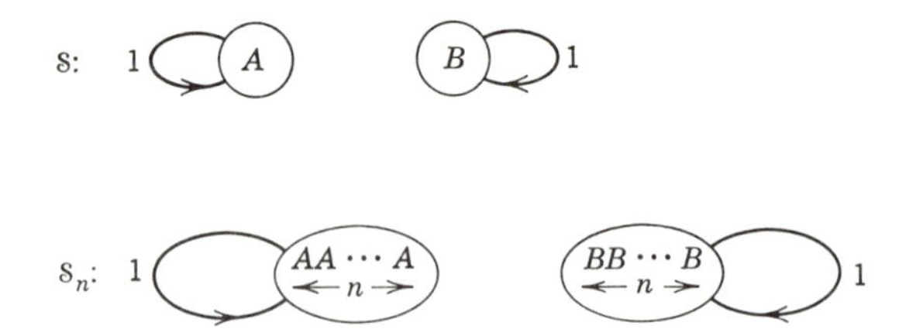

Problem 6.11. Any initial distribution $[w_A, w_B]$ is stationary; take $w_A > 0$, $w_B > 0$, $w_A + w_B = 1$.

6.12 A state diagram for the source $\{X_n\}$ is indicated in the accompanying figure. The states are a_{ij}, $j = 1, 2, \ldots, n_i$, $i = 1, 2, \ldots, M$, and the transition probabilities are $p(a_{ij}, a_{i,\,j+1}) = 1$ for $j = 1, 2, \ldots, n_i - 1$ and all i; $p(a_{in_i}, a_{k1}) = p_k$ for $k = 1, 2, \ldots, M$ and all i.

Since each code word begins with a different letter, we have a unifilar Markov source. Every state may be reached from every other state, and hence the source is indecomposable. Thus there is a unique stationary distribution with probabilities w_{ij} assigned to the states a_{ij}. By inspection of the state diagram, $w_{i1} = w_{i2} = \cdots = w_{in_i} =$ (say) b_i, and

$$b_k = \sum_{i=1}^{M} b_i p(a_{in_i}, a_{k1}) = p_k \sum_{i=1}^{M} b_i \quad \text{for all } k.$$

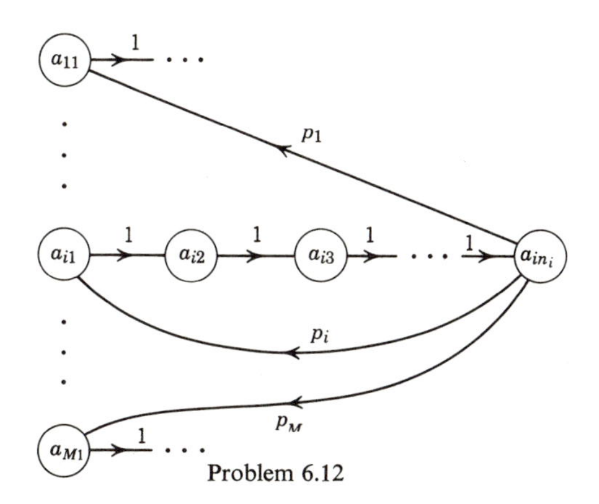

Problem 6.12

But $\sum_{i=1}^{M} b_i n_i = 1$, and hence $(\sum_{k=1}^{M} p_k n_k)(\sum_{i=1}^{M} b_i) = 1$. Consequently $b_i = p_i/\bar{n}$ for all i. It follows from Theorem 6.4.2 that

$$H\{\mathbf{X}\} = \sum_{i=1}^{M} w_{in_i} H(p_1, \ldots, p_M) = H(Y) \sum_{i=1}^{M} b_i = H(Y)/\bar{n}.$$

6.13. a. If we form a sequence of length $T + 1$ beginning at state s_i, the first transition must lead to some state s_j. If we are in state s_j after one transition, we may complete the sequence of length $T + 1$ by constructing a sequence of length T starting from s_j. Since the source is unifilar, all sequences formed in this way are distinct. Thus

$$N_i(T + 1) = \sum_{j=1}^{r} a_{ij} N_j(T), \qquad i = 1, \ldots, r.$$

This is a system of linear homogeneous difference equations with constant coefficients, and therefore the solution is a linear combination of exponentials λ^T. To find the particular λ's, we assume a solution of the form $N_i(T) = y_i \lambda^T$ to obtain $\lambda^T(\lambda y_i) = (\lambda^T) \sum_{j=1}^{r} a_{ij} y_j$, $i = 1, \ldots, r$, or $\lambda \mathbf{y} = A\mathbf{y}$. Thus the allowable λ's are the eigenvalues of A. The result (a) now follows from the observation that

$$N(T) \leq \sum_{i=1}^{r} N_i(T).$$

b. Since $H(X_1, \ldots, X_T)$ cannot exceed the logarithm of the total number of possible sequences of length T (Theorem 1.4.2), we have

$$T^{-1} H(X_1, \ldots, X_T) \leq T^{-1} \log N(T) \leq T^{-1} (\log K + T \log \lambda_0).$$

Using Theorem 6.4.1 we have

$$H\{\mathbf{X}\} = \lim_{T \to \infty} T^{-1} H(X_1, \ldots, X_T) \leq \log \lambda_0.$$

c. Since the components x_i are >0, $p_{ij} \geq 0$ for all i and j. Since $\mathbf{x}$ is an eigenvector for λ_0, $\sum_{j=1}^{r} p_{ij} = 1$, and hence the p_{ij} form a legitimate assignment of transition probabilities. Since $p_{ij} > 0$ if and only if $a_{ij} = 1$, the assignment is consistent with the given connection matrix. By Theorem 6.4.2, $H\{\mathbf{X}\} = \sum_{i=1}^{r} w_i H_i$, where H_i is the uncertainty of state s_i and $[w_1 \cdots w_r]$ is the unique stationary distribution for the source. (Note that in general, the stationary distribution will depend on the choice of the p_{ij}). Thus

$$H\{\mathbf{X}\} = -\sum_{i,j=1}^{r} w_i p_{ij} \log p_{ij} = \sum_{i,j} w_i p_{ij} (\log \lambda_0 + \log x_i - \log x_j).$$

(Note that only terms for which $a_{ij} = 1$ contribute to the summation.) Now

$$\sum_{i,j} w_i p_{ij} \log x_i = \sum_{i} w_i \log x_i$$

and

$$\sum_{i,j} w_i p_{ij} \log x_j = \sum_{j} \log x_j \sum_{i} w_i p_{ij} = \sum_{j} w_j \log x_j,$$

the last step following since the w_i form a stationary distribution. Thus

$$H\{\mathbf{X}\} = \sum_{i,j} w_i p_{ij} \log \lambda_0 = \log \lambda_0.$$

d. The connection matrix is

$$A = \begin{bmatrix} 1 & 1 \\ 1 & 0 \end{bmatrix}.$$

To find the eigenvalues, we set the determinant

$$\begin{vmatrix} 1 - \lambda & 1 \\ 1 & -\lambda \end{vmatrix}$$

equal to zero, to obtain $\lambda^2 - \lambda - 1 = 0$. The largest eigenvalue is $\lambda_0 = \frac{1}{2}(1 + \sqrt{5})$ and the maximum uncertainty is $\log \lambda_0$. An eigenvector with positive components may be formed by taking $x_1 = \frac{1}{2}(1 + \sqrt{5})$, $x_2 = 1$. The corresponding transition probabilities are given by

$$p_{11} = \lambda_0^{-1} = -\tfrac{1}{2} + \tfrac{1}{2}\sqrt{5}, \qquad p_{12} = [\tfrac{1}{4}(1 + \sqrt{5})^2]^{-1} = \tfrac{3}{2} - \tfrac{1}{2}\sqrt{5},$$
$$p_{21} = 1, \qquad p_{22} = 0.$$

Remark: Intuitively, we might expect that p_{ij} should be (roughly) the number of sequences of length $T + 1$ beginning at s_i and passing through s_j at the first transition [that is, $a_{ij}N_j(T)$], divided by the number of sequences of length $T + 1$ beginning at s_i [that is $N_i(T + 1)$]. If λ_0 is in fact greater than the magnitude of all other eigenvalues, then for large T, the term involving λ_0^T dominates the solution of the difference equation of (a). Thus asymptotically we have $N_i(T)$ proportional to $x_i \lambda_0^T$, hence

$$\frac{a_{ij}N_j(T)}{N_i(T + 1)} \sim \frac{a_{ij}x_j}{\lambda_0 x_i}.$$

7.1 a. Let the channel be in state s_{kij}, $k, i, j = 0$ or 1, at time $t = n$ if the ball labeled j is in the occupied slot at $t = n$, the ball labeled i was in the occupied slot at $t = n - 1$, and an input k was applied before the transition at $t = n$. The transition probability from s_{kij} to $s_{k_1i_1j_1}$ under the input k_1 is given by $M_{k_1}(s_{ij}, s_{i_1j_1})$, where the matrices M_{k_1} are given in Table 7.1.1. The transition probability from s_{kij} to $s_{k_1i_1j_1}$ under an input other than k_1 is zero. The output function g' is given by $g'(s_{kij}) = g(k, s_{ij})$ where g is the original output function of Table 7.1.1.

b. We proceed as in (a). Given a finite-state regular channel with states s_j, $j = 1, 2, \ldots, r$, and input alphabet $\Gamma = \{b_1, \ldots, b_t\}$, we define a new channel with the same input alphabet Γ, and states s_{ij}, $i = 1, 2, \ldots, t, j = 1, 2, \ldots, r$. The transition probability from s_{ij} to $s_{i_1j_1}$ under the input b_{i_1} is given by $M_{i_1}(s_j, s_{j_1})$; the transition probability under any other input is zero. If $g(b_i, s_j)$ is the original output function, the new output function is $g'(s_{ij}) = g(b_i, s_j)$. Since the original channel is regular, for any finite input sequence $b_{i_1} \cdots b_{i_k}$ there is a channel state s_{j_0} and an integer N such that for any initial channel state, s_{j_0} is reachable under the input sequence $b_{i_1} \cdots b_{i_k} \cdots b_{i_1} \cdots b_{i_k}$ (N times). In the new channel, the state $s_{i_kj_0}$ is reachable under the same input sequence, so the new channel is regular. If a source is connected to the new channel, the distribution of input and output sequences is (by construction) exactly the same as if the source were connected to the original channel, and thus the capacity is preserved.

7.2 First observe that

$$
\begin{aligned}
H(Y_{n+1} \mid X_1, \ldots, X_{n+1}, Y_1, \ldots, Y_n) &\le H(Y_{n+1} \mid X_2, \ldots, X_{n+1}, Y_2, \ldots, Y_n) \\
&\qquad \text{by Theorem 1.4.5 or Problem 1.4c} \\
&= H(Y_n \mid X_1, \ldots, X_n, Y_1, \ldots, Y_{n-1}) \\
&\qquad \text{by stationarity}
\end{aligned}
$$

Thus $\lim_{n\to\infty} H(Y_n \mid X_1, \ldots, X_n, Y_1, \ldots, Y_{n-1})$ exists. Now $n^{-1}H(Y_1, \ldots, Y_n \mid X_1, \ldots, X_n) = n^{-1}\sum_{i=1}^{n} g_{in}$, where $g_{in} = H(Y_i \mid X_1, \ldots, X_n, Y_1, \ldots, Y_{i-1})$. Since the channel is nonanticipatory, that is, the distribution of a given output letter does not depend on future inputs,

$$
g_{in} = g_i = H(Y_i \mid X_1, \ldots, X_i, Y_1, \ldots, Y_{i-1}).
$$

Since $\lim_{i\to\infty} g_i = H(\{\mathbf{Y}\} \mid \{\mathbf{X}\})$, it follows that

$$
\lim_{n\to\infty} n^{-1}H(Y_1, \ldots, Y_n \mid X_1, \ldots, X_n) = H(\{\mathbf{Y}\} \mid \{\mathbf{X}\}).
$$

Now by Theorem 6.4.1,

$$\begin{aligned} H\{\mathbf{X}, \mathbf{Y}\} &= \lim_{n\to\infty} n^{-1}H(X_1, \ldots, X_n, Y_1, \ldots, Y_n) \\ &= \lim_{n\to\infty} n^{-1}H(X_1, \ldots, X_n) + \lim_{n\to\infty} n^{-1}H(Y_1, \ldots, Y_n \mid X_1, \ldots, X_n) \\ &= H\{\mathbf{X}\} + H(\{\mathbf{Y}\} \mid \{\mathbf{X}\}). \end{aligned}$$

7.3 The results of Problem 7.2 do not hold in general. To see this, let $Z_0, Z_1, \ldots, Z_n, \ldots,$ be a sequence of independent, identically distributed random variables, and take $X_n = Z_{n-1}$, $Y_n = Z_n$, $n = 1, 2, \ldots$.

We then have $H(Y_n \mid X_1, \ldots, X_n, Y_1, \ldots, Y_{n-1}) = H(Z_n \mid Z_0, \ldots, Z_{n-1}) = H(Z_n) =$ a constant H for all n. But

$$\begin{aligned} H(Y_1, \ldots, Y_n \mid X_1, \ldots, X_n) &= H(Z_1, \ldots, Z_n \mid Z_0, \ldots, Z_{n-1}) \\ &= H(Z_n \mid Z_0) = H(Z_n) = H. \end{aligned}$$

Thus $H(\{\mathbf{Y}\} \mid \{\mathbf{X}\}) = H$, but

$$\lim_{n\to\infty} n^{-1}H(Y_1, \ldots, Y_n \mid X_1, \ldots, X_n) = \lim_{n\to\infty} n^{-1}H = 0 \neq H$$

in general.

However, if we make an assumption of "nonanticipatory" behavior, that is, $H(Y_i \mid X_1, \ldots, X_n, Y_1, \ldots, Y_{i-1}) = H(Y_i \mid X_1, \ldots, X_i, Y_1, \ldots, Y_{i-1})$ for all n, and all $i = 1, 2, \ldots, n$, then the argument of Problem 7.2 will go through intact.

References

Abramson, N. M. (1959), A Class of Systematic Codes for Non-Independent Errors, *IRE Trans. Inform. Theory*, **IT-5,** 150–157.

Abramson, N. M. (1961), Error-Correcting Codes from Linear Sequential Networks, *Proc., Fourth London Symposium on Information Theory*, C. Cherry, Ed., Butterworths, Washington, D.C.

Abramson, N. M. (1963), *Information Theory and Coding*, McGraw-Hill Book Co., New York.

Albert, A. A. (1956), *Fundamental Concepts of Higher Algebra*, Univ. of Chicago Press, Chicago, Ill.

Ash, R. B. (1963), Capacity and Error Bounds for a Time-Continuous Gaussian Channel, *Information and Control*, **6,** 14–27.

Ash, R. B. (1964), Further Discussion of a Time-Continuous Gaussian Channel, *Information and Control*, **7,** 78–83.

Ash, R. B. (1965), A Simple Example of a Channel for Which the Strong Converse Fails, *IEEE Trans. Inform. Theory*, in press.

Bethoux, P. (1962), Test et Estimations Concernant Certaines Functions Aleatoires en Particulier Laplaciennes, *Ann. Inst. Henri Poincaré*, **27,** 255–322.

Birkhoff, G., and S. MacLane (1953), *A Survey of Modern Algebra*, The Macmillan Co., New York.

Blackwell, D. (1961a), Information Theory, in *Modern Mathematics for the Engineer*, Second Series, E. F. Beckenbach, Ed., McGraw-Hill Book Co., New York.

Blackwell, D. (1961b), Exponential Error Bounds for Finite State Channels, *Proc. Fourth Berkeley Symposium on Math. Statistics and Probability*, Univ. of California Press, Berkeley, Calif., **1,** 57–63.

Blackwell, D., L. Breiman, and A. J. Thomasian (1958), Proof of Shannon's Transmission Theorem for Finite-State Indecomposable Channels, *Ann. Math. Stat.*. **29,** No. 4, 1209–1220.

Blackwell, D., L. Breiman, and A. J. Thomasian (1959), The Capacity of a Class of Channels, *Ann. Math. Stat.*, **30,** No. 4, 1229–1241.

Bose, R. C., and D. K. Ray-Chaudhuri (1960a), On a Class of Error Correcting Binary Group Codes, *Information and Control*, **3,** 68–79.

Bose, R. C., and D. K. Ray-Chaudhuri (1960b), Further Results on Error Correcting Binary Group Codes, *Information and Control*, **3,** 279–290.

Breiman, L. (1957), The Individual Ergodic Theorem of Information Theory, *Ann. Math. Stat.*, **28,** No. 3, 809–811; correction to this paper, 1960, *Ann. Math. Stat.*, **31,** No. 3, 809–810.

Chernoff, H. (1952), A Measure of Asymptotic Efficiency for Tests of a Hypothesis Based on the Sum of Observations, *Ann. Math. Stat.*, **23,** 493–507.

Chung, K. L. (1960), *Markov Chains with Stationary Transition Probabilities*, Springer-Verlag, Berlin.

Dobrushin, R. L. (1959), General Formulation of Shannon's Main Theorem in Information Theory, Usp. Math. Nauk, **14,** No. 6(90), 3–104, translated in *Am. Math. Soc. Translations*, **33,** Series 2, 323–438.

Doob, J. L. (1953), *Stochastic Processes*, John Wiley and Sons, New York.

Eisenberg, E. (1963), On Channel Capacity, *Technical Memorandum M-35*, Electronics Research Laboratory, Univ. of California, Berkeley, Calif.

Elias, P. (1955), Coding for Noisy Channels, *IRE Convention Record*, Part 4, pp. 37–46.

Elias, P. (1956), Coding for Two Noisy Channels, in *Information Theory*, Colin Cherry, Ed., Academic Press, New York, pp. 61–74.

Elspas, B. (1959), The Theory of Autonomous Linear Sequential Networks, *IRE Trans. Circuit Theory*, **CT-6,** 45–60.

Fadiev, D. A. (1956), On the Notion of Entropy of a Finite Probability Space (in Russian), *Usp. Math. Nauk*, **11,** No. 1 (67), 227–231.

Fano, R. M. (1961), *Transmission of Information*, MIT Press, Cambridge, Mass.

Feinstein, A. (1954), A New Basic Theorem of Information Theory, *IRE Trans. PGIT*, 2–22, Sept.

Feinstein, A. (1958), *Foundations of Information Theory*, McGraw-Hill Book Co., New York.

Feinstein, A. (1959), On the Coding Theorem and its Converse for Finite-Memory Channels, *Information and Control*, **2,** No. 1, 25–44.

Feller, W. (1950), *Introduction to Probability Theory*, John Wiley and Sons, New York.

Fire, P. (1959), A Class of Multiple Error Correcting Binary Codes for Non-Independent Errors, Sylvania Electric Products, Mountain View, Calif., *Report RSL-E-2*, March.

Fortet, R. (1961), Hypothesis Testing and Estimation for Laplacian Functions, *Fourth Berkeley Symposium on Mathematical Statistics and Probability*, **1,** 289–305, Univ. of California Press, Berkeley, Calif.

Friedland, B. (1959), Linear Modular Sequential Circuits, *IRE Trans. Circuit Theory*, **CT-6,** 61–68.

Friedland, B., and T. E. Stern (1959), On Periodicity of States in Linear Modular Sequential Circuits, *IRE Trans. Inform. Theory*, **IT-5,** 136–137.

Gallager, R. G. (1963), *Low Density Parity Check Codes*, MIT Press, Cambridge, Mass.

Gantmacher, F. R. (1959), *Applications of the Theory of Matrices*, Interscience Publishers, New York.

Gilbert, E. N. (1952), A Comparison of Signalling Alphabets, *Bell System Tech. J.*, **31,** 504–522.

Gilbert, E. N. (1960), Capacity of a Burst Noise Channel, *Bell System Tech. J.*, **39,** 1253–1265.

Goldberg, R. R. (1961), *Fourier Transforms*, Cambridge Univ. Press, London and New York.

Halmos, P. R. (1950), *Measure Theory*, D. Van Nostrand Co., Princeton, N.J.

Hamming, R. W. (1950), Error Detecting and Error Correcting Codes, *Bell System Tech. J.* **29,** 147–160.

Hocquenghem, A. (1959), Codes Correcteurs D'erreurs, *Chiffres*, **2,** 147–156.

Huffman, D. A. (1952), A Method for the Construction of Minimum Redundancy Codes, *Proc. IRE*, **40,** No. 10, 1098–1101.

Karush, J. (1961), A Simple Proof of an Inequality of McMillan, *IRE Trans. Inform. Theory*, **IT-7,** No. 2, 118.

Katz, M., and A. J. Thomasian (1961), A Bound for the Law of Large Numbers for Discrete Markov Processes, *Ann. Math. Stat.*, **32,** No. 1, 336–337.

Kelly, E. J., I. S. Reed, and W. L. Root (1960), The Detection of Radar Echoes in Noise , *J. Soc. Ind. Appl. Math.*, **8,** 309–341.

Kemeny, J. G., and J. L. Snell (1960), *Finite Markov Chains*, D. Van Nostrand Co., Princeton, N.J.

Khinchin, A. (1957), *Mathematical Foundations of Information Theory*, Dover Publications, New York.

Kraft, L. G. (1949), *A Device for Quantizing, Grouping and Coding Amplitude Modulated Pulses*, M. S. Thesis, Electrical Engineering Dept., MIT.

Landau, H. J., and H. O. Pollak (1961), Prolate Spheroidal Wave Functions, Fourier Analysis, and Uncertainty—II, *Bell System Tech. J.*, **40,** 65–84.

Landau, H. J., and H. O. Pollak (1962), Prolate Spheroidal Wave Functions, Fourier Analysis, and Uncertainty—III, *Bell System Tech. J.*, **41,** 1295–1336

Lee, P. M. (1964), On the Axioms of Information Theory, *Ann. Math. Stat.*, **35,** 415–418.

Liusternik, L. A. and V. J. Sobolev (1961), *Elements of Functional Analysis*, Frederick Ungar Publishing Co., New York.

Loève, M. (1955), *Probability Theory*, D. Van Nostrand Co., Princeton, N.J.

McMillan, B. (1953), The Basic Theorems of Information Theory, *Ann. Math. Stat.*, **24,** No.2, 196–219.

McMillan, B. (1956), Two Inequalities Implied by Unique Decipherability, *IRE Trans. Inform. Theory*, **IT-2,** 115–116.

Melas, C. M. (1960), A New Group of Codes for Correction of Dependent Errors in Data Transmission, *IBM J. Res. Development*, **4,** 58–65.

Moore, E. F. (1956), Gedanken Experiments on Sequential Machines, in *Automata Studies, Ann. Math. Studies*, **34,** Princeton Univ., Princeton, N.J.

Muroga, S. (1953), On the Capacity of a Discrete Channel, *J. Phys. Soc. Japan*, **8,** 484–494.

Peterson, W. W. (1960), Encoding and Error-Correction Procedures for the Bose-Chaudhuri Codes, *IRE Trans. Inform. Theory*, **IT-6,** 459–470.

Peterson, W. W. (1961), *Error Correcting Codes*, MIT Press, Cambridge, Mass.

Riesz, F., and B. Sz.-Nagy (1955), *Functional Analysis*, Frederick Ungar Publishing Co., New York.

Root, W. L., and T. S. Pitcher (1955), Some Remarks on Signal Detection, *IRE Trans. Inform. Theory*, **IT-1,** No. 3, 33–38.

Sacks, G. E. (1958), Multiple Error Correction by Means of Parity Checks, *IRE Trans. Inform. Theory*, **IT-4,** 145–147.

Sardinas, A. A., and G. W. Patterson (1950), A Necessary and Sufficient Condition for Unique Decomposition of Coded Messages, *Research Division Report* 50-27, Moore School of Electrical Engineering, University of Pennsylvania, Philadelphia, Pa.

Sardinas, A. A. and G. W. Patterson (1953), A Necessary and Sufficient Condition for the Unique Decomposition of Coded Messages, *IRE Convention Record*, Part 8, 104–108.

Shannon, C. E. (1948), A Mathematical Theory of Communication, *Bell System Tech. J.*, **27,** 379–423, 623–656. Reprinted in C. E. Shannon and W. Weaver, *The Mathematical Theory of Communication*, Univ. of Illinois Press, Urbana, Ill., 1949.

Shannon, C. E. (1957), Certain Results in Coding Theory for Noisy Channels, *Information and Control*, **1,** No. 1, 6–25.

Shannon, C. E. (1959), Probability of Error for Optimal Codes in a Gaussian Channel, *Bell System Tech. J.*, **38,** No. 3, 611–656.

Slepian, D. (1956), A Class of Binary Signaling Alphabets, *Bell System Tech. J.*, **35,** 203–234.

Slepian, D. (1965), Permutation Modulation, *Proc. IEEE*, **53,** No. 3, 228–236.

Slepian, D., and H. O. Pollak (1961), Prolate Spheroidal Wave Functions, Fourier Analysis, and Uncertainty—I, *Bell System Tech. J.*, **40,** 43–64.

Taylor, A. E. (1958), *Introduction to Functional Analysis*, John Wiley and Sons, New York.

Thomasian, A. J. (1960), An Elementary Proof of the AEP of Information Theory, *Ann. Math. Stat.*, **31,** 452–456.

Thomasian, A. J. (1961), Error Bounds for Continuous Channels, in *Fourth London Symposium on Information Theory*, C. Cherry, Ed., Butterworths, Washington, D.C., pp. 46–60.

Thomasian, A. J. (1963), A Finite Criterion for Indecomposable Channels, *Ann. Math. Stat.*, **34,** No. 1, 337–338.

Titchmarsh, E. C. (1937), *Introduction to the Theory of Fourier Integrals*, Oxford Univ. Press, London.

Varsharmov, R. R. (1957), Estimate of the Number of Signals in Error Correcting Codes, *Dokl. Akad. Nauk SSSR*, **117,** No. 5, 739–741.

Weiss, L. (1960), On the Strong Converse of the Coding Theorem for Symmetric Channels without Memory, *Quart. Appl. Math.*, **18,** No. 3, 209–214.

Wiener, N. (1949), *The Extrapolation, Interpolation, and Smoothing of Stationary Time Series*, John Wiley and Sons, New York.

Wolfowitz, J. (1960), Simultaneous Channels, *Arch. Rational Mech. Anal.*, **4,** No. 4, 371–386.

Wolfowitz, J. (1961), *Coding Theorems of Information Theory*, Prentice-Hall, Englewood Cliffs, N.J.

Wolfowitz, J. (1963a), On Channels without Capacity, *Information and Control*, **6,** 49–54.

Wolfowitz, J. (1963b), The Capacity of an Indecomposable Channel, *Sankhya*, **25,** Series A, 101–108.

Wolfowitz, J. (1964), *Coding Theorems of Information Theory*, second edition, Springer-Verlag, New York.

Wozencraft, J. M., and B. Reiffen (1961), *Sequential Decoding*, MIT Press, Cambridge, Mass.

Wyner, A. (1964), Improved Bounds for Minimum Distance and Error Probability in Discrete Channels, *Bell Telephone Laboratories Internal Report*, Murray Hill, N.J.

Wyner, A. (1965), Capabilities of Bounded Discrepancy Decoding, *Bell System Tech. J.*, **44,** pp. 1061–1122.

Yoshihara, K. (1964), Simple proof of the strong converse theorems in some channels, *Kodai Math. Sem. Rep.*, **16,** 213–222.

INDEX

A CATALOG OF SELECTED
DOVER BOOKS
IN SCIENCE AND MATHEMATICS

A CATALOG OF SELECTED

DOVER BOOKS

IN SCIENCE AND MATHEMATICS

Astronomy

BURNHAM'S CELESTIAL HANDBOOK, Robert Burnham, Jr. Thorough guide to the stars beyond our solar system. Exhaustive treatment. Alphabetical by constellation: Andromeda to Cetus in Vol. 1; Chamaeleon to Orion in Vol. 2; and Pavo to Vulpecula in Vol. 3. Hundreds of illustrations. Index in Vol. 3. 2,000pp. 6⅛ x 9¼.
23567-X, 23568-8, 23673-0 Pa., Three-vol. set $46.85

THE EXTRATERRESTRIAL LIFE DEBATE, 1750–1900, Michael J. Crowe. First detailed, scholarly study in English of the many ideas that developed between 1750 and 1900 regarding the existence of intelligent extraterrestrial life. Examines ideas of Kant, Herschel, Voltaire, Percival Lowell, many other scientists and thinkers. 16 illustrations. 704pp. 5⅜ x 8½. 40675-X Pa. $19.95

A HISTORY OF ASTRONOMY, A. Pannekoek. Well-balanced, carefully reasoned study covers such topics as Ptolemaic theory, work of Copernicus, Kepler, Newton, Eddington's work on stars, much more. Illustrated. References. 521pp. 5⅜ x 8½.
65994-1 Pa. $15.95

AMATEUR ASTRONOMER'S HANDBOOK, J. B. Sidgwick. Timeless, comprehensive coverage of telescopes, mirrors, lenses, mountings, telescope drives, micrometers, spectroscopes, more. 189 illustrations. 576pp. 5⅝ x 8¼. (Available in U.S. only.)
24034-7 Pa. $13.95

STARS AND RELATIVITY, Ya. B. Zel'dovich and I. D. Novikov. Vol. 1 of *Relativistic Astrophysics* by famed Russian scientists. General relativity, properties of matter under astrophysical conditions, stars, and stellar systems. Deep physical insights, clear presentation. 1971 edition. References. 544pp. 5⅝ x 8¼.
69424-0 Pa. $14.95

Chemistry

CHEMICAL MAGIC, Leonard A. Ford. Second Edition, Revised by E. Winston Grundmeier. Over 100 unusual stunts demonstrating cold fire, dust explosions, much more. Text explains scientific principles and stresses safety precautions. 128pp. 5⅜ x 8½. 67628-5 Pa. $5.95

THE DEVELOPMENT OF MODERN CHEMISTRY, Aaron J. Ihde. Authoritative history of chemistry from ancient Greek theory to 20th-century innovation. Covers major chemists and their discoveries. 209 illustrations. 14 tables. Bibliographies. Indices. Appendices. 851pp. 5⅜ x 8½. 64235-6 Pa. $24.95

CATALYSIS IN CHEMISTRY AND ENZYMOLOGY, William P. Jencks. Exceptionally clear coverage of mechanisms for catalysis, forces in aqueous solution, carbonyl- and acyl-group reactions, practical kinetics, more. 864pp. 5⅜ x 8½.
65460-5 Pa. $19.95

THE HISTORICAL BACKGROUND OF CHEMISTRY, Henry M. Leicester. Evolution of ideas, not individual biography. Concentrates on formulation of a coherent set of chemical laws. 260pp. 5⅜ x 8½. 61053-5 Pa. $8.95

A SHORT HISTORY OF CHEMISTRY, J. R. Partington. Classic exposition explores origins of chemistry, alchemy, early medical chemistry, nature of atmosphere, theory of valency, laws and structure of atomic theory, much more. 428pp. 5⅜ x 8½. (Available in U.S. only.) 65977-1 Pa. $12.95

GENERAL CHEMISTRY, Linus Pauling. Revised 3rd edition of classic first-year text by Nobel laureate. Atomic and molecular structure, quantum mechanics, statistical mechanics, thermodynamics correlated with descriptive chemistry. Problems. 992pp. 5⅜ x 8½. 65622-5 Pa. $19.95

Engineering

DE RE METALLICA, Georgius Agricola. The famous Hoover translation of greatest treatise on technological chemistry, engineering, geology, mining of early modern times (1556). All 289 original woodcuts. 638pp. 6¾ x 11. 60006-8 Pa. $21.95

FUNDAMENTALS OF ASTRODYNAMICS, Roger Bate et al. Modern approach developed by U.S. Air Force Academy. Designed as a first course. Problems, exercises. Numerous illustrations. 455pp. 5⅜ x 8½. 60061-0 Pa. $12.95

DYNAMICS OF FLUIDS IN POROUS MEDIA, Jacob Bear. For advanced students of ground water hydrology, soil mechanics and physics, drainage and irrigation engineering and more. 335 illustrations. Exercises, with answers. 784pp. 6⅛ x 9¼. 65675-6 Pa. $24.95

ANALYTICAL MECHANICS OF GEARS, Earle Buckingham. Indispensable reference for modern gear manufacture covers conjugate gear-tooth action, gear-tooth profiles of various gears, many other topics. 263 figures. 102 tables. 546pp. 5⅜ x 8½. 65712-4 Pa. $16.95

MECHANICS, J. P. Den Hartog. A classic introductory text or refresher. Hundreds of applications and design problems illuminate fundamentals of trusses, loaded beams and cables, etc. 334 answered problems. 462pp. 5⅜ x 8½. 60754-2 Pa. $13.95

MECHANICAL VIBRATIONS, J. P. Den Hartog. Classic textbook offers lucid explanations and illustrative models, applying theories of vibrations to a variety of practical industrial engineering problems. Numerous figures. 233 problems, solutions. Appendix. Index. Preface. 436pp. 5⅜ x 8½. 64785-4 Pa. $15.95

STRENGTH OF MATERIALS, J. P. Den Hartog. Full, clear treatment of basic material (tension, torsion, bending, etc.) plus advanced material on engineering methods, applications. 350 answered problems. 323pp. 5⅜ x 8½. 60755-0 Pa. $11.95

ANALYTICAL FRACTURE MECHANICS, David J. Unger. Self-contained text supplements standard fracture mechanics texts by focusing on analytical methods for determining crack-tip stress and strain fields. 336pp. 6⅛ x 9¼. 41737-9 Pa. $19.95

A HISTORY OF MECHANICS, René Dugas. Monumental study of mechanical principles from antiquity to quantum mechanics. Contributions of ancient Greeks, Galileo, Leonardo, Kepler, Lagrange, many others. 671pp. 5⅜ x 8½.
65632-2 Pa. $18.95

STATISTICAL MECHANICS: Principles and Applications, Terrell L. Hill. Standard text covers fundamentals of statistical mechanics, applications to fluctuation theory, imperfect gases, distribution functions, more. 448pp. 5⅜ x 8½.
65390-0 Pa. $14.95

THE VARIATIONAL PRINCIPLES OF MECHANICS, Cornelius Lanczos. Graduate level coverage of calculus of variations, equations of motion, relativistic mechanics, more. First inexpensive paperbound edition of classic treatise. Index. Bibliography. 418pp. 5⅜ x 8½. 65067-7 Pa. $14.95

THE VARIOUS AND INGENIOUS MACHINES OF AGOSTINO RAMELLI: A Classic Sixteenth-Century Illustrated Treatise on Technology, Agostino Ramelli. One of the most widely known and copied works on machinery in the 16th century. 194 detailed plates of water pumps, grain mills, cranes, more. 608pp. 9 x 12.
28180-9 Pa. $24.95

ORDINARY DIFFERENTIAL EQUATIONS AND STABILITY THEORY: An Introduction, David A. Sánchez. Brief, modern treatment. Linear equation, stability theory for autonomous and nonautonomous systems, etc. 164pp. 5⅜ x 8¼.
63828-6 Pa. $6.95

ROTARY WING AERODYNAMICS, W. Z. Stepniewski. Clear, concise text covers aerodynamic phenomena of the rotor and offers guidelines for helicopter performance evaluation. Orignially prepared for NASA. 537 figures. 640pp. 6⅛ x 9¼.
64647-5 Pa. $16.95

INTRODUCTION TO SPACE DYNAMICS, William Tyrrell Thomson. Comprehensive, classic introduction to space-flight engineering for advanced undergraduate and graduate students. Includes vector algebra, kinematics, transformation of coordinates. Bibliography. Index. 352pp. 5⅜ x 8½. 65113-4 Pa. $10.95

HISTORY OF STRENGTH OF MATERIALS, Stephen P. Timoshenko. Excellent historical survey of the strength of materials with many references to the theories of elasticity and structure. 245 figures. 452pp. 5⅜ x 8½. 61187-6 Pa. $14.95

CONSTRUCTIONS AND COMBINATORIAL PROBLEMS IN DESIGN OF EXPERIMENTS, Damaraju Raghavarao. In-depth reference work examines orthogonal Latin squares, incomplete block designs, tactical configuration, partial geometry, much more. Abundant explanations, examples. 416pp. 5⅜ x 8¼.
65685-3 Pa. $10.95

Math–Decision Theory, Statistics, Probability

ELEMENTARY DECISION THEORY, Herman Chernoff and Lincoln E. Moses. Clear introduction to statistics and statistical theory covers data processing, probability and random variables, testing hypotheses, much more. Exercises. 364pp. 5⅜ x 8½. 65218-1 Pa. $12.95

STATISTICS MANUAL, Edwin L. Crow et al. Comprehensive, practical collection of classical and modern methods prepared by U.S. Naval Ordnance Test Station. Stress on use. Basics of statistics assumed. 288pp. 5⅜ x 8½. 60599-X Pa. $8.95

SOME THEORY OF SAMPLING, William Edwards Deming. Analysis of the problems, theory and design of sampling techniques for social scientists, industrial managers and others who find statistics important at work. 61 tables. 90 figures. xvii +602pp. 5⅜ x 8½. 64684-X Pa. $16.95

STATISTICAL ADJUSTMENT OF DATA, W. Edwards Deming. Introduction to basic concepts of statistics, curve fitting, least squares solution, conditions without parameter, conditions containing parameters. 26 exercises worked out. 271pp. 5⅜ x 8½. 64685-8 Pa. $9.95

LINEAR PROGRAMMING AND ECONOMIC ANALYSIS, Robert Dorfman, Paul A. Samuelson and Robert M. Solow. First comprehensive treatment of linear programming in standard economic analysis. Game theory, modern welfare economics, Leontief input-output, more. 525pp. 5⅜ x 8½. 65491-5 Pa. $17.95

DICTIONARY/OUTLINE OF BASIC STATISTICS, John E. Freund and Frank J. Williams. A clear concise dictionary of over 1,000 statistical terms and an outline of statistical formulas covering probability, nonparametric tests, much more. 208pp. 5⅜ x 8½. 66796-0 Pa. $8.95

PROBABILITY: An Introduction, Samuel Goldberg. Excellent basic text covers set theory, probability theory for finite sample spaces, binomial theorem, much more. 360 problems. Bibliographies. 322pp. 5⅜ x 8½. 65252-1 Pa. $11.95

GAMES AND DECISIONS: Introduction and Critical Survey, R. Duncan Luce and Howard Raiffa. Superb nontechnical introduction to game theory, primarily applied to social sciences. Utility theory, zero-sum games, n-person games, decision-making, much more. Bibliography. 509pp. 5⅜ x 8½. 65943-7 Pa. $14.95

FIFTY CHALLENGING PROBLEMS IN PROBABILITY WITH SOLUTIONS, Frederick Mosteller. Remarkable puzzlers, graded in difficulty, illustrate elementary and advanced aspects of probability. Detailed solutions. 88pp. 5⅜ x 8½. 65355-2 Pa. $5.95

PROBABILITY THEORY: A Concise Course, Y. A. Rozanov. Highly readable, self-contained introduction covers combination of events, dependent events, Bernoulli trials, etc. 148pp. 5⅜ x 8¼. 63544-9 Pa. $8.95

STATISTICAL METHOD FROM THE VIEWPOINT OF QUALITY CONTROL, Walter A. Shewhart. Important text explains regulation of variables, uses of statistical control to achieve quality control in industry, agriculture, other areas. 192pp. 5⅜ x 8½. 65232-7 Pa. $8.95

THE COMPLEAT STRATEGYST: Being a Primer on the Theory of Games of Strategy, J. D. Williams. Highly entertaining classic describes, with many illustrated examples, how to select best strategies in conflict situations. Prefaces. Appendices. 268pp. 5⅜ x 8½. 25101-2 Pa. $9.95

Math–Geometry and Topology

ELEMENTARY CONCEPTS OF TOPOLOGY, Paul Alexandroff. Elegant, intuitive approach to topology from set-theoretic topology to Betti groups; how concepts of topology are useful in math and physics. 25 figures. 57pp. 5⅜ x 8½. 60747-X Pa. $4.95

COMBINATORIAL TOPOLOGY, P. S. Alexandrov. Clearly written, well-organized, three-part text begins by dealing with certain classic problems without using the formal techniques of homology theory and advances to the central concept, the Betti groups. Numerous detailed examples. 654pp. 5⅜ x 8½. 40179-0 Pa. $18.95

EXPERIMENTS IN TOPOLOGY, Stephen Barr. Classic, lively explanation of one of the byways of mathematics. Klein bottles, Moebius strips, projective planes, map coloring, problem of the Koenigsberg bridges, much more, described with clarity and wit. 43 figures. 210pp. 5⅜ x 8½. 25933-1 Pa. $8.95

CONFORMAL MAPPING ON RIEMANN SURFACES, Harvey Cohn. Lucid, insightful book presents ideal coverage of subject. 334 exercises make book perfect for self-study. 55 figures. 352pp. 5⅝ x 8¼. 64025-6 Pa. $11.95

THE GEOMETRY OF RENÉ DESCARTES, René Descartes. The great work founded analytical geometry. Original French text, Descartes's own diagrams, together with definitive Smith-Latham translation. 244pp. 5⅜ x 8½. 60068-8 Pa. $9.95

THE THIRTEEN BOOKS OF EUCLID'S ELEMENTS, translated with introduction and commentary by Sir Thomas L. Heath. Definitive edition. Textual and linguistic notes, mathematical analysis. 2,500 years of critical commentary. Unabridged. 1,414pp. 5⅜ x 8½.. Three-vol. set. Vol. I: 60088-2 Pa. $11.95
Vol. II: 60089-0 Pa. $11.95
Vol. III: 60090-4 Pa. $12.95

GEOMETRY OF COMPLEX NUMBERS, Hans Schwerdtfeger. Illuminating, widely praised book on analytic geometry of circles, the Moebius transformation, and two-dimensional non-Euclidean geometries. 200pp. 5⅝ x 8¼. 63830-8 Pa. $8.95

DIFFERENTIAL GEOMETRY, Heinrich W. Guggenheimer. Local differential geometry as an application of advanced calculus and linear algebra. Curvature, transformation groups, surfaces, more. Exercises. 62 figures. 378pp. 5⅜ x 8½. 63433-7 Pa. $11.95

CURVATURE AND HOMOLOGY: Enlarged Edition, Samuel I. Goldberg. Revised edition examines topology of differentiable manifolds; curvature, homology of Riemannian manifolds; compact Lie groups; complex manifolds; curvature, homology of Kaehler manifolds. New Preface. Four new appendixes. 416pp. 5⅜ x 8½.
40207-X Pa. $14.95

TOPOLOGY, John G. Hocking and Gail S. Young. Superb one-year course in classical topology. Topological spaces and functions, point-set topology, much more. Examples and problems. Bibliography. Index. 384pp. 5⅝ x 8¼. 65676-4 Pa. $13.95

LECTURES ON CLASSICAL DIFFERENTIAL GEOMETRY, Second Edition, Dirk J. Struik. Excellent brief introduction covers curves, theory of surfaces, fundamental equations, geometry on a surface, conformal mapping, other topics. Problems. 240pp. 5⅜ x 8½. 65609-8 Pa. $9.95

Math–History of

A SHORT ACCOUNT OF THE HISTORY OF MATHEMATICS, W. W. Rouse Ball. One of clearest, most authoritative surveys from the Egyptians and Phoenicians through 19th-century figures such as Grassman, Galois, Riemann. Fourth edition. 522pp. 5⅜ x 8½. 20630-0 Pa. $13.95

THE HISTORICAL ROOTS OF ELEMENTARY MATHEMATICS, Lucas N. H. Bunt, Phillip S. Jones, and Jack D. Bedient. Fundamental underpinnings of modern arithmetic, algebra, geometry and number systems derived from ancient civilizations. 320pp. 5⅜ x 8½. 25563-8 Pa. $9.95

GAMES, GODS & GAMBLING: A History of Probability and Statistical Ideas, F. N. David. Episodes from the lives of Galileo, Fermat, Pascal, and others illustrate this fascinating account of the roots of mathematics. Features thought-provoking references to classics, archaeology, biography, poetry. 1962 edition. 304pp. 5⅜ x 8½. (Available in U.S. only.) 40023-9 Pa. $9.95

HISTORY OF MATHEMATICS, David E. Smith. Nontechnical survey from ancient Greece and Orient to late 19th century; evolution of arithmetic, geometry, trigonometry, calculating devices, algebra, the calculus. 362 illustrations. 1,355pp. 5⅜ x 8½. Two-vol. set. Vol. I: 20429-4 Pa. $13.95
Vol. II: 20430-8 Pa. $14.95

A CONCISE HISTORY OF MATHEMATICS, Dirk J. Struik. The best brief history of mathematics. Stresses origins and covers every major figure from ancient Near East to 19th century. 41 illustrations. 195pp. 5⅜ x 8½. 60255-9 Pa. $8.95

THE HISTORY OF THE CALCULUS AND ITS CONCEPTUAL DEVELOPMENT, Carl B. Boyer. Origins in antiquity, medieval contributions, work of Newton, Leibniz, rigorous formulation. Treatment is verbal. 346pp. 5⅜ x 8½.
60509-4 Pa. $9.95

Physics

OPTICAL RESONANCE AND TWO-LEVEL ATOMS, L. Allen and J. H. Eberly. Clear, comprehensive introduction to basic principles behind all quantum optical resonance phenomena. 53 illustrations. Preface. Index. 256pp. 5⅜ x 8½.
65533-4 Pa. $10.95

ULTRASONIC ABSORPTION: An Introduction to the Theory of Sound Absorption and Dispersion in Gases, Liquids and Solids, A. B. Bhatia. Standard reference in the field provides a clear, systematically organized introductory review of fundamental concepts for advanced graduate students, research workers. Numerous diagrams. Bibliography. 440pp. 5⅜ x 8½. 64917-2 Pa. $11.95

QUANTUM THEORY, David Bohm. This advanced undergraduate-level text presents the quantum theory in terms of qualitative and imaginative concepts, followed by specific applications worked out in mathematical detail. Preface. Index. 655pp. 5⅜ x 8½. 65969-0 Pa. $16.95

ATOMIC PHYSICS (8th edition), Max Born. Nobel laureate's lucid treatment of kinetic theory of gases, elementary particles, nuclear atom, wave-corpuscles, atomic structure and spectral lines, much more. Over 40 appendices, bibliography. 495pp. 5⅜ x 8½. 65984-4 Pa. $14.95

AN INTRODUCTION TO HAMILTONIAN OPTICS, H. A. Buchdahl. Detailed account of the Hamiltonian treatment of aberration theory in geometrical optics. Many classes of optical systems defined in terms of the symmetries they possess. Problems with detailed solutions. 1970 edition. xv + 360pp. 5⅜ x 8½.
67597-1 Pa. $10.95

THIRTY YEARS THAT SHOOK PHYSICS: The Story of Quantum Theory, George Gamow. Lucid, accessible introduction to influential theory of energy and matter. Careful explanations of Dirac's anti-particles, Bohr's model of the atom, much more. 12 plates. Numerous drawings. 240pp. 5⅜ x 8½. 24895-X Pa. $8.95

ELECTRONIC STRUCTURE AND THE PROPERTIES OF SOLIDS: The Physics of the Chemical Bond, Walter A. Harrison. Innovative text offers basic understanding of the electronic structure of covalent and ionic solids, simple metals, transition metals and their compounds. Problems. 1980 edition. 582pp. 6⅛ x 9¼.
66021-4 Pa. $19.95

HYDRODYNAMIC AND HYDROMAGNETIC STABILITY, S. Chandrasekhar. Lucid examination of the Rayleigh-Benard problem; clear coverage of the theory of instabilities causing convection. 704pp. 5⅜ x 8¼. 64071-X Pa. $17.95

INVESTIGATIONS ON THE THEORY OF THE BROWNIAN MOVEMENT, Albert Einstein. Five papers (1905–8) investigating dynamics of Brownian motion and evolving elementary theory. Notes by R. Fürth. 122pp. 5⅜ x 8½.
60304-0 Pa. $7.95

THE PHYSICS OF WAVES, William C. Elmore and Mark A. Heald. Unique overview of classical wave theory. Acoustics, optics, electromagnetic radiation, more. Ideal as classroom text or for self-study. Problems. 477pp. 5⅜ x 8½.
64926-1 Pa. $14.95

PHYSICAL PRINCIPLES OF THE QUANTUM THEORY, Werner Heisenberg. Nobel Laureate discusses quantum theory, uncertainty, wave mechanics, work of Dirac, Schroedinger, Compton, Wilson, Einstein, etc. 184pp. 5⅜ x 8½.
60113-7 Pa. $8.95

ATOMIC SPECTRA AND ATOMIC STRUCTURE, Gerhard Herzberg. One of best introductions; especially for specialist in other fields. Treatment is physical rather than mathematical. 80 illustrations. 257pp. 5⅜ x 8½. 60115-3 Pa. $11.95

AN INTRODUCTION TO STATISTICAL THERMODYNAMICS, Terrell L. Hill. Excellent basic text offers wide-ranging coverage of quantum statistical mechanics, systems of interacting molecules, quantum statistics, more. 523pp. 5⅜ x 8½.
65242-4 Pa. $14.95

THEORETICAL PHYSICS, Georg Joos, with Ira M. Freeman. Classic overview covers essential math, mechanics, electromagnetic theory, thermodynamics, quantum mechanics, nuclear physics, other topics. First paperback edition. xxiii + 885pp. 5⅜ x 8½. 65227-0 Pa. $24.95

PROBLEMS AND SOLUTIONS IN QUANTUM CHEMISTRY AND PHYSICS, Charles S. Johnson, Jr. and Lee G. Pedersen. Unusually varied problems, detailed solutions in coverage of quantum mechanics, wave mechanics, angular momentum, molecular spectroscopy, more. 280 problems plus 139 supplementary exercises. 430pp. 6½ x 9¼. 65236-X Pa. $14.95

THEORETICAL SOLID STATE PHYSICS, Vol. 1: Perfect Lattices in Equilibrium; Vol. II: Non-Equilibrium and Disorder, William Jones and Norman H. March. Monumental reference work covers fundamental theory of equilibrium properties of perfect crystalline solids, non-equilibrium properties, defects and disordered systems. Appendices. Problems. Preface. Diagrams. Index. Bibliography. Total of 1,301pp. 5⅜ x 8½. Two volumes. Vol. I: 65015-4 Pa. $16.95
Vol. II: 65016-2 Pa. $16.95

A TREATISE ON ELECTRICITY AND MAGNETISM, James Clerk Maxwell. Important foundation work of modern physics. Brings to final form Maxwell's theory of electromagnetism and rigorously derives his general equations of field theory. 1,084pp. 5⅜ x 8½. Two-vol. set. Vol. I: 60636-8 Pa. $14.95
Vol. II: 60637-6 Pa. $14.95

OPTICKS, Sir Isaac Newton. Newton's own experiments with spectroscopy, colors, lenses, reflection, refraction, etc., in language the layman can follow. Foreword by Albert Einstein. 532pp. 5⅜ x 8½. 60205-2 Pa. $13.95

THEORY OF ELECTROMAGNETIC WAVE PROPAGATION, Charles Herach Papas. Graduate-level study discusses the Maxwell field equations, radiation from wire antennas, the Doppler effect and more. xiii + 244pp. 5⅜ x 8½.
65678-0 Pa. $9.95